AQA Geography A

GCSE

Judith Canavan

Alison Rae

Simon Ross

Editor

Simon Ross

Nelson Thornes

Published in 2009 by:
Nelson Thornes Ltd
Delta Place
27 Bath Road
CHELTENHAM
GL53 7TH
United Kingdom

11 12 13 / 10 9 8 7 6 5

A catalogue record for this book is available from the British
Library

ISBN 978 1 4085 0271 6

Cover photograph by Getty/Gazimal

Illustrations by David Russell Illustration, Tim Jay,
Peters & Zabransky and GreenGate Publishing

Page make-up by GreenGate Publishing, Tonbridge, Kent

Printed and bound in Egypt by Sahara Printing Company

Contents

Nelson Thornes and AQA

Nelson Thornes has worked in partnership with AQA to ensure this book and the accompanying online resources offer you the best support for your GCSE course.

All resources have been approved by senior AQA examiners so you can feel assured that they closely match the specification for this subject and provide you with everything you need to prepare successfully for your exams.

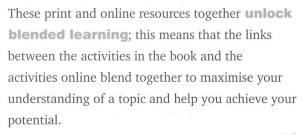

These print and online resources together **unlock blended learning**; this means that the links between the activities in the book and the activities online blend together to maximise your understanding of a topic and help you achieve your potential.

These online resources are available on *kerboodle!* which can be accessed via the internet at **www.kerboodle.com/live**, anytime, anywhere. If your school or college subscribes to *kerboodle!* you will be provided with your own personal login details. Once logged in, access your course and locate the required activity.

For more information and help on how to use *kerboodle!* visit **www.kerboodle.com**.

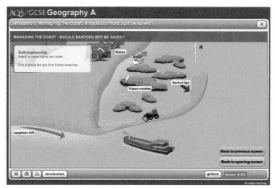

How to use this book

Visit **www.nelsonthornes.com/aqagcse** for more information.

AQA examination-style questions are reproduced by permission of the Assessment and Qualifications Alliance.

This specification focuses on the physical and human processes and factors that have shaped the environment in which we live. You will learn about the interdependence of physical environments and how human activity affects the environment, and vice versa. You will also gain an understanding of the need for sustainable management of both physical and human environments. You will learn to appreciate the differences and similarities between people, places and cultures and improve your understanding of societies and economies.

Your full GCSE geography course has three units (Unit 1: Physical Geography; Unit 2: Human Geography, and Unit 3: Local Fieldwork Investigation, which will be based on a topic covered in the first two units). Units 1 and 2 each have two sections, A and B, and the chapters of this book take their titles from the topic titles in these sections. If you are taking the short course, you have to study one Physical Geography topic and one Human Geography topic, and complete a local fieldwork investigation as for the full course.

◼ Unit 1 Physical Geography

In the exam, you must answer three questions. One question should be taken from Section A, one from Section B and the third question from either section.

Section A

- The Restless Earth
- Rocks, Resources and Scenery
- Challenge of Weather and Climate
- Living World

Section B

- Water on the Land
- Ice on the Land
- The Coastal Zone

These topics are intended to provide you with a solid foundation in physical geography to enable you to fully appreciate the physical world in which you live.

A great deal of emphasis has been placed on topicality, using up-to-date case studies and examples to explore themes and concepts, such as the 2004 Indian Ocean tsunami and recent floods in Boscastle and Tewkesbury.

All of the physical geography topics have a link to human activities. For example, in *The Restless Earth* you will study the impact of tectonic hazards on people's lives. The impacts of climate change are also discussed in *Challenge of Weather and Climate*.

Sustainable development is a key concept in all topics. For example, in *The Coastal Zone*, strategies for coastal defence and coastal zone management are underpinned by the need for a sustainable approach. In *Living World*, sustainability is at the heart of ecosystem management, whether it is hot deserts or tropical rainforest.

◼ Unit 2 Human Geography

In the exam, you must answer three questions. One question should be taken from Section A, one from Section B and the third question from either section.

Section A

- Population Change
- Changing Urban Environments
- Changing Rural Environments

Section B

- The Development Gap
- Globalisation
- Tourism

These topics have been chosen to reflect current thinking and interests in geography in the 21st century. They all involve the study of recent issues and extensive use is made of up-to-date case studies and examples. For example, in *Population Change*, detailed consideration is made of the implications of China's controversial 'one child policy'. The current (and future) issue of ageing populations in Europe is also considered.

Sustainable development is a significant consideration in human geography. For example, in *Tourism*, you will discuss the issues surrounding mass tourism as well as examining recent trends such as ecotourism and adventure tourism. In *Globalisation*, topics such as renewable energy and pollution control are also tackled.

As you study human geography there will be many opportunities for you to discuss and debate real world issues, such as fair trade in *The Development Gap* and organic farming in *Changing Rural Environments*. You will be able to formulate and share your own opinions and, hopefully, you will be able to make a positive difference to the world around you.

1 The restless earth

1.1 Why is the earth's crust unstable?

The structure of the earth

The **crust** – the outer layer of the earth – is relatively thin (diagram **A**). The crust is not one single piece of skin, like that of an apple. Instead, it is split into **plates** of varying size and at **plate margins** it is liable to move. This is because the slabs of crust float on the semi-molten upper **mantle**. **Convection currents** within the mantle determine the direction of plate movement. Therefore, in some cases the plates are moving together and sometimes they are moving apart. There are two types of crust: oceanic and continental (diagram **B**). The location of the plates, plate boundaries and direction of movement of the plates is shown in map **C**.

In this section you will learn

about the structure of the earth and the difference between oceanic and continental crust

how and why destructive, constructive and conservative plate margins are different

how the earth's crust is unstable, especially at plate margins.

Key terms

Crust: the outer layer of the earth.

Plate: a section of the earth's crust.

Plate margin: the boundary where two plates meet.

Mantle: the dense, mostly solid layer between the outer core and the crust.

Convection currents: the circular currents of heat in the mantle.

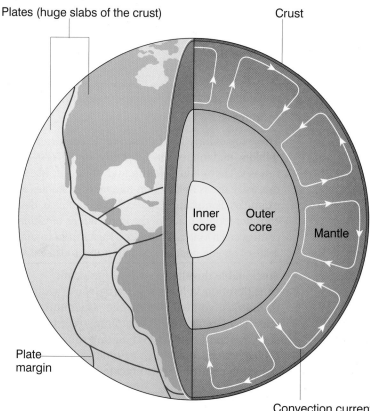

Plates (huge slabs of the crust)

Crust

Inner core

Outer core

Mantle

Plate margin

Convection currents in the mantle

A *The structure of the earth*

Did you know ??????

The earth's crust is divided into 14 major plates and 38 minor plates, making a total of 52 jigsaw pieces altogether.

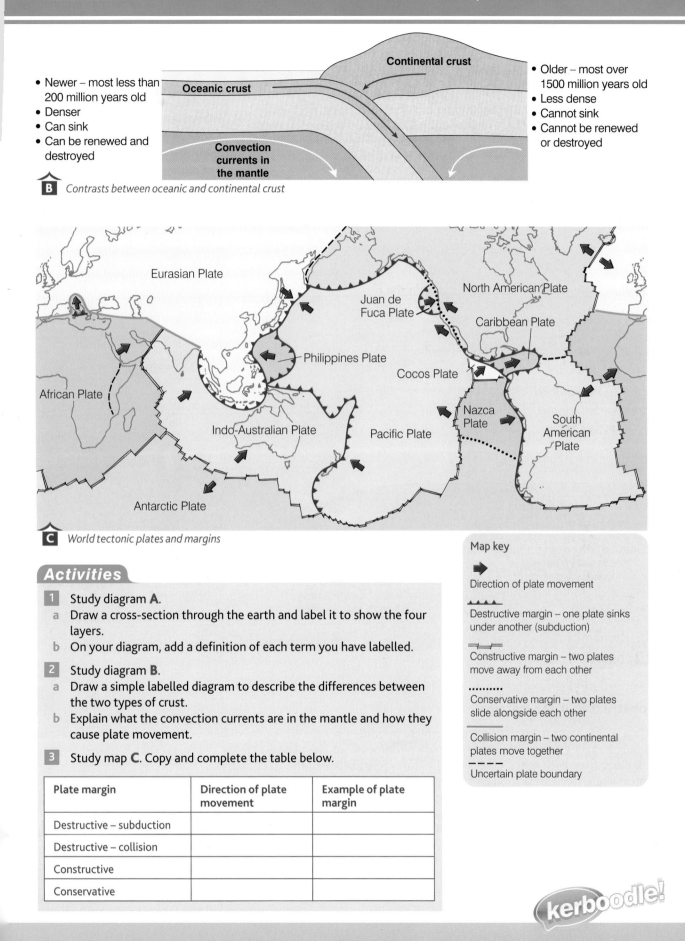

Oceanic crust
- Newer – most less than 200 million years old
- Denser
- Can sink
- Can be renewed and destroyed

Continental crust
- Older – most over 1500 million years old
- Less dense
- Cannot sink
- Cannot be renewed or destroyed

Convection currents in the mantle

B   Contrasts between oceanic and continental crust

C   World tectonic plates and margins

Map key

➡ Direction of plate movement

▲▲▲ Destructive margin – one plate sinks under another (subduction)

➔⌐ Constructive margin – two plates move away from each other

········· Conservative margin – two plates slide alongside each other

──── Collision margin – two continental plates move together

── ── Uncertain plate boundary

Activities

1   Study diagram **A**.
   a   Draw a cross-section through the earth and label it to show the four layers.
   b   On your diagram, add a definition of each term you have labelled.

2   Study diagram **B**.
   a   Draw a simple labelled diagram to describe the differences between the two types of crust.
   b   Explain what the convection currents are in the mantle and how they cause plate movement.

3   Study map **C**. Copy and complete the table below.

Plate margin	Direction of plate movement	Example of plate margin
Destructive – subduction		
Destructive – collision		
Constructive		
Conservative		

kerboodle!

Types of plate margin

Destructive plate margins

Convection currents in the mantle cause the plates to move together. If one plate is made from oceanic crust and the other from continental crust, the denser oceanic crust sinks under the lighter continental crust in a process known as **subduction**. Great pressure is exerted and the oceanic crust is destroyed as it melts to form magma.

If two continental plates meet each other, they collide rather than one sinking beneath the other. This **collision** boundary is a different type of destructive margin.

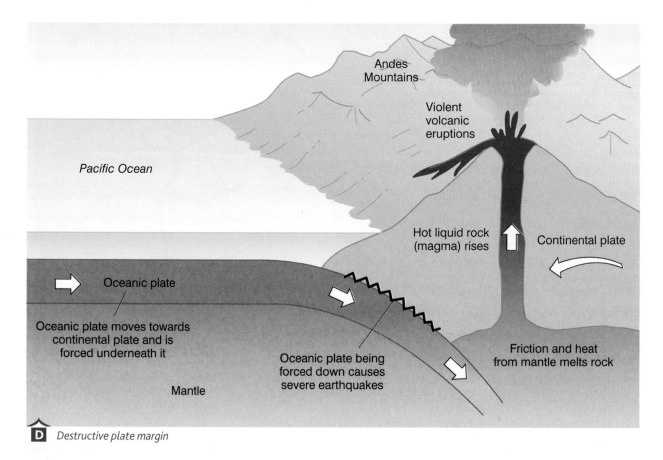

Andes Mountains

Violent volcanic eruptions

Pacific Ocean

Hot liquid rock (magma) rises

Continental plate

Oceanic plate

Oceanic plate moves towards continental plate and is forced underneath it

Oceanic plate being forced down causes severe earthquakes

Friction and heat from mantle melts rock

Mantle

D   *Destructive plate margin*

Constructive plate margins

When plates move apart, a constructive plate boundary results. This usually happens under the oceans, as shown in map **C**. As the plates pull away from each other, cracks and fractures form between the plates where there is no solid crust. Magma forces its way into the cracks and makes its way to the surface to form volcanoes. In this way new land is formed as the plates gradually pull apart.

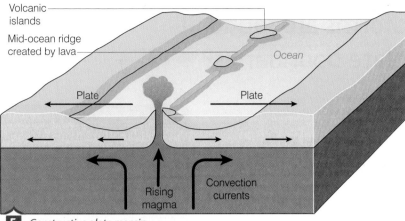

E *Constructive plate margin*

Conservative plate margins

At conservative plate margins, the plates are sliding past each other. They are moving in a similar (though not the same) direction, at slightly different angles and speeds. As one plate is moving faster than the other and in a slightly different direction, they tend to get stuck. Eventually, the build-up of pressure causes them to be released. This sudden release of pressure causes an earthquake. At a conservative margin, crust is being neither destroyed nor made.

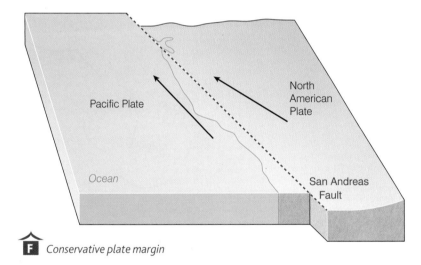

F *Conservative plate margin*

AQA *Examiner's tip*

Ensure that you can draw and label a diagram of an oceanic plate being subducted beneath a continental plate. This could occur in several types of exam question.

Activity

4 Study diagrams **D**, **E** and **F**.
a Read the task you have to complete in b below. Before starting the task, discuss with a partner what information the diagrams ought to have on them, and how they might be drawn and presented to ensure the information is clear.

b For each plate margin, draw a labelled cross-section to describe the characteristics of the boundary and the processes that are occurring there.

⃝links

You can find out more about plates and the different types of crust at http://en.wikipedia.org/wiki/List_of_tectonic_plates.

What landforms are found at different plate boundaries?

Fold mountains and ocean trenches

Young **fold mountains** (those that have been formed in the last 65 million years) are the highest areas in the world. All peaks over 7,000 m are in central Asia, including Mt Everest at 8,850 m. This dwarfs the highest mountain in England – Scafell Pike at 978 m. Young fold mountains include ranges such as the Himalayas, the Rockies, the Andes and the Alps. Older ranges of fold mountains that are less high due to erosion include the Cambrian mountains and the Cumbrian mountains in the UK. **Ocean trenches** form some of the deepest parts of the ocean. The distribution of young fold mountains and ocean trenches is shown in map **A**. Look back at map **C** on page 9 to see the link between these landforms and plate margins.

Look back at map **C** on page 9

> **In this section you will learn**
>
> why fold mountains and ocean trenches form at destructive plate margins
>
> the differences between composite volcanoes, which are associated with destructive plate margins, and shield volcanoes, which are associated with constructive plate margins.

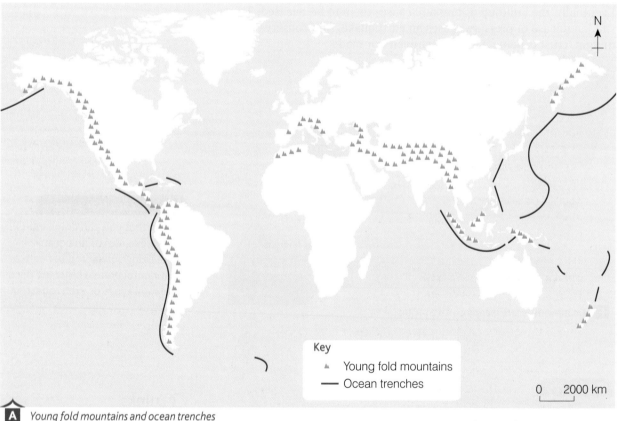

A *Young fold mountains and ocean trenches*

Key
▲ Young fold mountains
— Ocean trenches

0   2000 km

N

Both fold mountains and ocean trenches result from plates moving together. If both landforms occur in the same area, they are found in association with subduction. If fold mountains occur by themselves, they are in areas where collision is taking place. Either way, the sequence relating to their formation is similar, as shown in diagram **B**.

> AQA **Examiner's tip**
>
> To access the highest levels on questions relating to fold mountains, the correct sequence of formation is essential.

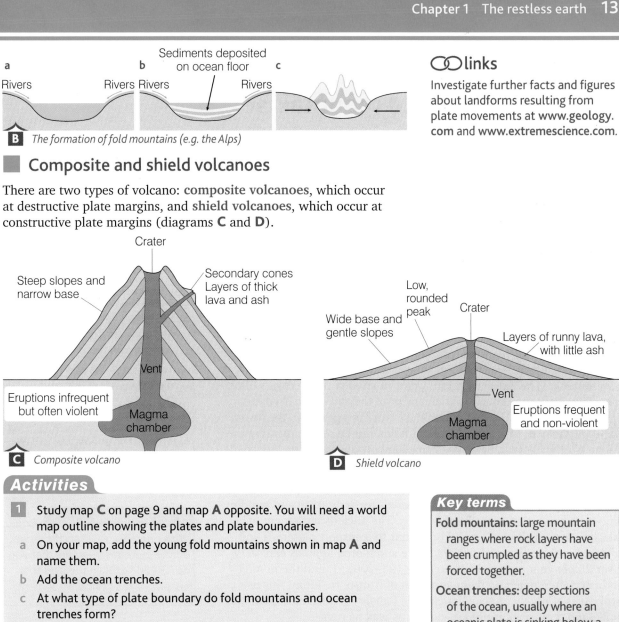

B *The formation of fold mountains (e.g. the Alps)*

GD **links**

Investigate further facts and figures about landforms resulting from plate movements at www.geology. com and www.extremescience.com.

Composite and shield volcanoes

There are two types of volcano: **composite volcanoes**, which occur at destructive plate margins, and **shield volcanoes**, which occur at constructive plate margins (diagrams **C** and **D**).

C *Composite volcano*

D *Shield volcano*

Activities

1   Study map **C** on page 9 and map **A** opposite. You will need a world map outline showing the plates and plate boundaries.

a   On your map, add the young fold mountains shown in map **A** and name them.

b   Add the ocean trenches.

c   At what type of plate boundary do fold mountains and ocean trenches form?

2   Study diagram **B**.

a   Make a copy of each part of the diagram.

b   Use the key terms on the right to help you label each diagram to explain the formation of fold mountains.

c   Draw another diagram to explain the formation of ocean trenches (see diagram **D** on page 10).

3   Study diagrams **C** and **D**.

a   On your own, write five questions you would ask to find out about the contrasts between a composite volcano and a shield volcano.

b   Swap questions with a partner and answer their questions.

c   What were the good points about the questions you have just answered? How might they be improved?

d   Use an internet search engine to find images of composite and shield volcanoes. Draw a sketch of each image and label them to show the characteristics of each type of volcano.

e   Summarise how and why composite and shield volcanoes are different.

Key terms

Fold mountains: large mountain ranges where rock layers have been crumpled as they have been forced together.

Ocean trenches: deep sections of the ocean, usually where an oceanic plate is sinking below a continental plate.

Composite volcano: a steep-sided volcano that is made up of a variety of materials, such as lava and ash.

Shield volcano: a broad volcano that is mostly made up of lava.

How do people use an area of fold mountains?

The Andes and its people

The Andes is a range of young fold mountains formed some 65 million years ago (map **A**, page 12). It is the longest range of fold mountains in the world at 7,000 km and extends the length of South America. The Andes are about 300 km in width and have an average height of 4,000 m.

Farming

Despite the high altitude of the Andes, the mountain slopes are used for farming. In Bolivia, many **subsistence** farmers grow a variety of crops on the steep slopes, including potatoes which are a main source of food. The use of **terraces** (photo **A**) creates areas of flat land on the slopes. Terracing offers other advantages in trying to farm in this harsh environment. The flat areas retain water in an area that receives little. They also limit the downward movement of the soil in areas where the soils are thin in the first place. Most crops are grown in the lower valleys (photo **B**) and a patchwork of fields can be seen indicating the range of crops grown. Some cash crops are produced such as soybeans, rice and cotton.

Llamas are synonymous with the Andes. For hundreds of years they have been pack animals, carrying materials for **irrigation** and buildings into inhospitable and inaccessible areas. The ancient settlement of Machu Picchu (photo **C**) relied on llamas to transport materials and goods due to its remote location. These surefooted animals can carry over 25 per cent of their body weight (125–200 kg); and the mining industry often relied on them as a form of transport. Today, this use still exists – largely of male llamas. The females are used for meat and milk, and their wool is used in clothes as well as rugs.

> **In this section you will learn**
>
> how people use areas of fold mountains
>
> how people adapt to the difficult conditions within them.

> **Key terms**
>
> **Subsistence:** farming to provide food and other resources for the farmer's own family.
>
> **Terraces:** steps cut into hillsides to create areas of flat land.
>
> **Irrigation:** artificial watering of the land.
>
> **Hydroelectric power:** the use of flowing water to turn turbines to generate electricity.

> AQA **Examiner's tip**
>
> With reference to a specific range of fold mountains, ensure that you can list and explain the uses that people make of this environment.

A Farming in the Andes

B The lower valleys

Mining

The Andes has a range of important minerals and the Andean countries rank in the top 10 for tin (Peru and Bolivia), nickel (Colombia), silver (Peru and Chile) and gold (Peru). More than half of Peru's exports are from mining. The Yanacocha gold mine (map **D**) is the largest gold mine in the world. It is a joint venture between a Peruvian mining company and a US-based one that has a 51 per cent share. It is an open pit and the gold-bearing rock is loosened by daily dynamite blasts. The rock is then sprayed with cyanide and the gold extracted from the resulting solution. This can lead to contamination of water supplies.

The nearby town of Cajamarca has grown from 30,000 inhabitants (when the mine began) to 240,000 in 2005. This brings with it alternative sources of jobs. However, this growth also brings many problems, including a lack of services and an increased crime rate.

C   *Machu Picchu*

D   *The location of Yanacocha gold mine*

Hydroelectric power

The steep slopes and narrow valleys that limit farming are an advantage for **hydroelectric power** (HEP). They can be more easily dammed than wider valleys and the relief encourages the rapid fall (flow) of water needed to ensure the generation of electricity. The melting snow in spring increases the supply of water, but the variation throughout the year is a disadvantage rather than an advantage. The Yuncan project dams the Paucartambo and Huachon rivers in north-east Peru, while the El Platinal project is due to begin construction in 2009. This will be the second largest in Peru and will dam the Cañete River.

Tourism

There are many natural attractions in the Andes such as mountain peaks, volcanoes, glaciers and lakes. Some tourist attractions show how people settled in these inhospitable areas, such as the remains of early settlements built by the Incas like Machu Picchu (photo **C**). The Inca Trail combines both (extract **E**).

The Inca Trail is South America's best-known trek

The Inca Trail is South America's best-known trek. It combines a stunning mix of Inca ruins, mountain scenery, lush forest and tropical jungle. Over 250 species of orchid exist in the Machu Picchu historic sanctuary, as well as numerous species of birds. The sanctuary is an important natural and archaeological reserve – it is only one of 23 UNESCO world heritage sites to be classified as important both culturally and naturally.

The Inca Trail is a hike that finishes at Machu Picchu, the sacred mysterious 'Lost City of the Incas'. The 45 km trek is usually covered in four days, arriving at Machu Picchu at daybreak on the final day before returning to Cusco by train in the afternoon. The trek is best undertaken from April to October, when the weather is drier. Any fit person should be able to cover the route. It is fairly challenging and altitudes of 4,200 m are reached, so it is important to be well acclimatised.

E *The Inca Trail*

Did you know ??????

The Inca Trail consists of three routes that meet at Machu Picchu. The trails were made for walkers and llamas rather than wheeled transport. It is closed for conservation every February. Strict regulations have been introduced, with only 200 trekkers allowed to start the trail each day. Tickets must be bought at least a month in advance.

∞ links

Investigate further facts and discussion about gold mining by searching on www.google.co.uk.

For more on the Inca Trail, go to www.llamatravel.com/about-peru.

Activities

1. Study map **C** on page 9 and an atlas map of South America. Locate the Andes on the map of South America. Using these resources, produce a fact file about the Andes to provide the following information:
   - location
   - countries
   - plates responsible for mountains
   - direction of plate movement
   - type of plate margin
   - dimensions
   - highest peak and its location.

2. Study photos **A** and **B**.
   a. Describe how the mountains are used for crops.
   b. Outline the role of the llama in the life of Andean residents.
   c. Explain how difficulties of steep relief and poor soils are overcome.

3. Study map **D**.
   a. Work in pairs to produce a short presentation to include:
   - a map showing the location of the Yanacocha gold mine
   - the advantages of the area for the mine and for Peru as a whole
   - the disadvantages of the area for the mine and for Peru as a whole.
   b. Show your presentation to another group.

4. Research HEP and specifically that produced in the Andes. On one side of A4 paper, define HEP, explain why the Andes is suitable for the development of HEP and give details of one HEP project. Include at least one map and one photo to illustrate your information.

5. Read extract **E**. Imagine you have visited Peru and part of your holiday included the Inca Trail. Write a postcard to send home, which includes the following:
   - a map showing the trail
   - a description of how you got to the trail
   - a description of the attractions (natural and human) that you saw. Explain why these were so awesome and different to what you had seen previously.

1.4   How do volcanoes affect people?

The distribution of volcanoes

Volcanoes are an example of a **natural hazard**. Their spread relates closely to plate margins (map **A** and map **C**, page 9). The area around the Pacific Ocean is especially prone to volcanoes and is known as 'the Pacific Ring of Fire'. Occasionally, active volcanoes are found away from plate margins. The two volcanoes we will focus on represent the two different types described on page 13; Nyiragongo is a shield volcano found at a constructive margin, whereas Mt St Helens is a composite volcano found at a destructive plate margin.

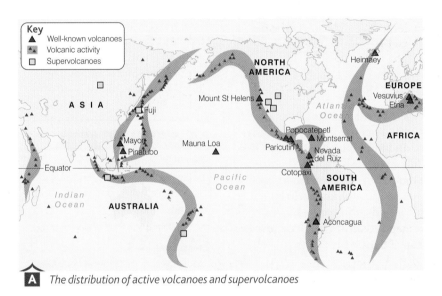

A *The distribution of active volcanoes and supervolcanoes*

The eruption of Nyiragongo, Africa

On 17 January 2002, Nyiragongo volcano in the Democratic Republic of Congo was disturbed by the movement of plates along the East African rift valley. This led to lava spilling southwards in three streams. The speed of the lava reached 60 kph, which is especially fast. The lava flowed across the runway at Goma airport (photo **C**) and through the town, splitting it in half (map **B**). The lava destroyed many homes as well as roads and water pipes, set off explosions in fuel stores and power plants, and killed 45 people. These were the **primary effects**. In addition, there were many **secondary effects**. Half a million people fled from Goma into neighbouring Rwanda to escape the lava. They spent the night sleeping on the streets of Gisenyi. Here, there was no shelter, electricity or clean water as the area could not cope with the influx. Diseases such as cholera were a real risk. People were frightened of going back. However, looting was a problem in Goma and many residents returned within a week in the hope of receiving **aid**. In the aftermath of the eruption, water had to be supplied in tankers. Aid agencies, including Christian Aid and Oxfam, were involved in the distribution of food, medicine and blankets.

In this section you will learn

where volcanoes are found

how the primary and secondary effects of volcanoes can be positive as well as negative

about immediate and long-term responses to eruptions, including monitoring and prediction.

Key terms

Natural hazard: an occurrence over which people have little control, which poses a threat to people's lives and possessions. This is different from a natural event as volcanoes can erupt in unpopulated areas without being a hazard.

Primary effects: the immediate effects of the eruption, caused directly by it.

Secondary effects: the after-effects that occur as an indirect effect of the eruption on a longer timescale.

Aid: money, food, training and technology given by richer countries to poorer ones, either to help with an emergency or to encourage long-term development.

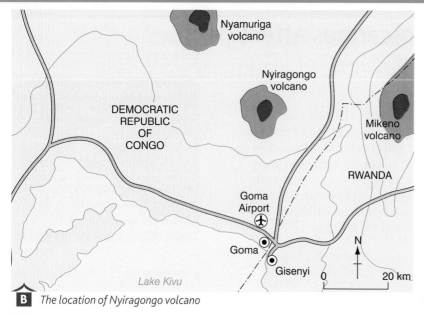

B *The location of Nyiragongo volcano*

C *Lava on the runway at Goma airport*

The eruption of Mt St Helens, USA

Mt St Helens is one of five volcanoes in the Cascade Range in Washington State, USA. Like the others, it had been dormant for many years. In March 1980 there were signs of an impending eruption, as first **earthquakes** occurred and then steam filled with ash exploded onto the pristine white glacial summit of the mountain. Residents had been told to leave and visitors were not allowed inside an 8 km exclusion zone around the crater. The vulcanologist David Johnston, who was monitoring the volcano, said 'This is like standing next to a dynamite keg and the fuse is lit, but you don't know how long the fuse is.'

The expected, yet unpredictable, eruption happened at 8.32am on 18 May (photos in **D**). An earthquake measuring 5.1 on the Richter scale caused a landslide on the north-east side of the mountain. This was the biggest landslide ever recorded and the sideways blast of pulverised rock, glacier ice and ash wiped out all living things up to 27 km north of the crater. In the outer 'blow-down' zone, the trees (some 500-year-old cedars) were uprooted and tossed around like matchsticks in the swirling volcanic fallout. Fifty-seven people died, one of them David Johnston, who realised his prediction of a lateral blast had come true. He radioed his last message to the Observatory in Oregon: 'Vancouver! Vancouver! This is it!'

AQA Examiner's tip

Be clear about the differences between the terms 'effects' and 'responses' when you write about volcanic activity.

Key terms

Earthquake: a sudden and often violent shift in the rocks forming the earth's crust, which is felt at the surface.

Immediate responses: how people react as the disaster happens and in the immediate aftermath.

Long-term responses: later reactions that occur in the weeks, months and years after the event.

D *Mt St Helens before and after the eruption*

The aftermath

The **immediate responses** involved mobilising helicopters to search and rescue those in the vicinity of the catastrophic blast. It became clear to Vancouver that those caught in the path of the immediate blast could not have survived. The remains of a small yellow car, belonging to geologist Jim Fitzgerald, and the disappearance of David Johnston's jeep and trailer, verified this. Tourists within the exclusion zone were caught up in the mudflows; others in the lateral blast. Rescuing survivors was a priority, followed by emergency treatment at nearby towns. The ash clogged air-conditioning systems and blocked roads with drifts a metre deep in places, and stranded tourists had to be found shelter. The ash had to be cleared to allow traffic to flow, which was done within three days of the eruption. A shortage of masks resulted and US President Carter promised to send two million extra immediately.

The ash that had been such a disaster initially would have a positive impact on the quality of the soil by increasing its fertility – one reason why, in some areas, people choose to live near volcanoes. The **long-term responses** began to take effect. Buildings and bridges needed rebuilding. The drainage in the area had to be looked at to see that flooding would not occur as a result of all the debris. The forest in the area to the north began to be replanted by the forest service following removal of fallen timber. Roads had to be rebuilt and attempts to bring tourists back became a consideration.

There have also been natural changes occurring in the area, which continue today. The apparently sterile moonscape has been transformed into one bursting with life, as photo **E** shows. Within 10 years of the eruption, small traces of green had already appeared. Insects, birds and animals, including elk, have all returned as the area recovers.

Mt St Helens is a better-known area today than before the 1980 eruption; the books at the visitor centres bear signatures from all over the world. The tourist industry has recovered over a long period of time. The area was designated a national monument in 1982, which led to the spending of $1.4 million to transform the area and allow access for the 3 million visitors it receives each year. The road to the north of the volcano was reopened in 1990 and the Johnston Ridge Observatory in 1997. Photos **F** and **G** show the area today. At Spirit Lake, the trees felled by the eruption still cover a third of the lake surface although many have sunk. The new cone in the crater grew between 2004 and February 2008. What will happen next?

E *Regeneration of vegetation near Johnston Ridge Observatory, 2006*

F *Spirit Lake today, with Mt St Helens in the background*

G *The crater today*

Monitoring and predicting volcanoes

The eruption of Mt St Helens was expected and prepared for, but in spite of the monitoring that had taken place, the exact date and time of the eruption took people by surprise. Earthquakes are a frequent sign of an impending eruption and their frequency and strength can be recorded. The bulge that appeared on the northern flank of Mt St Helens was evidence of the movement of magma below. This 'swelling' was obvious and could be monitored easily. Tiltmeters can identify small, subtle changes in the landscape (diagram **H**). Global positioning systems (GPS) use satellites to detect movement of as little as 1 mm (photo **K**). The change in temperature at the surface can be seen on satellite images. Digital cameras placed on the rim of craters can photograph events relatively safely (photo **J**). Gases being emitted from the **vent** change before an eruption – there is an increasing amount of sulphur dioxide. The night before the Mt St Helens eruption, samples were collected by David Johnston entering the crater. Today, robots called 'spiders' are often used to monitor changes. The past frequency of eruptions; the gap between them; and the pattern of lava flows, ash movement and **lahars** can tell us about how the volcano is likely to behave. This technology allows people to prepare for an eruption by organising the evacuation of people and arranging supplies.

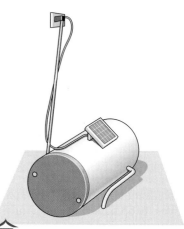

H *Tiltmeters detect a change in slope caused by shifting magma beneath the surface*

Tectonic-like earthquakes

Shallow volcanic earthquakes

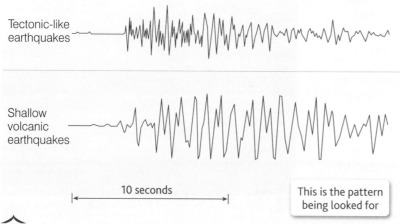

10 seconds

This is the pattern being looked for

I *Seismograms indicate what to look out for*

J *Time-lapse cameras in the crater allow geologists to make safe observations*

Activities

1 Study map **A** on page 17.

a Describe the distribution of the world's active volcanoes.

b Compare the distribution with map **C** on page 9. Provide evidence to support the statement that 'volcanoes usually occur at plate boundaries'.

c Explain the cause of either the Nyiragongo eruption or the Mt St Helens eruption.

2 Study map **B** and photo **C** on page 18.

a Draw a sketch map based on the map to show the location of the volcano, the countries and the main towns. Use an arrow to show the flow of the lava towards Goma.

b Label your sketch map to show the primary and secondary effects of the eruption at Nyiragongo.

3 Study the photos in **D**. Produce a front-page newspaper report detailing the eruption of Mt St Helens. Your article should be illustrated with a map showing the location and at least one labelled photo. The text should make clear the effect of the eruption on the volcano itself, the surrounding area, and the residents and visitors to the area.

4 Read the text.

a Summarise the differences in the immediate responses to the eruptions at Nyiragongo and Mt St Helens.

b Describe the long-term responses at Mt St Helens.

5 Study photos **E**, **F** and **G**. Using the internet, work in pairs to produce an information board entitled 'Recovery at Mt St Helens'. Show the changes that have occurred in the natural environment, which will be displayed at Johnston Ridge Observatory. Illustrate your board with photos of some of the changes. Make sure your board is clear, quick to read, interesting and able to catch people's attention.

6 Study photo **E** and use the internet. Tourism is one positive effect of volcanoes.

a Summarise the natural and human attractions of Mt St Helens in a leaflet advertising the area. Include a map showing the location of the area and describe the variety of natural and human attractions on offer. You could provide labelled photos or sketches of these.

b Research two other positive effects of volcanoes. Present your information to describe each positive effect and provide an example of where this occurs.

7 Study the text together with diagram **H**, photos **J** and **K** and graph **I**.

a Make a list of the different means of monitoring and predicting volcanoes.

b Describe in detail two methods of monitoring volcanoes. Explain the reasons for each form of monitoring and how it indicates that an eruption is likely.

c How successful do you think these methods are in predicting volcanoes?

K   *GPS use satellites to detect minute movement*

∞links

Investigate the eruption of Mt St Helens by visiting www.olywa.net/ radu/valerie/StHelens.html and the National Monument website at www.fs.fed.us/gpnf./mshnvm.

What is a supervolcano?

Imagine an eruption 1,000 times bigger than that of Mt St Helens in 1980. This was the scale of an eruption at Yellowstone, Montana in the USA 630,000 years ago. Yellowstone is a **supervolcano**. The distribution of these is shown in map **A** on page 17.

■ Characteristics of a supervolcano

Supervolcanoes are on a much bigger scale than volcanoes. They emit at least 1,000 km³ of material – compare this with an eruption on the magnitude of Mt St Helens, which emitted some 1 km³. Supervolcanoes do not look like a volcano with its characteristic cones. Instead, they are large depressions called **calderas**, often marked by a rim of higher land around the edges (diagram **A**).

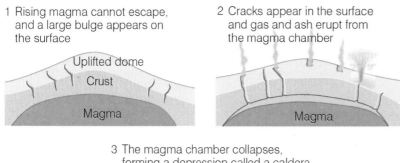

1 Rising magma cannot escape, and a large bulge appears on the surface

2 Cracks appear in the surface and gas and ash erupt from the magma chamber

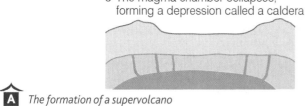

3 The magma chamber collapses, forming a depression called a caldera

A *The formation of a supervolcano*

■ The Yellowstone supervolcano

Many visitors stand in awe looking at Old Faithful and the **geothermal** features of the Norris **Geyser** basin without realising the vulnerability of where they are standing. The very forces that created such a unique area could be responsible for its destruction and threaten the existence of people – at the very least in North America, if not globally. Map **B** shows the area of Yellowstone and some of its attractions.

There is evidence that the magma beneath Yellowstone is shifting. The caldera is bulging up beneath Lake Yellowstone. There are signs of increasing activity at Norris and the ground has risen 70 cm in places. Is this just part of a natural cycle? The magma chamber beneath Yellowstone is believed to be 80 km long, 40 km wide and 8 km deep. It is not known whether the magma is on top of other materials, which would be necessary for an eruption.

Eruptions have occurred at this **hot spot** 2 million years ago, 1.3 million years ago and 630,000 years ago. An eruption today would have a catastrophic effect. It is potentially five times the minimum size for a supervolcanic eruption by the size of the magma chamber.

In this section you will learn

what a supervolcano is and how it differs from a volcano

the potential impact of a supervolcano eruption in contrast to a volcano eruption.

Key terms

Supervolcano: a mega colossal volcano that erupts at least 1,000 km³ of material.

Caldera: the depression of the supervolcano marking the collapsed magma chamber.

Fissures: extended openings along a line of weakness that allow magma to escape.

Geothermal: water that is heated beneath the ground, which comes to the surface in a variety of ways.

Geyser: a geothermal feature in which water erupts into the air under pressure.

Hot spot: a section of the earth's crust where plumes of magma rise, weakening the crust. These are away from plate boundaries.

Did you know ??????

The last supervolcano eruption occurred on Sumatra when Toba erupted 74,000 years ago. It is thought world temperatures fell by between 3 and 5°C. Toba exuded 3,000 km³ of magma and covered India in 15 cm of ash.

AQA Examiner's tip

Make sure that you can describe three differences between a supervolcano and a volcano.

B   *The location of the Yellowstone caldera*

An eruption is likely to destroy 10,000 km^2 of land, kill 87,000 people, 15 cm of ash would cover buildings within 1,000 km and 1 in 3 people affected would die. The ash would affect transport, electricity, water and farming. Lahars are a probability. The UK would await the arrival of the ash some five days later. Global climates would change, crops would fail and many people would die.

Activities

1   Study map **A** on page 17.

a   Describe the distribution of supervolcanoes.

b   How is the distribution of supervolcanoes different from that of volcanoes?

2   Study diagram **A**.

a   Draw simple diagrams of a volcano and a supervolcano.

b   Label your diagrams to show the characteristics and contrasts between them.

3   Study map **B**.

a   Measure the size of the caldera at its widest point north to south and west to east.

b   Describe the attractions in the area.

4   Produce a front-page report for a newspaper published in the UK describing the local, national, international and global impacts as 'Yellowstone supervolcano erupts'.

⊂⊃ **links**

Investigate the Yellowstone supervolcano further at **www. discovery.com** and **www.bbc.co.uk**. Enter Yellowstone supervolcano into the search box in both websites.

1.6 What are earthquakes and where do they occur?

Characteristics of earthquakes

The place where **earthquakes** begin, deep within the earth's crust, is called the **focus** (diagram **A**). Deep-focus earthquakes cause less damage and are felt less than shallow-focus ones. This is why the earthquake at Market Rasen in Lincolnshire on 27 February 2008, measuring 5.2 on the **Richter scale**, was felt so widely but not severely. The focus was 18.6 km from the surface. The point above the focus where the earthquake is most strongly felt is called the **epicentre**. Radiating out from this point are **shock waves**. Initially, there are primary waves that are relatively weak and cause the surface to move forward and backward in the direction of the wave movement. These warn of worse to come – the stronger secondary waves come next, at right angles to the outward movement of the main wave. Both these waves travel well below the surface. The final waves to arrive are those that travel near the surface and are, therefore, most powerful. These involve two types: those causing an up-and-down wave-like movement (like ripples in a pond) and those moving at right angles, which cause most damage.

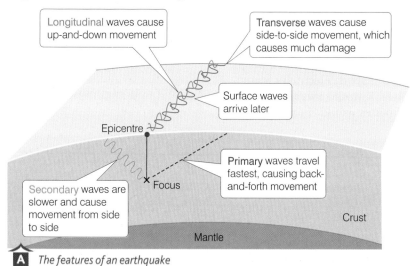

Longitudinal waves cause up-and-down movement

Transverse waves cause side-to-side movement, which causes much damage

Surface waves arrive later

Epicentre

Primary waves travel fastest, causing back-and-forth movement

Secondary waves are slower and cause movement from side to side

Focus

Crust

Mantle

A *The features of an earthquake*

Measuring earthquakes

When earthquakes occur, seismomographs record the extent of the shaking by a pen identifying the trace of the movement on a rotating drum. The line graph produced is called a seismogram (graph **B**).

The Richter scale

The strength of earthquakes is generally given according to the Richter scale. There is no upper limit to this scale. The logarithmic nature of the scale means that there is a 10-fold increase every time the scale increases by 1. So a scale 2 earthquake on the Richter scale is 10 times more powerful than a scale 1; a scale 3 earthquake is 10 times more powerful than a scale 2 and 100 times more powerful than a scale 1.

In this section you will learn

the features of an earthquake and how they are measured

where volcanoes occur and why they are found at constructive, destructive and conservative plate margins.

Key terms

Earthquake: a sudden and brief period of intense ground shaking.

Focus: the point in the earth's crust where the earthquake originates.

Richter scale: a logarithmic scale used for measuring earthquakes, based on scientific recordings of the amount of movement.

Epicentre: the point at the earth's surface directly above the focus.

Shock waves: seismic waves generated by an earthquake that pass through the earth's crust.

Mercalli scale: a means of measuring earthquakes by describing and comparing the damage done, on a scale of I to XII.

Did you know ??????

The most powerful earthquake ever recorded hit Valdivia in Chile in 1960 and measured 9.5 on the Richter scale. The two quakes that caused the most deaths occurred in China, with 830,000 people dying in Shensi in 1556 (over 8.0 on the Richter scale) and 255,000 (official figure) or 655,000 (unofficial figure) dying in Tangshan in 1976. The most powerful earthquake in the UK was 6.0 at Dogger Bank in 1931.

The Mercalli scale

The **Mercalli scale** measures the effects of earthquakes using a scale from I to XII. It uses subjective descriptions of the resulting damage (table **C**).

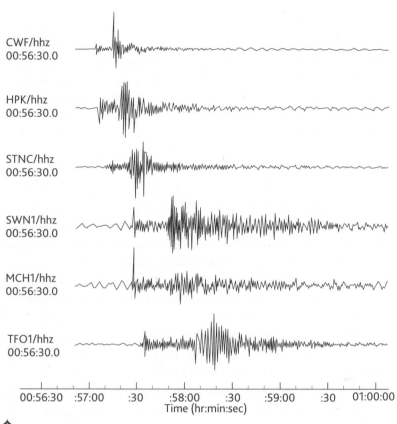

B   Seismogram of Market Rasen earthquake, 27 February 2008

◼ Where and why do earthquakes occur?

Map **D** shows the location of areas prone to earthquakes. Look back at map **C** on page 9, and you will see a close link between plate margins and where earthquakes occur. The friction and pressures that build up where the plates meet are the causes of earthquakes.

- At destructive plate margins, the pressure resulting from the sinking of the subducting plate and its subsequent melting can trigger strong earthquakes as this pressure is periodically released.

- Earthquakes that occur at constructive plate boundaries tend to be less severe than those at destructive or conservative plate margins. The friction and pressure caused by the plates moving apart is less intense than at destructive plate margins.

- Earthquakes tend to be of greater strength at conservative plate margins. Here, where the plates slide past each other, the plates tend to stick for periods of time. This causes stresses and pressure to build. The release of the pressure occurs in a sudden, quick release of the plates and the result is an earthquake.

C   *The Mercalli scale*

I	Barely felt
II	Felt by a few sensitive people; some suspended objects may swing
III	Slightly felt indoors as though a large truck were passing
IV	Felt indoors by many people; most suspended objects swing; windows and dishes rattle; standing cars rock
V	Felt by almost everyone; sleeping people are awakened; dishes and windows break
VI	Felt by everyone; some are frightened and run outside; some chimneys break; some furniture moves; slight damage
VII	Considerable damage in poorly built structures; felt by people driving; most are frightened and run outside
VIII	Slight damage to well-built structures; poorly built structures are heavily damaged; walls, chimneys and monuments fall
IX	Underground pipes break; foundations of buildings are damaged and buildings shift off foundations; considerable damage to well-built structures
X	Few structures survive; most foundations destroyed; water moved out of banks of rivers and lakes; avalanches and rockslides; railroads are bent
XI	Few structures remain standing; total panic; large cracks in the ground
XII	Total destruction; objects thrown into the air; the land appears to be liquid and is visibly rolling like waves

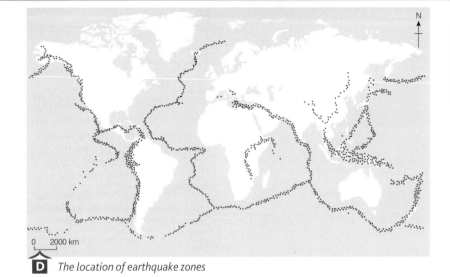

0 2000 km

D *The location of earthquake zones*

AQA **Examiner's tip**

Ensure that you can describe and explain the distribution of earthquakes – where they occur and why.

∞ **links**

There are many facts and figures available at **www.usgs.com** and **www.earthquakes.bgs.co.uk**.

Activities

1 Study diagram **A**.

a Draw simple diagrams to illustrate:

- the focus

- the epicentre

- the different types of shock wave.

b Include a definition of the terms next to your diagrams.

2 a Make a bulleted list of the main points about the Richter scale.

b Illustrate the logarithmic scale by drawing a bar graph for 1 to 4 on a simple diagram of the scale.

c How many times more powerful is an earthquake measuring 7 than an earthquake measuring 5 on the Richter scale?

d How many times more powerful is an earthquake measuring 9 than an earthquake measuring 5 on the Richter scale?

3 Describe the seismogram for the Market Rasen earthquake (graph **B**).

4 Study table **C**.

a Select two scores on the Mercalli scale that are at least four apart and draw simple diagrams to illustrate the damage.

b What are the advantages and the disadvantages of the two methods of measuring earthquakes?

5 Study map **D**, above, and map **C** on page 9.

a On an outline map of the world, shade in the areas that experience earthquakes.

b Label your map to describe the distribution of earthquakes. Use an atlas to help you with place names.

c Add the locations of the following five recent earthquakes to your map:

- Sichuan, China (2008)

- Kashmir, Pakistan (2005)

- Bam, Iran (2003)

- Gujarat, India (2001)

- Kobe, Japan (1995).

Add any additional recent earthquakes.

d Provide evidence to support the hypothesis that earthquakes mostly occur at plate boundaries.

6 Study diagrams **D**, **E** and **F** on pages 10 and 11, and the text on these pages.

a Draw simple, cross-sectional diagrams of each type of plate margin.

b Label your diagrams to explain where and why earthquakes occur at each plate margin.

c Give an example of an earthquake that has happened at each plate margin and name the plates responsible for it.

1.7 How do the effects of earthquakes differ in countries at different stages of development?

■ The Kobe earthquake, Japan

At 5.46am on 17 January 1995, the Philippines Plate shifted uneasily beneath the Eurasian Plate along the Nojima fault line that runs beneath Kobe (diagram **A**). This collision of plates led to an earthquake measuring 7.2 on the Richter scale, with tremors lasting 20 seconds.

1 Philippines Plate moves towards Eurasian Plate.
2 Philippines Plate is forced down as it is oceanic crust.
3 Plates jam together and pressure builds up.
4 Pressure is suddenly released and plate jerks forward.
5 Earthquake shockwaves travel outwards.

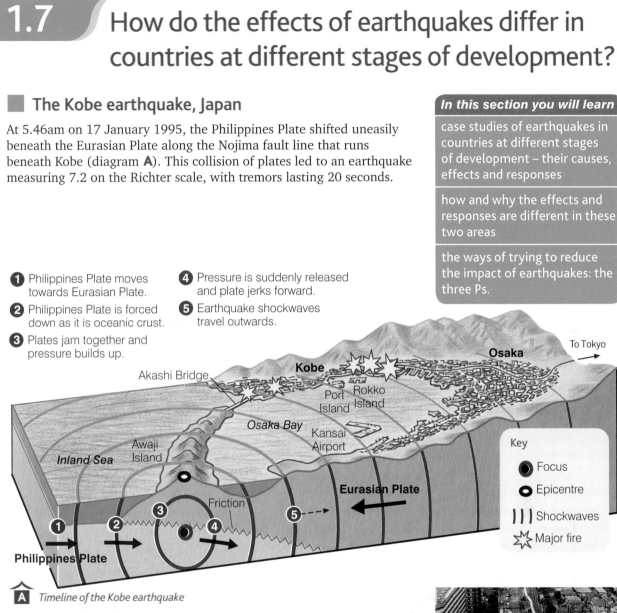

A Timeline of the Kobe earthquake

The effects of the Kobe earthquake

In this short time, the earthquake claimed the lives of 6,434 people (4,600 of them Kobe residents); seriously injured over 40,000; made 300,000 homeless; and wreaked havoc with the city's infrastructure. Gas mains ruptured, water pipes fractured, sections of elevated roads collapsed (photo **B**), railway lines buckled and only 30 per cent of the Osaka to Kobe tracks were usable. Two million homes were without electricity and one million had to cope without water for 10 days.

Fires engulfed parts of the city, especially to the west of the port, devouring the wooden structures (photo **C**). Damage to roads and water supply made attempts to extinguish them impossible. People huddled in blankets on the streets and in tented shelters in parks in fear of returning to buildings damaged by the earthquake. The damage caused was in excess of $220 billion and the economy suffered. Companies such as Panasonic had to close temporarily.

B The Great Hanshin Expressway

Responses to the Kobe earthquake

Friends and neighbours searched through the rubble for survivors, joined by the emergency services when access was possible. Hospitals struggled to cope with the injured, treating people and operating in corridors. Major retailers such as 7-Eleven helped to provide essentials and Motorola maintained telephone connections free of charge. The railways were 80 per cent operational within a month. It took longer to restore the road network – most was operational by July, although it was not until September 1996 that the Hanshin Expressway was fully open again. A year later, the port was 80 per cent operational, but much of the container shipping business had been lost.

C Fire in the area west of Kobe port

Buildings and structures that had survived the earthquake had been built to a 1981 code, whereas those that had complied with earlier 1960s practices had collapsed. This led to changes. New buildings were built further apart, to prevent the domino effect. High-rise buildings had to have flexible steel frames; others were built of concrete frames reinforced with steel instead of wood. Rubber blocks were put under bridges to absorb shocks.

The Sichuan earthquake, China

On 12 May 2008 at 2.28pm, the pressure resulting from the Indian Plate colliding with the Eurasian Plate was released along the Longmenshan fault line that runs beneath Sichuan (map **D**). This led to an earthquake measuring 7.9 on the Richter scale, with tremors lasting 120 seconds.

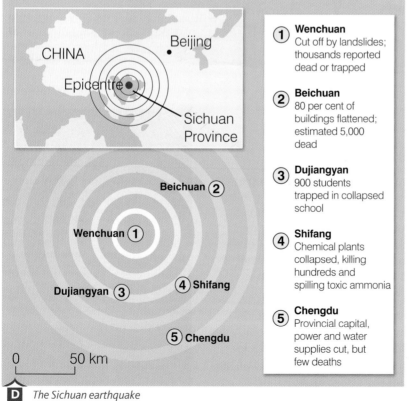

(1) Wenchuan
Cut off by landslides; thousands reported dead or trapped

(2) Beichuan
80 per cent of buildings flattened; estimated 5,000 dead

(3) Dujiangyan
900 students trapped in collapsed school

(4) Shifang
Chemical plants collapsed, killing hundreds and spilling toxic ammonia

(5) Chengdu
Provincial capital, power and water supplies cut, but few deaths

D The Sichuan earthquake

Key terms

Debt: money owed to others, to a bank or to a global organisation such as the World Bank.

The effects of the Sichuan earthquake

Initial reports placed the death toll at 8,700 but this rose to 55,000 eleven days later and 69,000 two months later, with a further 18,000 missing. Some 374,000 were injured and at least 5 million (though some said up to 11 million) people were homeless. In rural areas such as Beichuan county near the epicentre, 80 per cent of the buildings collapsed (photo **E**). These were built before building codes were introduced.

Around a total of 5 million buildings collapsed, including a number of schools; one – Juyuan middle school in Dujiangyan city – killed 900 pupils (photo **F**). There was anger among parents about the cheap building materials used, which led to their collapse.

Communications were brought to a halt – neither land nor mobile phones worked in Wenchuan. Roads were blocked by landslides and some landslides also blocked rivers, leading to fears of flooding. The cost was predicted to be $75 million.

E   *Collapsed buildings at Beichuan*

Responses to the Sichuan earthquake

There was dismay that areas in Wenchuan had not been reached within 30 hours of the quake, but access was impossible. Twenty helicopters were assigned to rescue and relief efforts immediately after the disaster. Troops began parachuting in to assess the situation, while others hiked on foot. Thousands of army troops were deployed after the earthquake. Large-scale efforts were made to free trapped survivors from collapsed buildings (photo **G**). Immediate needs were clean water, food supplies and tents to shelter people from the spring rains. Calls were made to increase the production of tents to meet the 3.3 million needed. Areas of land were flattened to allow them to be erected.

On 14 May, China requested international help. Teams from Japan, Russia and South Korea joined the rescue effort, but cash donations were the preferred option. Donations to the Red Cross exceeded £100 million in the fortnight after the earthquake. Much of this went into running the camps – ensuring that food, medicine and doctors were available, that tents had mattresses and blankets, and that there were volunteers available.

One million temporary small homes will be built to house those made homeless. The vice governor of Sichuan hoped that rebuilding would be complete in three years. The Chinese government pledged a $10 million rebuilding fund and banks wrote off **debts** owed by survivors who did not have insurance.

F   *Relatives gather at the collapsed Juyuan middle school*

G   *Rescue workers search the rubble for survivors*

Prediction, protection and preparation

The three Ps provide the key to trying to reduce the impact of earthquakes. **Prediction** involves trying to forecast when an earthquake will happen. Japan tries to monitor earth tremors with a belief that warning can be given, but this did not happen at Kobe. Foreshocks do occur, but not on a timescale useful to evacuation. Experts know *where* earthquakes are likely to happen, but struggle to establish *when*. Even looking at the time between earthquakes in a particular area does not seem to work. Similarly, experts struggle to pinpoint exactly where along a plate margin they will occur. Animal behaviour has been used in the east, but it is viewed sceptically in the USA. China evacuated the city of Haicheng (population 1 million) in 1975, partly due to the strange, unexplained behaviour of animals. Days later an earthquake struck, measuring 7.3 on the Richter scale. There were relatively few deaths, but it is estimated that 150,000 would have died without the evacuation.

Building to an appropriate standard and using designs to withstand movement is the main way of ensuring **protection** (photo **H** and diagram **I**). **Preparation** involves hospitals, emergency services and inhabitants practising for major disasters, including having drills in public buildings and a code of practice so that people know what to do to reduce the impact and increase their chance of survival.

H *The Transamerica Pyramid, San Francisco*

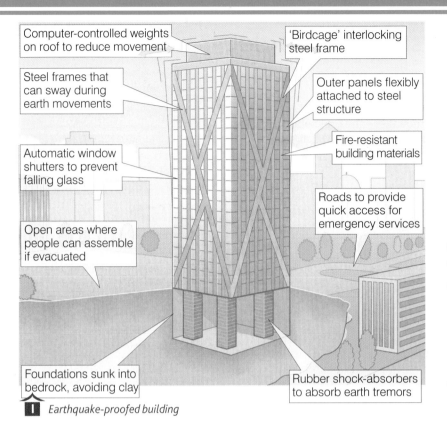

Computer-controlled weights on roof to reduce movement

'Birdcage' interlocking steel frame

Steel frames that can sway during earth movements

Outer panels flexibly attached to steel structure

Fire-resistant building materials

Automatic window shutters to prevent falling glass

Roads to provide quick access for emergency services

Open areas where people can assemble if evacuated

Foundations sunk into bedrock, avoiding clay

Rubber shock-absorbers to absorb earth tremors

I *Earthquake-proofed building*

∞ links

To research Kobe, go to **www.seismo.unr.edu**; for Sichuan go to **http://earthquake.usgs.gov**.

For information on earthquake preparation and drills, visit **www.bbc.co.uk** and **www.sfgate.com**. Enter 'earthquake drill' into the search box.

Activities

1 Study diagram **A** and photos **B** and **C**.

a Produce a fact file to summarise the main points about the Kobe earthquake such as the date, time, focus and epicentre.

b Imagine you are the science editor on a newspaper. Explain the specific causes of the Kobe earthquake.

c Imagine you are a resident of Kobe. Describe the effects of the earthquake, distinguishing between primary and secondary effects in your account. Make sure what you write relates specifically to Kobe.

d Summarise the immediate and long-term responses to the earthquake in Kobe. Do this as a piece of writing, a table or a list of bullet points.

2 Copy the table below to summarise key facts about the two earthquakes.

Feature	Kobe earthquake	Sichuan earthquake
Cause		
Primary effects		
Secondary effects		
Immediate responses		
Long-term responses		

3 Describe how and explain why the effects of an earthquake differ in countries at different stages of development.

4 For each of prediction, protection and preparation, answer the following questions.

a Explain the meaning of the term.

b Give an example of the method for reducing the impact of earthquakes from either Kobe or Sichuan.

c Give one advantage and one disadvantage of the method.

d Which of the three Ps do you think is the most useful in reducing the impact of earthquakes? Justify your choice.

5 Draw an illustrated earthquake code for a school in an earthquake-prone area.

How tsunamis form

Tsunamis are usually triggered by earthquakes. The crust shifting is the primary effect; a knock-on (secondary) effect of this is the displacement of water above the moving crust. This is the start of a tsunami.

A normal, wind-driven wave may have a length of 100 m from crest to crest, but a tsunami may be 200 km in length. The heights also greatly differ: 2 m for a normal wave versus 1 m for a tsunami out at sea. Tsunamis move at speeds of around 800 kph, rapidly approaching the coast almost unnoticed. As they near land they slow, reduce in length and gain in height (diagram **A**).

Key term

Tsunami: a special type of wave where the entire depth of the sea or ocean is set in motion by an event, often an earthquake, which displaces the water above it and creates a huge wave.

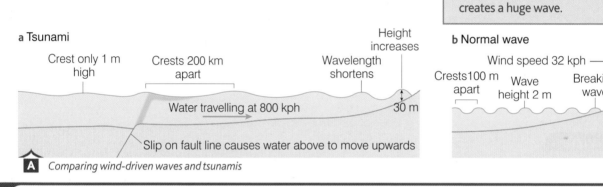

a Tsunami

Crest only 1 m high · Crests 200 km apart · Wavelength shortens · Height increases · Water travelling at 800 kph · 30 m · Slip on fault line causes water above to move upwards

b Normal wave

Wind speed 32 kph → · Crests 100 m apart · Wave height 2 m · Breaking wave

A *Comparing wind-driven waves and tsunamis*

The Indian Ocean tsunami

The Indian Ocean tsunami that occurred on 26 December 2004 was the result of the Indo-Australian Plate subducting beneath the Eurasian Plate (diagram **B**). The earthquake measured 9.1 on the Richter scale according to the US Geological Survey, but others have estimated it at 9.3.

The earthquake set in motion a sequence of events that were to devastate first the coast of Sumatra, especially the province of Banda Aceh, and then Sri Lanka and Thailand (map **C**). The highest wave to come ashore was over 25 m. Estimates suggested more than 220,000 died, 650,000 were seriously injured and up to 2 million made homeless. Public buildings including schools and hospitals were wiped out in some areas. Many people posted photos in affected areas in the vain hope that a loved one had survived. Identification of the dead on such a massive scale was a real issue. Some 1,500 settlements are believed to have been wiped out completely in Banda Aceh, where the scene was one of widespread devastation.

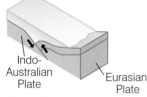

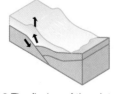

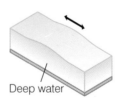

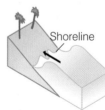

1 If one tectonic plate is dragged beneath another, stress on the boundary causes the edges of the plates to flex and deform.

Indo-Australian Plate · Eurasian Plate

2 The flexing of the plates displaces the entire column of water vertically.

3 Quickly the water column splits into two with one wave travelling out to sea and the other towards the coast.

Deep water

4 The tsunami comes ashore and can surge far inland. Often secondary waves are far more powerful than the initial one.

Shoreline

B *What triggered the Indian Ocean tsunami?*

Responses to the disaster

Rescue services and emergency teams were swamped by the scale of the disaster. Injured people were untreated for days as wounds turned gangrenous and conditions worsened. Bodies littered the street before being buried in mass graves.

There was an immediate response from the international community. Fresh water, water purification tablets, food, sheeting and tents all poured in as aid. Medical teams and forensic scientists arrived. The UK government promised £75 million and public donations of £100 million followed.

A year later £372 million had been donated by the British public, but only £128 million had been spent by the Disasters Emergency Committee (DEC). There were organisational issues following the collection of such large sums of money. However, rebuilding was progressing and DEC has spent £40 million on projects in Sri Lanka and Indonesia. There were plans to spend a further £190 million in the second year, building 20,000 houses for 100,000 homeless people. There were concerns about the number still homeless one year on.

Before the tsunami there had been no early-warning system in the Indian Ocean, so there was a real push to rectify this (there has been one in the Pacific Ocean since 1949). The Indian Ocean Tsunami Warning System was set up in June 2006. Ensuring people know how to respond and that local authorities have plans in place are essential for its success.

⛓links

There are many websites that give much information on the tsunami, so **www.google.co.uk** is worth researching.

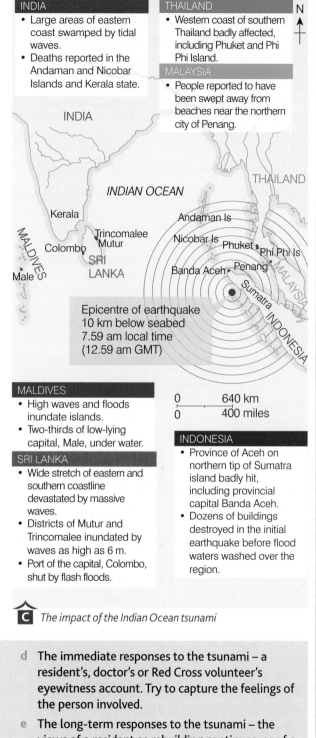

C   *The impact of the Indian Ocean tsunami*

INDIA
- Large areas of eastern coast swamped by tidal waves.
- Deaths reported in the Andaman and Nicobar Islands and Kerala state.

THAILAND
- Western coast of southern Thailand badly affected, including Phuket and Phi Phi Island.

MALAYSIA
- People reported to have been swept away from beaches near the northern city of Penang.

Epicentre of earthquake
10 km below seabed
7.59 am local time
(12.59 am GMT)

0 — 640 km
0 — 400 miles

MALDIVES
- High waves and floods inundate islands.
- Two-thirds of low-lying capital, Male, under water.

SRI LANKA
- Wide stretch of eastern and southern coastline devastated by massive waves.
- Districts of Mutur and Trincomalee inundated by waves as high as 6 m.
- Port of the capital, Colombo, shut by flash floods.

INDONESIA
- Province of Aceh on northern tip of Sumatra island badly hit, including provincial capital Banda Aceh.
- Dozens of buildings destroyed in the initial earthquake before flood waters washed over the region.

Activity

Working in small groups of three or four, produce a report or presentation on the Indian Ocean tsunami to include the following:

a   A definition of 'tsunami', supported by a labelled diagram.

b   The cause of the tsunami from a science editor's viewpoint, supported by a labelled diagram.

c   The effects of the tsunami from a resident's or tourist's point of view. Include a labelled photograph showing the impact and a map showing the areas affected.

d   The immediate responses to the tsunami – a resident's, doctor's or Red Cross volunteer's eyewitness account. Try to capture the feelings of the person involved.

e   The long-term responses to the tsunami – the views of a resident as rebuilding continues, or of a government or aid agency worker.

f   A summary of 10 important facts about the tsunami.

It must be clear that you are writing about the Indian Ocean tsunami. You should try to convey the scale of the disaster and the human suffering.

2.1 How were rocks formed?

What is the geological timescale?

Although we measure time in minutes and hours, geologists are seldom interested in anything less than hundreds of thousands of years! Scientists believe that the earth is about 4,600 million years old. For much of that time the earth was cooling and forming a basic atmosphere. Life on earth only became abundant some 542 million years ago. It is from this time that geologists have divided time into eras and periods to form the **geological timescale** (diagram **A**). The boundaries between the different periods represent critical stages in the earth's history, such as periods of mountain building or widespread sea-level change.

Era		Millions of years ago	Period	Major UK events
Cenozoic		23	Quaternary* / Neogene	Ice Age
		65	Palaeogene	Formation of Alps caused folding of rocks in UK
Mesozoic		145	Cretaceous	Much of England covered by sea
		199	Jurassic	Limestone deposited that now forms the Cotswold Hills
		251	Triassic	Much of the UK would have been desert
Palaeozoic	Late	299	Permian	
		359	Carboniferous	Tropical conditions affected the UK. Coal formed
		416	Devonian	
	Early	443	Silurian	
		488	Ordovician	Volcanoes active in Wales. Great Glen Fault formed in Scotland
		542	Cambrian	
			Pre-Cambrian	* The Quaternary period is divided into the PLEISTOCENE (ICE AGE) and the most recent period (from 10,000 years ago until the present day) called the HOLOCENE

A The geological timescale

In this section you will learn

the concept of the geological timescale

that the earth's crust is made up of three types of rock: igneous, sedimentary and metamorphic.

Key terms

Geological timescale: the period of geological time since life became abundant 542 million years ago, which geologists have divided into eras and periods.

Crust: the outer layer of the earth.

Igneous rocks: rocks formed from the cooling of molten magma.

Sedimentary rocks: most commonly, rocks formed from the accumulation of sediment on the sea floor.

Metamorphic rocks: rocks that have undergone a change in their chemistry and texture as a result of heating and/or pressure.

AQA Examiner's tip

When studying rocks and landscapes, it is important to appreciate that processes have been operating for an extremely long period of time, often measured in millions of years.

Did you know ??????

If the whole of geological time were to be represented by a single calendar year, humans started to walk on the earth at 1 minute to midnight on 31 December.

As methods of dating have improved in recent years, the boundary dates on the geological timescale have changed slightly. The geological timescale in diagram **A** was published in 2004 and is the latest version.

In this chapter we will be focusing on four rocks that are abundant in the UK. They were formed at different times on the geological timescale:

- granite – formed about 280 million years ago
- Carboniferous limestone – formed some 340 million years ago
- chalk – formed during the Cretaceous period
- clay – formed on many occasions throughout geological time but especially during the Jurassic, Cretaceous and Tertiary periods. We will look at landscapes formed on clay deposited during the Cretaceous period.

Rock types

The earth's **crust** is the thin outer layer of the earth, surrounding the mantle and the core. Look at diagram **A** on page 8 to see the structure of the earth and plate margins.

Three types of rock make up the earth's crust: **igneous**, **sedimentary** and **metamorphic**. Table **B** summarises the formation and characteristics of these three rock groups.

Activity

Study diagram **A**.

a   Use a ruler to draw the geological timescale to scale – a scale of 4 cm = 100 million years will work well. Note that the youngest periods are at the top and oldest at the bottom.

b   Write the names of the eras and periods. Write the dates alongside the boundary lines.

c   Use arrows to locate the times when granite, Carboniferous limestone, chalk and clay were formed.

d   Select a few 'Major UK events' to add to your diagram.

⊖⊖ links

The British Geological Survey has a good website at www.bgs.ac.uk/education/britstrat/home.html.

B   *Formation and characteristics of igneous, sedimentary and metamorphic rocks*

Rock type	Formation	Characteristics	Examples	Photo
Igneous	Formed by the cooling of molten magma either underground (intrusive) or on the ground (extrusive) by volcanic activity.	Igneous rocks are composed of interlocking crystals (they are said to be crystalline). They are generally tough rocks and are resistant to erosion.	Basalt, andesite and rhyolite are examples of extrusive lavas. Granite, gabbro and dolerite are intrusive rocks.	
Sedimentary	Formed by the compaction and cementation of sediments; usually deposited in the sea. Also includes organic material (e.g. coal) and rocks precipitated from solutions (e.g. limestone).	Sedimentary rocks usually form layers called beds. They often contain fossils. Although some rocks can be tough (e.g. limestone), most are weaker than igneous and metamorphic rocks.	Common sedimentary rocks include sandstone, limestone, shale, clay and mudstone. The rock chalk is a form of limestone.	
Metamorphic	Formed by the alteration of pre-existing igneous, sedimentary or metamorphic rocks by heat and/or pressure.	Metamorphic rocks are also crystalline. They often exhibit layering (not beds) called cleavage (as with the rock slate) and banding. Metamorphic rocks tend to be very tough and resistant to erosion.	Slate is one of the most common metamorphic rocks. Other examples include gneiss (pronounced 'nice') and schist.	

The rock cycle

Rocks are constantly being recycled. For example, igneous rocks are broken down by weathering and transported to the sea as sediment. On the sea bed the sediment is turned into sedimentary rock. When uplifted to form a new mountain range, the sedimentary rock is put under enormous pressure. Some of it is transformed into metamorphic rock. Some might be completely melted to form magma and, on cooling, brand-new igneous rock.

The connections between the main rock groups can be described as the **rock cycle** (diagram **A**).

Key term

Rock cycle: connections between the three rock types shown in the form of a diagram.

Did you know ??????

What is thought to be the oldest rock on earth was discovered encased in gneiss (metamorphic rock) in Canada. It is over 4,000 million years old.

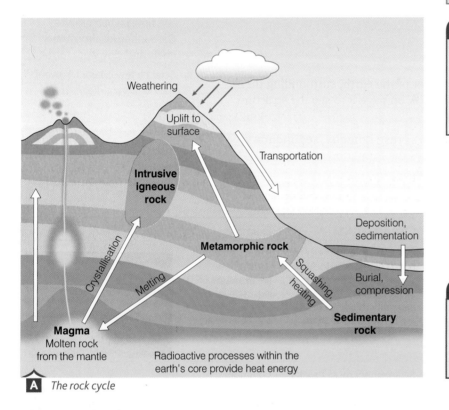

A The rock cycle

AQA **Examiner's tip**

Ensure that you learn the three definitions of rock types and practise labelling the various parts of the rock cycle.

Activity

Study diagram **A**.

a   Make a large copy of the rock cycle diagram.

b   Draw lava coming out of the volcano and add the label 'extrusive igneous rock'.

c   What is the difference between extrusive and intrusive igneous rock?

d   Where does magma come from?

e   What provides the heat to melt rocks deep within the earth?

f   How do metamorphic rocks become exposed on the ground surface?

g   In what environment do most sedimentary rocks form?

h   Describe in your own words how a metamorphic rock may become a sedimentary rock.

2.3   What is weathering and how does it operate?

What is weathering?

Weathering is the disintegration or decay of rocks in their original place at or close to the ground surface. As the name suggests, it is largely caused by elements of the weather such as rainfall and changes in temperature.

Weathering affects natural outcrops of rock as well as manmade structures such as churches, bridges and schools. Close inspection of a wall often reveals small pits, flaking of the outer surface or discolouration. This is weathering.

Types of weathering

There are three types of weathering:

1 **Mechanical weathering** – also known as physical weathering, this involves the disintegration (breakup) of rocks without any chemical changes taking place. It often results in piles of angular rock fragments called **scree** found at the foot of bare rocky outcrops.

2 **Chemical weathering** – here, a chemical change occurs when weathering takes place. Rainwater, being slightly acidic, can slowly dissolve certain rocks and minerals. Those minerals or particles unaffected by chemical weathering are usually left behind to form a fine clay deposit.

3 **Biological weathering** – this involves the actions of flora and fauna. Plant roots are effective at growing and expanding in cracks in the rocks (photo **A**). Rabbits can be effective in burrowing into weak rocks such as sands.

Mechanical weathering

Freeze–thaw weathering

Freeze–thaw weathering, or frost shattering as it is sometimes known, involves the action of water as it freezes and thaws in a crack or hole in the rock. It is a common process and operates wherever there is plenty of water and where temperatures fluctuate repeatedly above and below freezing point.

Look at diagram **B**. The process of freeze–thaw starts with liquid water collecting in cracks or holes (**pores**) in a rock. At night, this water freezes and expands by approximately 9 per cent. If the water is in a confined space, the expansion creates stresses within the rock, widening any cracks that already exist. When the temperature rises and the ice thaws, the water seeps deeper into the rock along newly formed cracks. After repeated cycles of freezing and thawing, fragments of rock may become detached and fall to the foot of the slope to collect as scree.

> **In this section you will learn**
>
> the meaning of and different types of weathering that affect rocks
>
> the main weathering processes associated with mechanical and chemical weathering.

A *Biological weathering involving the expansion of tree roots*

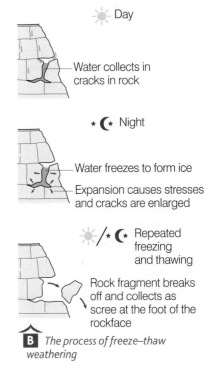

☀ Day

Water collects in cracks in rock

☽ Night

Water freezes to form ice

Expansion causes stresses and cracks are enlarged

☀/☽ Repeated freezing and thawing

Rock fragment breaks off and collects as scree at the foot of the rockface

B *The process of freeze–thaw weathering*

Exfoliation

Look at photo **C**, which shows an outcrop of rock that has been affected by **exfoliation** over time. Note how the outer 'skin' of the rock appears to be flaking away from the rest of the rock. You can see why exfoliation is sometimes known as onion-skin weathering.

C   *Exfoliation in rocks exposed in Antrim, Northern Ireland*

Exfoliation is most commonly associated with large fluctuations in temperature, which occur particularly in hot deserts. Rock is a poor conductor of heat. This means that only the outer part of a rock warms and cools in response to changes in temperature. As it warms during the day it expands and as it cools at night it contracts. Repeated cycles of expansion (heating) and contraction (cooling) can ultimately lead to the outer skin peeling away from the rest of the rock. The presence of water is important for exfoliation to take place as it weakens the rock, making it more vulnerable to flaking.

Chemical weathering

Solution

In the same way that sugar dissolves in tea, some minerals and rocks dissolve in rainwater. This dissolving process is called **solution**. Rock salt is a sedimentary rock that formed under desert conditions from the accumulation of salt crystals on a dried-up lakebed. Deposits of rock salt are found in Cheshire and it is extracted and used as a de-icer on roads and pavements in winter. Just like table salt, rock salt dissolves in water and is therefore vulnerable to the weathering process of solution.

Carbonation

Carbonation is similar to solution in that it involves dissolving. It affects rocks that are made up of calcium carbonate ($CaCO_3$) such as limestone and chalk. Look at diagram **D** to see how the process of carbonation works. Carbonation is responsible for forming some of the landforms associated with limestone landscapes.

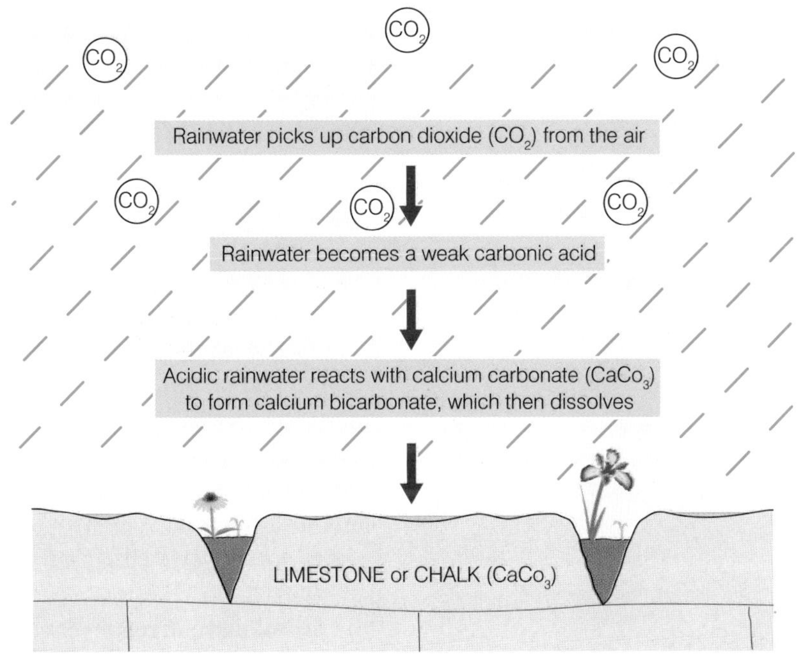

CO2
Rainwater picks up carbon dioxide (CO2) from the air
Rainwater becomes a weak carbonic acid
Acidic rainwater reacts with calcium carbonate (CaCo3) to form calcium bicarbonate, which then dissolves
LIMESTONE or CHALK (CaCo3)

D *The process of carbonation*

Did you know ??????

Rock salt put on roads as a de-icer is a powerful weathering agent and can cause considerable damage to steel structures, for example those supporting bridges. At the coast, salt from the sea causes exterior paint to flake and cars to rust.

AQA Examiner's tip

Learn the definition of weathering and of the various processes. Using diagrams to describe weathering processes is effective and makes them easier to learn.

Activities

1 Study photo **A**.
  - a What is the meaning of the term 'weathering'?
  - b How is the tree root breaking up the rock?
  - c Apart from tree roots and rabbits, can you think of any other forms of biological weathering?
  - d How does chemical weathering differ from biological weathering?

2 Study diagram **B**. Draw your own series of diagrams to describe the process of freeze–thaw.

3 Make a sketch of photo **C**. Add annotations or a written commentary to explain how the process of exfoliation operates.

4 Work in pairs to produce a PowerPoint presentation to describe one of the processes of weathering. Freeze–thaw, exfoliation and carbonation lend themselves particularly well to this. Use labelled photos and diagrams to describe your chosen process. Do not use more than six slides.

links

Further information and links can be found at www.geographypages.co.uk/weathering.htm.

What are the characteristics of granite landscapes?

The formation and distribution of granite

Granite is an intrusive igneous rock that was formed deep underground. Following uplift and the erosion of the overlying rocks, the granite has become exposed on the surface.

In south-west England the apparently isolated exposures of granite (map **A**) are actually all part of the same enormous igneous rock mass called a **batholith**. The irregular upper surface is gradually being exposed by erosion. One day, south-west England will appear to be almost completely granite.

The distribution of granite in the UK is patchy (map **A**). Apart from the exposures in south-west England, granite is only found in parts of north-west England, Scotland and Ireland.

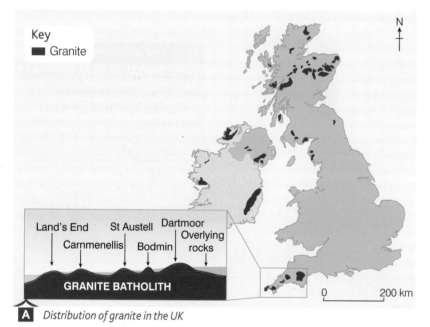

Key
■ Granite

N

Land's End   St Austell   Dartmoor
Carnmenellis   Bodmin   Overlying rocks

GRANITE BATHOLITH

0   200 km

A *Distribution of granite in the UK*

The characteristics of granite

Granite is a tough rock resistant to the processes of weathering and erosion. This explains why it forms upland areas such as Dartmoor and Bodmin in south-west England. Granite consists of three main minerals: grey-coloured quartz, black mica and pink feldspar. Feldspar is a mineral that is vulnerable to chemical weathering and will decay to form white clay called china clay or kaolin.

Granite contains cracks or **joints**. Vertical joints formed when the granite cooled and contracted. Horizontal joints in granite resulted from pressure release as the overlying rocks were removed by erosion. The rock expanded as the pressure was released, causing the joints to form roughly parallel to the ground surface. The presence of these joints makes granite vulnerable to freeze–thaw weathering and, in some places, exfoliation.

In this section you will learn

the formation and characteristics of granite

the distinctive landforms associated with granite.

Key terms

Batholith: a huge irregular-shaped mass of intrusive igneous rock that only reaches the ground surface when the overlying rocks are removed.

Joints: cracks that may run vertically or horizontally through rock.

Impermeable rock: a rock that does not allow water to pass through it.

Tor: an isolated outcrop of rock on a hilltop, typically found in granite landscapes.

Mass movement: the downhill movement of material under the influence of gravity.

Did you know ??????

Hard, durable and criss-crossed with joints, granite is a popular rock with climbers. Several popular peaks in the Mont Blanc range in France are made of granite.

AQA Examiner's tip

Tors are the main features associated with granite, so you need to be able to describe their features and suggest how they were formed. Practise using diagrams to help your explanation. Label your diagrams – this is what the examiners will award marks for.

Despite the presence of the joints, granite is an **impermeable rock** – it does not allow water to pass through it. Therefore, granite landscapes are often wet and marshy with plenty of rivers. As photo **B** shows, granite moorlands tend to be bleak, wet and windswept. As upland areas, they experience heavy rainfall and snow in winter. Low grasses with a few stunted trees mostly cover the moors, which are deeply dissected by rivers.

Granite landforms

The most distinctive granite landform is an isolated outcrop of rock called a **tor** (photo **C**). Tors are found on the tops of hills and initially look like a pile of rocks dumped on the ground. They are, however, part of the solid geology.

There are several theories about the formation of these desolate rocky outcrops. The most widely accepted theory was suggested by Linton in 1955 (diagram **D**).

B   *Granite landscape (with Hound Tor in the distance)*

Horizontal joints formed by pressure release when the overlying rocks were removed

Rounded edges caused by chemical weathering

Vertical joints formed when the granite cooled

Broken rocks at the foot of the tor resulting from freeze–thaw weathering

Enlarged joints caused by freeze–thaw weathering

C   *Haytor, one of Dartmoor's most impressive tors*

Linton suggested that the spacing of the vertical joints in granite varied across an area and that this influenced the effectiveness of weathering.

While underground and in a previously warmer and wetter climate, the closely spaced joints were weathered rapidly compared with the zone of more widely spaced joints.

As the granite became exposed on the surface during the Ice Age, erosion and **mass movement** (slumping) removed the broken-up granite, leaving behind the largely unweathered jointed granite to form a tor. Under current conditions, the tor has continued to be weathered by physical and chemical processes.

a Horizontal cracks caused by pressure release

Soil

Zone of widely spaced joints

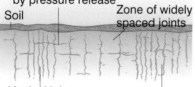

Vertical joints caused by cooling

Zone of closely spaced joints

b Deep chemical weathering under warm and wet conditions

Weathered granite

Zone of widely spaced joints is weathered less rapidly than where the joints are closely spaced

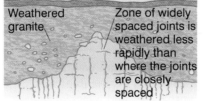

c Weathered granite removed by surface processes to leave the tor exposed

Tor

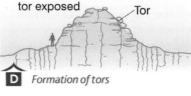

D   *Formation of tors*

Activities

1   Study map **A**.

  a   On a blank outline map of the UK, draw the areas of granite shown on the map.

  b   Using an atlas, locate and name the moors of south-west England.

  c   Add locational labels to other outcrops of granite on your map.

  d   Write a few sentences describing the distribution of granite in the UK.

2   Study photo **C**.

  a   Draw a sketch of Haytor from the photo.

  b   Use arrows to locate the features listed in the photo.

  c   Describe the main features of the tor.

  d   Is there evidence that the tor is being weathered under present-day climatic conditions?

3   Study diagram **D**. Draw a series of annotated diagrams to show the formation of tors as suggested by Linton in 1955.

⬤⬤ **links**

Further information about granite and Dartmoor can be found at **www.dartmoor-npa.gov.uk**. Enter 'geology and landforms' into the search box.

What are the characteristics of chalk and clay landscapes?

Chalk and clay are two examples of sedimentary rocks. They were both formed under the sea and then uplifted by tectonic activity to form land. Although they are similar in terms of their formation, they are rocks with very different characteristics.

Map **A** shows the distribution of chalk in the UK. It is found in the south and east of England, and tends to form bands rather than the isolated outcrops that we saw with granite (map **A**, page 40). This is because, being a sedimentary rock, chalk forms beds (layers) that cover large areas. When exposed at the surface, these beds appear as bands. Clay is such a common rock, formed throughout geological time, that it is not really feasible to plot every outcrop on a map.

In this section you will learn

the formation and characteristics of chalk and clay

the distinctive landforms associated with chalk and clay.

Key
■ Chalk

N

0   200 km

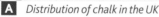
A *Distribution of chalk in the UK*

Key terms

Cliff: a steep or vertical face of rock at the coast.

Permeable rock: a rock that allows water to pass through it.

Water table: the upper surface of underground water.

Spring: water re-emerging from the rock onto the ground surface. Springs often occur as a line of springs (springline) at the base of a scarp slope.

Vale: in the landscape, a flat plain typically formed on clay.

▮ Characteristics of chalk and clay

Chalk, familiar to most people as forming the 'white **cliffs** of Dover', is physically a tough rock although it is not as strong as granite. It does form upland areas such as the Chilterns, Yorkshire Wolds and the South Downs, but this is partly because chalk is a **permeable rock**. This means that it does not readily support rivers, and a lack of rivers means a lack of erosion.

Chalk is permeable because it is heavily jointed and porous (it contains holes or pores). Rainwater soaks through the joints and pores until it reaches the **water table** (diagram **B**). This is the upper surface of underground water. Water stored within the chalk is a valuable resource, as you will discover later. Where the water table reaches the ground surface, **springs** are formed.

Did you know ? ? ? ? ? ?

Classroom chalk is not the same as rock chalk. It is made of a much softer mineral called gypsum, which is the same as the substance used to make plaster boards in buildings and the plaster used to mend broken bones.

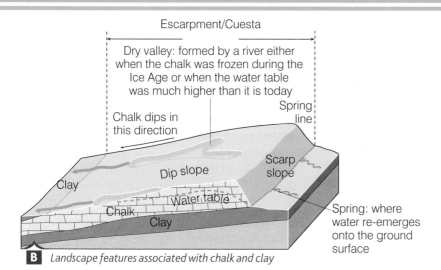

Escarpment/Cuesta

Dry valley: formed by a river either when the chalk was frozen during the Ice Age or when the water table was much higher than it is today

Chalk dips in this direction

Spring line

Dip slope

Scarp slope

Clay

Water table

Chalk

Clay

Spring: where water re-emerges onto the ground surface

B   *Landscape features associated with chalk and clay*

Chalk is a pure form of limestone and rich in calcium carbonate ($CaCO_3$). This means that it is vulnerable to the chemical weathering process of carbonation (page 39). As it contains many joints and pores, chalk is also vulnerable to freeze–thaw weathering, which has been held responsible for some notable recent cliff collapses (photo **A**, page 160).

In contrast, clay is a weak and impermeable rock. Rivers easily erode clay, which explains why it mainly forms low, flat ground called **vales**.

Chalk and clay landscapes

Look at diagram **B**, which describes the main landscape features associated with chalk and clay. The key thing to note is that the rocks are exposed at the ground at a slight angle rather than being horizontal. This is due to the rocks being folded by tectonic activity centred at the Alps in southern Europe. As you can see, this has had a profound effect on the landscape features.

> ### Remember
>
> **Permeability** is the ability of a rock to allow water to pass through it.
>
> **Permeable** means water can pass through a rock (via joints and/or pores).
>
> **Impermeable** means water cannot pass through a rock.
>
> **Porosity** is a description of the relative amount (percentage) of pore spaces (holes) in a rock.
>
> **Porous** means there is high proportion of pores.
>
> **Non-porous** means there is a low proportion of pores.
>
> **Granite** is impermeable and non-porous.
>
> **Chalk** is permeable and porous.
>
> **Clay** is impermeable and can be porous, but the pores are so tiny that it tends to become waterlogged.
>
> **Carboniferous limestone** is permeable but non-porous (water passes through joints in the rock).

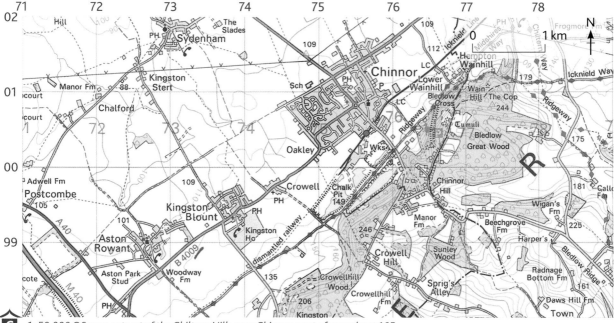

C   *1 : 50,000 OS map extract of the Chiltern Hills near Chinnor, part of map sheet 165*

1 Study map **A**.

a On a blank outline map of the UK, draw the areas of chalk shown on the map.

b Using an atlas, locate and name the upland areas formed by chalk.

c Write a few sentences describing the distribution of chalk in the UK.

d Explain why chalk tends to form bands of rock when exposed at the surface, whereas granite tends to form 'blobs'.

2 Study map **C**, which is a 1:50,000 OS map extract of part of the Chiltern Hills. The chalk **escarpment/cuesta** is the hilly area to the south-east of the map extract. The steeper scarp slope is at the north-western edge of the chalk. The clay vale is the flatter area to the north-west of the map extract.

a What is the highest point and what is its six-figure grid reference?

b What is the lowest point on the clay and what is its six-figure grid reference?

c Give an example of a springline settlement.

d At roughly what height do you think the junction between the chalk and the clay lies? Support your answer with evidence from the map.

e In which grid square is the scarp slope clearly evident? With the aid of a simple sketch map, describe how the scarp slope can be recognised on the map.

f Locate grid square 7799. In the centre of the square, a **dry valley** can be seen clearly by the shape of the contours. Draw a simple sketch (include a scale) to show the shape of these contours and label the dry valley.

g Compare the land uses between the chalk and the clay. Refer to place names and grid references in your answer.

D The Chilterns

Key terms

Escarpment/cuesta: an outcrop of chalk comprising a steep scarp slope and a more gentle dip slope.

Dry valley: a valley formed by a river during a wetter period in the past but now without a river.

links

Further information about the Chiltern Hills can be found at **www.chilternsaonb.org**.

The White Cliffs Countryside Project has information on chalk at **www.whitecliffscountryside.org.uk**. Click on the 'Chalk' link.

2.6   What are the characteristics of Carboniferous limestone landscapes?

Carboniferous limestone is a sedimentary rock so named because it was formed during the Carboniferous geological period about 340 million years ago. It was formed by the accumulation of calcium carbonate ($CaCO_3$) in warm tropical seas, rather like the present-day Caribbean. The corals and shellfish that were present in these ancient seas are often preserved in the limestone as fossils.

Map **A** shows the distribution of Carboniferous limestone. Like chalk (map **A**, page 42), limestone forms bands across the UK. One of the main exposures of Carboniferous limestone runs down the spine of England to form the Pennine Hills.

The characteristics of carboniferous limestone

Carboniferous limestone is a tough and resistant rock. It forms upland areas in the UK such as the Pennine Hills and the Mendips. When exposed at the coast, as it is in parts of south Wales, it forms dramatic towering cliffs.

Despite being physically strong, it is chemically weak. As it is composed of calcium carbonate, it is vulnerable to the slow dissolving action associated with carbonation. It is well jointed, with horizontal joints called bedding planes between the layers (beds) of limestone and regularly spaced vertical joints. These joints promote freeze–thaw weathering. They also account for limestone being a permeable rock.

Carboniferous limestone landscapes

Carboniferous limestone tends to form high ground, often with exposures of bare rock and steep-sided valleys or **gorges** (photo **B**). Weathering produces thin soils that support grass, used for grazing by sheep, and just a few isolated trees.

In this section you will learn

about the formation and characteristics of Carboniferous limestone

about the distinctive landforms associated with Carboniferous limestone.

AQA Examiner's tip

Draw a large labelled diagram to help you learn and recognise the main limestone features.

∞ links

Find out more at **www.bbc.co.uk**. Enter 'upland limestone' into the search box.

Key terms

Gorge: steep-sided deep valley that may be formed by cavern collapse.

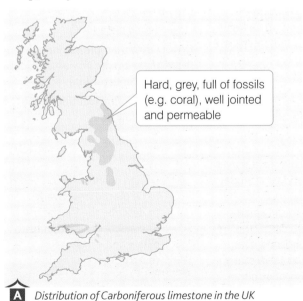

Hard, grey, full of fossils (e.g. coral), well jointed and permeable

A *Distribution of Carboniferous limestone in the UK*

B *Limestone landscape near Malham, Yorkshire*

There are several distinctive features associated with Carboniferous limestone landscapes (diagram **C**). Some features are found on the surface whereas others are underground.

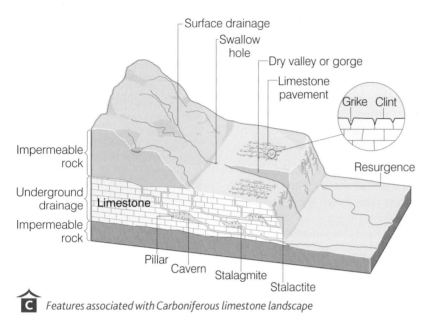

C *Features associated with Carboniferous limestone landscape*

Surface features

Bare rocky surfaces called **limestone pavements** are common features. The pattern of jointing results in a blocky appearance, with blocks called clints separated by enlarged joints called grikes. Chemical weathering causes the surface of the limestone pavement to be relatively smooth. Water flowing over adjacent impermeable rocks disappears down holes in the limestone called **swallow holes**. These are found at the intersection of the joints and are enlarged by weathering processes and river erosion.

Dry valleys are common features in limestone areas. They are formed in much the same way as the dry valleys in chalk (page 44). Most were probably formed by meltwater at the end of the last ice advance, when some of the limestone was still frozen and impermeable and water tables were much higher than today. Occasionally very narrow gorges can be found. These may result from the collapse of roofs of underground **caverns**. Springs or **resurgences** occur where water flowing underground emerges onto the ground surface, often from a small cave.

Underground features

Unlike granite, chalk or clay, limestone exhibits a number of underground features. As water flows through the joints in the limestone, weathering and erosion enlarge the joints to create tunnels and caverns. When water, rich in dissolved calcium carbonate, drips from the roofs of caverns it leaves a minute deposit of calcite as it evaporates. Over hundreds and thousands of years an icicle-like **stalactite** forms hanging down from the roof. The drips on the floor of the cavern also deposit calcite but, due to the splatter effect, the resultant **stalagmite** is shorter and stubbier. In rare cases, the two features join together to form a **pillar**. Calcite can be deposited over a wider surface where water flows over a rock face or drips occur in many places along a crack in a wall. This can result in the formation of a sheet-like **curtain** rather than an individual stalactite.

Activities

1 Study photo **B** and diagram **C**.

a Describe the landscape in the photo.

b What limestone features in diagram **C** can you identify in the photo?

c What evidence is there that weathering is active?

2 Study diagram **C**.

a Draw a simple diagram to describe the characteristics of a limestone pavement.

b Explain the location of the swallow hole.

c Describe in your own words how dry valleys were formed.

3 Design an information board to be placed in a cavern to describe the formation of stalactites, stalagmites, pillars and curtains for tourists. Use diagrams to help explain these features.

2.7   What are the uses of rocks?

Rocks have many uses to people. They can be extracted to provide raw materials for industry or used as building stone or for **cement**. They provide opportunities for water supply, either by storing water underground in cracks and pores or, if impermeable, by enabling surface reservoirs to be constructed. Soils that develop from rocks enable farming to take place. Rocks also create distinctive landscapes and scenery, providing a tremendous breadth of opportunity for leisure and recreation.

Table **A** outlines some of the main uses for the four types of rock that you have been studying in this chapter.

> **In this section you will learn**
>
> the opportunities for human use of granite, chalk, clay and Carboniferous limestone.

> **Key terms**
>
> **Cement:** mortar used in building, made from crushed limestone and shale.

A *Uses of granite, chalk, clay and Carboniferous limestone*

Rock type	Resource for extraction	Farming	Water supply	Scenery
Granite	Building stone used throughout Cornwall. Aberdeen is known as the 'city of granite'.\n\nCommonly used for kitchen surfaces.\n\nIn the past, granite contained valuable veins of tin and other metals. Kaolin (china clay) – used in industry as a whitener – originated as granite.	Mainly extensive sheep farming on poor pastures and in harsh environmental conditions.\n\nThis is because granite forms upland areas.	Impermeable rock. Several reservoirs have been constructed in steep valleys, such as the Burrator Reservoir which supplies Plymouth.	Bleak and windswept, granite forms wild and attractive moorland scenery.\n\nAttractive for outdoor activities especially walking, bird watching, mountain biking and climbing.\n\nWater sports (fishing, sailing) on the reservoirs.
Chalk	Quarried to be manufactured into cement.\n\nSource of lime for industry and farming, to neutralise acidic soils.	Reasonably fertile land used for sheep farming and some arable crops such as wheat and barley.	Important store of underground water (aquifers). Supplies large parts of the south-east of England including London.	Characterised by rolling hills. Popular with naturalists due to rich wildlife, particularly flowers and birds. Opportunities for walking and horse riding.
Clay	Used in making bricks and for pottery.	Fertile soils but with a tendency to become waterlogged. Mostly used as pasture for sheep and dairy cattle.	Impermeable rock. Some reservoirs have been constructed but flat land is not ideal.	Featureless landscape is not especially attractive.
Carboniferous limestone	Quarried to be manufactured into cement.\n\nSource of lime for industry and farming, to neutralise acidic soils.\n\nUsed as a building stone and in dry-stone walls as field boundaries.\n\nPopular stone for gardens, which has led to some destruction of limestone pavements.	Generally thin, upland soils (most of the limestone dissolves when weathered) so mostly used for sheep.	Spring water flowing out of the limestone can be a source of water.	Attractive upland scenery is popular with tourists. A number of National Parks and Areas of Outstanding Natural Beauty are on limestone areas, e.g. Peak District and Yorkshire Dales.\n\nMany opportunities for walking, mountain biking, climbing and potholing.

Farming on Dartmoor

Over the last 5,000 years farming has been the main land use on Dartmoor. Over 90 per cent of the land within the National Park today is farmed; half of this is open moorland, which is used for grazing livestock, and the rest is made up of fringe farmland and improved grassland.

The whole of Dartmoor is now designated as an Environmentally Sensitive Area and farmers enter into management agreements, for payment, to carry out agricultural practices that conserve the upland landscape and wildlife habitats. This may include reducing the numbers of livestock grazing sensitive areas, but they must also restrict the use of fertilisers and pesticides. Farmers are paid to maintain stone walls and hedgerows, to develop hay meadows and to adopt agricultural practices that help to protect the area's archaeological and historic interest.

B *A sheep farm on Dartmoor*

The London Basin chalk aquifer

The rocks underneath London form a basin called a **syncline** (diagram **C**). Water soaks into the chalk where it is exposed on either side of London and then percolates through the chalk to form a giant underground reservoir called an **aquifer**. For hundreds of years this water has supplied London with its water. The aquifer is carefully managed by the Environment Agency to ensure that its use is sustainable. In the 1960s, industrial use caused the water table to drop to 88 m below sea level, which resulted in some seawater contamination. Following careful management and reduced demand from industry since the 1990s, the water table has risen by as much as 3 m a year.

> **Key terms**
>
> **Syncline:** the lower arc of the fold in fold mountains.
>
> **Aquifer:** an underground reservoir of water stored in pores and/or joints in a rock, e.g. chalk.

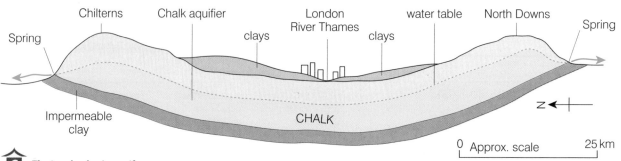

C *The London basin aquifer*

Limestone scenery: recreation in the Yorkshire Dales

The Yorkshire Dales is a National Park largely made up of Carboniferous limestone. The landscape is spectacular, with steep valleys, cliffs and extensive grassy plateaux (photo **B**, page 45).

The Yorkshire Dales offers many opportunities for leisure and recreation. The area is criss-crossed with footpaths including the long-distance footpath known as the Pennine Way. There are opportunities for nature tourism (birds and wild flowers), as well as more adventurous outdoor pursuits such as climbing (limestone is an excellent rock for climbing with its many joints), mountain biking and caving. The area around Malham (map **D**) is a particular focus for tourism. For this reason it is referred to as a honeypot site (think bees and honeypot!).

> AQA **Examiner's tip**
>
> Table **A** is probably the best way of learning this topic. You could add to it from your own research and from additional articles.

Tourism brings both benefits and costs to the Yorkshire Dales. Tourists bring money into the area, which they spend in shops, cafés and hotels. Many local people benefit from employment opportunities, working in restaurants and hotels or acting as guides. Local craft industries and farms also benefit from the visitors. However, there are costs associated with tourism. Traffic jams form in the narrow roads and cause additional pollution. People leave litter, which spoils the area and can be harmful to wildlife. Farm gates may be left open and animals worried by dogs. Shop prices may be higher than elsewhere as people cash in on the tourists, which is bad news for local residents. House prices may also be higher than elsewhere due to a demand for holiday homes. Thoughtful and sustainable management is vital if the area is to retain its special beauty in the future.

∞ links

The Yorkshire Dales National Park website is a useful source of information at **www.yorkshiredales.org.uk**.

A number of excellent farm case studies are available at the Farming and Countryside Education (FACE) site at **www.face-online.org.uk**.

Further information about Malham can be found at **www.malhamdale.com**.

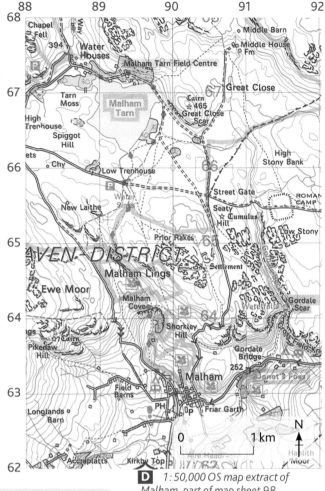

D 1 : 50,000 OS map extract of Malham, part of map sheet 98

Activities

1   Study map **D**.

a   Using grid references or place names, suggest some activities available for visitors in this area.

b   What amenities have been provided for tourists?

c   The Field Studies Council runs a popular Field Centre at Malham Tarn. In which grid square is the Field Centre located?

d   Why do you think the Field Studies Council decided to run a centre in this area?

e   Look closely at the roads. What challenges does the landscape create for travelling in the area?

f   Malham is popular with visitors and it gets extremely busy at peak times. When do you think the 'peak times' are and why?

g   Suggest some problems that might occur in the area resulting from the high number of visitors.

h   Suggest some benefits that tourism might bring to the area.

2   Use the Internet to find out more about recreational activities available in the Yorkshire Dales. A good starting point is the Yorkshire Dales National Park website at **www.yorkshiredales.org.uk**. Present your information in the form of a single-page advertisement encouraging people to visit the Yorkshire Dales. You might like to focus on just one or two activities, such as climbing and caving.

What are the issues associated with quarrying?

Limestone quarrying in the Peak District

As early as Roman times, limestone was quarried in the Peak District. It has many uses including building stone, cement, lime for use in farming, and as **aggregate** (crushed stone) for road building and the construction industry. Today there are 12 active quarries in the Peak District. The amount of limestone extracted from the area has increased from 1.5m tonnes in 1951 to 7.8m tonnes in 2001. This mainly reflects increasing demand for aggregate.

Hope Quarry, Castleton

Hope Quarry is one of the largest limestone quarries in the Peak District. It is located on the outskirts of Castleton in the Peak District National Park (map **A**). It supplies 2m tonnes of limestone a year to the nearby Hope Cement Works, which produces 1.3m tonnes of cement a year. The cement works was opened in 1929 to exploit the nearby reserves of limestone and shale, both of which are required to make cement.

The quarry and cement works, now owned by the Lafarge Group, employs 182 local people, many of whom live in nearby Hope. The local economy enjoys huge benefits, as the locally employed people support nearby shops and services. This is called the multiplier effect.

The quarry is estimated to have reserves for another 30 to 35 years and is working to a planned programme of extraction and restoration.

Quarrying and the environment

In 1951 the Peak District National Park was designated. It is the responsibility of the Park Authority to strike a balance between the conservational needs of the environment and the economic and social needs of the area.

Hope Quarry is a huge hole in the ground (photo **B**) and it has a massive visual impact on the landscape. Several initiatives have been undertaken to minimise the effects of the quarry and the cement works on the environment:

Key terms

Aggregate: crushed stone made from tough rocks such as limestone, used in the construction industry and in road building.

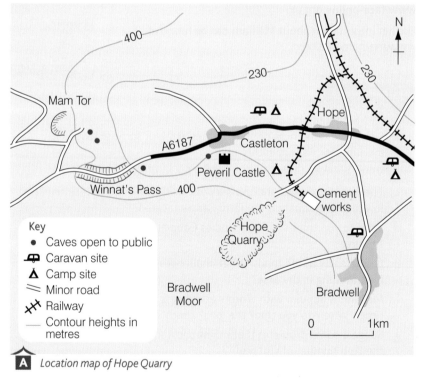

Key
- • Caves open to public
- ⌂ Caravan site
- △ Camp site
- ≈ Minor road
- ╳╳ Railway
- — Contour heights in metres

A *Location map of Hope Quarry*

Did you know ??????

In 2003 Hope Cement Works started to use chipped tyres as a fuel to preserve fossil fuels and recycle materials.

- Landscaping and tree planting have reduced the visual impact of the quarry.
- Efforts have been made to reduce dust.
- Some £15m has been spent to improve transport. Rail is used rather than road to reduce the impact of heavy traffic. A single train is equivalent to 57 lorries.
- Hope Cement Works produces 1m tonnes of carbon dioxide a year but, in 2003 in an attempt to offset some of this, 7,000 trees were planted.
- One old quarry area is now managed as a wetland reserve.

AQA **Examiner's tip**

You need to learn some facts and figures about Hope Quarry to act as a case study. Be aware of the various advantages and disadvantages of quarrying.

B Hope Quarry

C *Main uses of limestone in the Peak District*

Use	%
Aggregate	56
Cement	23
Chemicals	17
Iron and steel	4

Activities

1 Study table **C**.

a Draw a pie graph to present the information in the table.

b What is aggregate and what is it used for?

c Why has the demand for aggregate increased in recent years?

d Why is limestone extracted from Hope Quarry?

e What factors will determine how long limestone continues to be extracted from the quarry?

2 Work in pairs or small groups for this activity.

a Draw a table with two columns. At the top of one column write the heading 'Benefits of quarrying' and at the top of the other 'Problems caused by quarrying'.

b With reference to Hope Quarry and the cement works, complete the table using the text and the resources in this section. Consider the economic (money), social (people) and environmental impacts of quarrying.

c On balance, do you think quarrying should be allowed to continue at Hope Quarry or should it be stopped? Give reasons for your answer.

links

The Virtual Quarry site offers a great deal of related information at www.virtualquarry.co.uk/text/textlimestonelandscapes.htm.

The Peak District National Park has a quarrying factsheet available at www.peakdistrict.org/helpshapethefuture/quarrying.pdf.

How can quarries be restored?

Quarrying has the potential to create tremendous environmental damage. Apart from the visual impact, quarrying can cause pollution of rivers and underground aquifers and can destroy habitats when trees and vegetation are removed.

There are now strict environmental controls on quarrying, both during the operation phase and after the resource is exhausted. Quarrying companies are expected to restore or improve on the original environmental qualities of the area. This is called **quarry restoration**.

Quarry restoration can begin while work is still in progress. Parts of a quarry that have been exhausted can be restored while other parts continue to be worked. However, most of the environmental repair work takes place when quarrying operations have ceased completely.

There are many uses for exhausted quarries. They can be restored to farmland by having the topsoil replaced. They can create interesting undulating courses for motocross or mountain bikes. Waste tips can be used as dry ski slopes. Often quarries contain lakes and these are ideal for wildlife reserves, fishing or water sports.

Restoration during extraction

Drayton Sand and Gravel Quarries are located near Chichester in West Sussex. There are two quarries: Drayton North and Drayton South. Drayton North has been worked for some time and the site is now being extended to incorporate Drayton South.

In this section you will learn

the options for quarry restoration during extraction and after the resource has been exhausted.

Key terms

Quarry restoration: restoring or improving the environmental quality of a quarry, either during its operation or afterwards.

Did you know ??????

The National Watersports Centre near Nottingham is the site of an old quarry.

A *Quarry restoration at Drayton Quarries, Chichester*

Even before extraction started on the new site, restoration had begun. This involved planting hedgerows and creating an avenue of oak trees in between the two quarries. Much of the site is waterlogged and it will be worked wet to retain and deepen the existing lake.

When quarrying has been completed, an extensive lake covering some 15 ha will have been created. It will have reed beds, deep and shallow areas to provide a range of habitats and small islands. The edges of the lake will be grassland and woods (photo **A**). Nesting boxes will be sited to encourage birds into the area. The operators expect to have increased biodiversity in an area that was previously species-poor intensive farmland.

Restoration after extraction

Hollow Banks Quarry is a 20 ha quarry near Catterick, North Yorkshire where sand and gravel were extracted between 1999 and 2003. Soon after the quarry closed in 2003, the owners Tarmac Ltd began restoration (photo **B**):

- The site was contoured to create a gently undulating landscape with small ponds bordered by grass and woodland.
- After soil had been added, it was loosened and had stones removed. It was then divided into separate parcels of land, some of which were sown with grasses for pasture (farming) and others established as woodland. By planting a variety of plants and trees, a number of different habitats were created for animals and birds.
- Woodland areas have been fenced to prevent trees being damaged by browsing farm animals.
- Over 20,000 trees and shrubs raised locally were planted during 2004 and 2005.
- Aquatic plants have been planted at the margins of the ponds.
- Footpaths have been established to provide public access to the woods and ponds.

links

The British Cement Association has some good case studies at **www.cementindustry.co.uk**. Look under Sustainability, Our sustainability agenda, Quarry restoration.

Further information on Drayton Quarries is at the West Sussex County Council website at **www.westsussex.gov.uk**. Type 'Drayton quarries' into the search box.

For more information on Tarmac Ltd, go to **www.tarmac.co.uk**.

B   *Restoration of Hollow Banks Quarry, Catterick*

Activities

1   Study photo **B**.

a   What is meant by 'quarry restoration'?

b   What is the evidence that restoration has taken place here?

c   Describe the different habitats that will exist in a few years' time.

d   Do you think it is important to provide public access to this site? Explain your answer.

e   For what reasons do you think people might wish to visit this site in the future?

f   Can you suggest any other improvements that could be made?

2   Study photo **B** on page 51. Suggest how Hope Quarry could be restored over the remaining 30 to 35 years of its productive life. Consider options that will restore the environment and provide amenity land for local people and visitors. Include a plan or simple sketch to illustrate your proposed restoration.

3 Challenge of weather and climate

3.1 What is the climate of the UK?

The **climate** of the UK is largely influenced by its global position (diagram **A**). It lies in an atmospheric battleground between warm tropical air and cold polar air. As the two distinctly different types of air battle one another for control over the mid-latitudes, the UK experiences contrasting and changeable **weather** conditions as a result.

The global position of the UK also affects the relative position of the sun in the sky and the impact of the seasons. Unlike places nearer to the Equator, we experience significant differences in the relative position of the sun during the year. During the summer, the sun is high in the sky bringing long periods of potential warm sunshine. Occasionally the high temperatures can trigger **convectional rainfall**, resulting in thunderstorms. During the winter, our global position results in a much lower-angled sun. This is less powerful and accounts for much lower winter temperatures.

A number of other factors control our climate:

- Ocean currents – a warm ocean current called the **North Atlantic Drift** is a major driving force, particularly in the winter when it results in much milder conditions than would otherwise be expected for our latitude.

- **Prevailing winds** – the North Atlantic Drift coincides with the south-westerly prevailing winds in the UK. This accounts for our relatively mild but damp climate, with winds mostly approaching the UK from a warm ocean.

- **Maritime influence** – the sea has a significant effect on the climate of the UK. Being an island, the air over the UK tends to be humid. This explains the relatively high amounts of rainfall together with the predominantly cloudy weather we tend to experience.

- **Continentality** – inland areas away from the influence of the sea tend to be drier than the coast. In summer, energy from the sun (**insolation**) heats up the land more rapidly than the sea. This explains the higher summer temperatures inland. In winter, the land loses heat more rapidly and this accounts for colder winters inland.

- Altitude – upland areas tend to record higher amounts of precipitation, as air is forced to rise up and over them (diagram **B**). Temperatures decrease with altitude, which is why it is often colder in upland areas.

A Factors affecting the climate of the UK

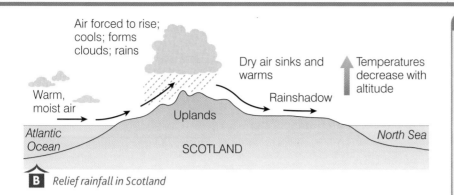

B Relief rainfall in Scotland

Characteristics

The maps in **C** show some of the characteristics of the climate of the UK. Take time to study the key for each of the maps and look closely at the overall patterns and some of the anomalies (exceptions to the general rule). Note the features listed overleaf.

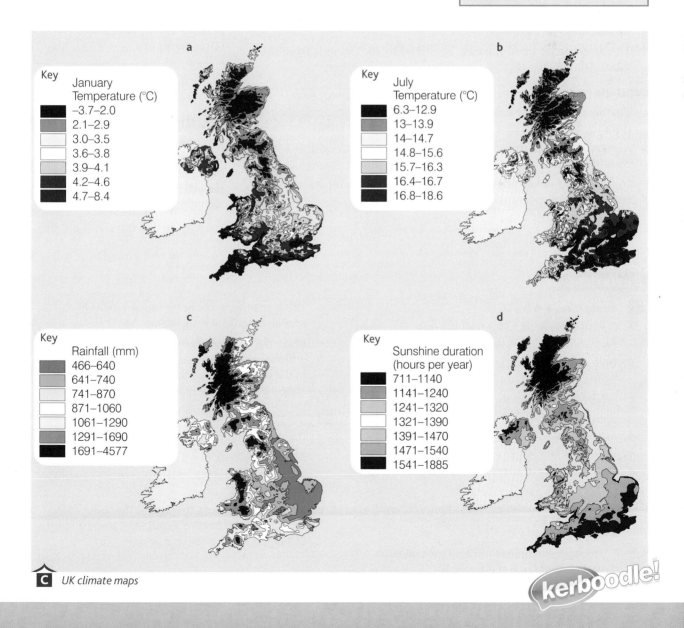

C UK climate maps

Temperature in January (map a):

- In winter, the warmest parts of the UK are to the west and the south where the North Atlantic Drift has a big influence.

- The coldest areas are in the north where the sun is at a lower angle in the sky. It is particularly cold over the mountains due to the effect of altitude.

- Cold polar air moving south is likely to have a greater influence in the north of the UK than in the south.

- The sea warms up the coastal fringe of the UK.

Temperature in July (map b):

- In summer, the warmest places are in the south. Here the sun is more powerful than in the north.

- Warm air from southern Europe often nudges into the south of the UK during the summer.

Precipitation, or rainfall (map c):

- The prevailing south-westerly winds explain the relatively high precipitation values in western areas. Note also the effect of the upland areas on rainfall.

- The driest areas are in the east, sheltered by the uplands from the moist Atlantic winds.

Sunshine (map d):

- The highest sunshine values are in the south and east, again sheltered from the relatively moist and cloudy south-westerly winds.

- Anticyclones from the south that bring stable and cloud-free conditions often affect coastal areas in the south.

- Upland areas tend to have less sunshine than lowland regions as air rising over the mountains cools to form clouds.

Did you know ???????

The highest temperature ever recorded in the UK was 38.5°C at Faversham in Kent on 10 August 2003 during the European heatwave. The lowest recorded temperature was −27.2°C, which was recorded at Altnaharra in the Scottish highlands on 30 December 1995.

∞ links

Visit the Met Office education section on UK climate at www.metoffice.gov.uk/education/secondary/students. Select the 'Climate of the UK' section.

Activities

1 Study maps **a** & **b** in **C**.

a Describe and account for the lowest temperatures recorded in the UK in January.

b Why are the highest January temperatures in Scotland recorded along the west coast?

c Where in the UK are the highest January temperatures recorded?

d Suggest reasons for the location of the highest January temperatures.

e Describe the pattern of UK temperatures in July.

2 Study map **c** in **C**.

a Compare the amounts of rainfall between western and eastern Scotland.

b Try to suggest reasons for the differences you have described (look at diagram **B**!).

c Using an atlas, account for the variations in rainfall in south-west England.

d Most of the rainfall in the UK falls in the north and the west, whereas most demand for water is in the south and the east. How does this present challenges for water management in the UK?

3 Study map **d** in **C**.

a Locate your home town or city. How many hours of sunshine do you receive on average each year?

b Why do upland areas tend to receive less sunshine than lowlands?

c Many people choose to travel to the south coast for holidays or to retire. Explain this trend and try to suggest reasons for your answer.

3.2   What causes the weather in the UK?

▉ The formation and development of depressions

A **depression** is an area of low atmospheric pressure. In a depression, air is rising and this leads to the formation of cloud and rain. In the UK most of our weather is associated with the passage of depressions which, having formed over the Atlantic, pass across the UK from west to east driven by the prevailing winds. Winds in a depression in the northern hemisphere circulate in an anticlockwise direction.

Depressions are isolated storms that form at the boundary of cold polar air moving south and warm tropical air moving north. This boundary is called the polar **front** (diagram **A**). Once formed, the depression intensifies and starts to rotate, with air spiralling upwards to form cloud and rain.

In this section you will learn

the formation and development of a depression

the sequence of weather associated with a depression

the characteristics of an anticyclone and the contrasting weather conditions between winter and summer.

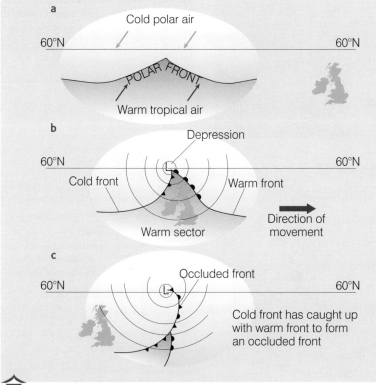

Key terms

Depression: an area of low atmospheric pressure.

Front: a boundary between warm and cold air.

Warm front: a boundary with cold air ahead of warm air.

Cold front: a boundary with warm air ahead of cold air.

Occluded front: a front formed when the cold front catches up with the warm front.

A   *The formation and development of a depression*

The boundaries between the warm and the cold air are called fronts (diagram **A**). The **warm front** marks the front of warmer air and the **cold front** represents the front of colder air. At these fronts air is forced to rise, often forming distinct bands of cloud and rain. Over time the cold front catches up with the warm front to form a single boundary called an **occluded front**. Eventually the depression fizzles out and dies.

Look at satellite photo **B**. It shows a satellite image of an intense depression over the UK in 1998. Look closely at the labels to see the main features of the depression.

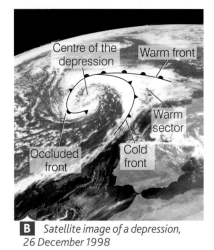

B   *Satellite image of a depression, 26 December 1998*

Weather associated with depressions

When a depression passes over the UK it brings a sequence of weather conditions. Look at satellite photo **C**. There is a depression centred off Northern Ireland to the west coast of Scotland. The warm front stretches through the North Sea and the cold front runs through the centre of England. Note on chart **D** that weather symbols have been used to describe the weather at a number of locations. The symbols are explained in the key (**E**).

Ahead of the warm front, conditions slowly become cloudier before a prolonged period of steady rain sets in (diagram **F** and table **G**). Behind the warm front the temperature increases in the **warm sector**. It remains cloudy with patchy rain and drizzle. The cold front brings a short period of heavy rain, often accompanied by strong and gusty winds. Behind the cold front the temperature drops sharply. The rain stops but heavy showers may occur. A frontal system similar to the one in diagram **D** would usually take between 12 and 18 hours to cross the UK.

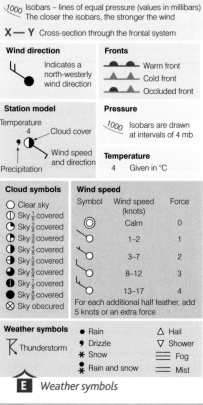

C Satellite image, 1800 hours, 4 January 2008

D Depression off Northern Ireland: 1800 hours, 4 January 2008

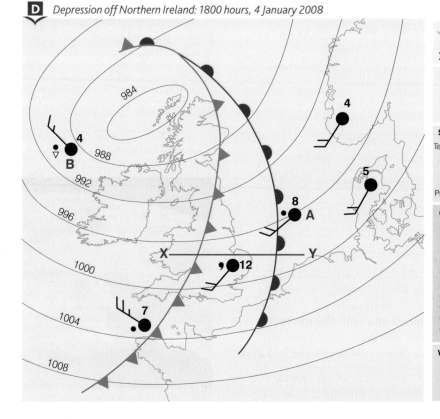

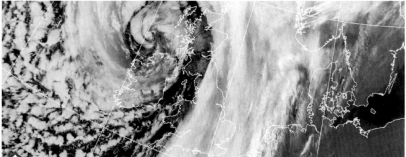

1000 Isobars – lines of equal pressure (values in millibars) The closer the isobars, the stronger the wind

X — Y Cross-section through the frontal system

Wind direction

Indicates a north-westerly wind direction

Fronts

⬤⬤⬤ Warm front
▲▲▲ Cold front
⬤▲⬤ Occluded front

Station model

Temperature
4
⬤ Cloud cover
Wind speed and direction
Precipitation

Pressure

1000 Isobars are drawn at intervals of 4 mb

Temperature

4 Given in °C

Cloud symbols

○ Clear sky
◐ Sky $\frac{1}{8}$ covered
◓ Sky $\frac{2}{8}$ covered
◔ Sky $\frac{3}{8}$ covered
◑ Sky $\frac{4}{8}$ covered
◕ Sky $\frac{5}{8}$ covered
◕ Sky $\frac{6}{8}$ covered
◕ Sky $\frac{7}{8}$ covered
● Sky $\frac{8}{8}$ covered
⊗ Sky obscured

Wind speed

Symbol	Wind speed (knots)	Force
◎	Calm	0
○	1–2	1
○	3–7	2
○	8–12	3
○	13–17	4

For each additional half feather, add 5 knots or an extra force

Weather symbols

● Rain
❟ Drizzle
✳ Snow
✳ Rain and snow
Қ Thunderstorm

△ Hail
▽ Shower
≡ Fog
≡ Mist

E Weather symbols

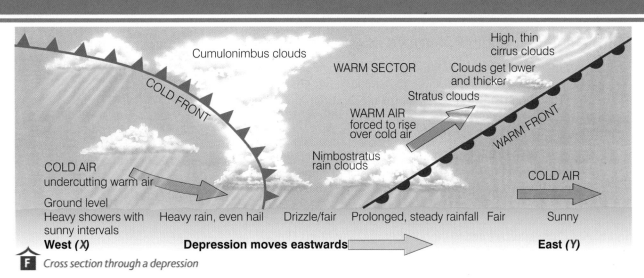

High, thin
cirrus clouds

Cumulonimbus clouds

WARM SECTOR

Clouds get lower
and thicker

COLD FRONT

Stratus clouds

WARM AIR
forced to rise
over cold air

WARM FRONT

Nimbostratus
rain clouds

COLD AIR
undercutting warm air

COLD AIR

Ground level

Heavy showers with   Heavy rain, even hail   Drizzle/fair   Prolonged, steady rainfall   Fair   Sunny
sunny intervals

West (X)   **Depression moves eastwards**   **East (Y)**

F *Cross section through a depression*

G *Weather associated with the passage of a depression*

	Cold sector	Passage of the cold front	Warm sector	Passage of the warm front	Ahead of the warm front
Pressure	Continues to rise	Starts to rise	Steadies	Continues to fall	Starts to fall steadily
Temperature	Remains cold	Sudden drop	Quite mild	Continues to rise	Quite cold, starts to rise
Cloud cover	Clouds thin with some cumulus	Clouds thicken (sometimes with large cumulonimbus)	Cloud may thin and break	Cloud base is low and thick (nimbostratus)	Cloud base drops and thickens (cirrus and altostratus)
Wind speed and direction	Winds are squally	Speeds increase, sometimes to gale force, sharp change in wind direction	Remain steady	Wind changes direction and becomes blustery with strong gusts	Speeds increase
Precipitation	Showers	Heavy rain, sometimes hail, thunder or sleet	Rain turns to drizzle or stops	Continues, and sometimes heavy rainfall	None at first, rain closer to front, sometimes snow on leading edge

Activities

1 Study diagram **D**.

a Describe the weather experienced at weather station **A**.

b Why is it raining at **A**?

c Describe the weather experienced at **B**.

d Why is it much colder at **B** than at **A**?

e What is the weather like in the warm sector?

2 Draw simple sketches to show what the following weather features would look like on a satellite image (photo **C**):

■ a front

■ the centre of a depression

■ showers.

3 Place the following statements in their correct order to describe the weather associated with the passage of a depression (diagram **F**):

■ short period of heavy rain and gusty winds

■ cloud gradually thickens and sun becomes hazy

■ cold and showery

■ warm and cloudy with light patchy rain

■ prolonged period of steady rain.

AQA **Examiner's tip**

In an exam you could be asked to label features on a weather chart or satellite image. You should learn the sequence of weather associated with the passage of a depression, including the cross section.

∞links

The Met Office has regular updated weather charts and satellite images at www.metoffice.gov.uk.

The characteristics of anticyclones

An **anticyclone** is an area of high atmospheric pressure caused by air sinking towards the ground surface. The sinking air stops pockets of air from rising. Rainfall is unlikely and for much of the time the conditions are clear and sunny. Winds in an anticyclone in the northern hemisphere circulate in a clockwise direction.

In the winter the clear, cloudless conditions result in sunny and crisp days (photo **H**). At night, heat from the ground is lost rapidly due to the absence of a blanket of cloud. This can lead to **frosty** nights. Surrounded by sea, the air over the UK is often moist. Under cold conditions this moist air readily condenses to form low cloud and **fog**. Sometimes these overcast and rather depressing conditions (known as 'anticyclonic gloom') can last for several days at a time.

In the summer, the sun is more powerful and it is able to burn off any low cloud and fog. This often leads to warm and sunny days, typical of summer (photo **I**).

Key terms

Anticyclone: an area of high atmospheric pressure.

Frost: results from the temperature of the ground or of the air dropping below 0°C.

Fog: water that has condensed close to the ground to form a dense low cloud with poor visibility.

H *Typical winter's day under the influence of an anticyclone*

I *Warm sunny day during a summer anticyclone*

Case study

Winter anticyclone, 2 February 2006

At the end of January and into the start of February 2006, a large anticyclone became established over the UK (chart **K**). It contained a great deal of moisture and, with the low winter temperatures and little wind (see how the isobars are far apart), it resulted in dull and overcast conditions with mist and fog (satellite photo **J**). If the air had been drier continental air, conditions would have been different with cold, crisp and sunny days and frosty nights.

J

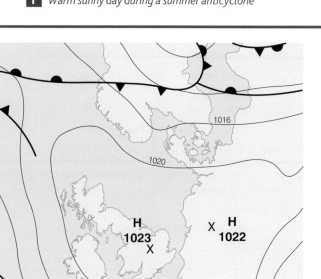

K *Synoptic chart of winter anticyclone, 2 February 2006*

Summer anticyclone, 24 July 2008

In mid-July 2008 an anticyclone settled to the east of the UK (satellite photo **L**). With air circulating around the anticyclone in a clockwise direction, this introduced warm easterly winds to much of the UK. Map **M** shows the weather experienced in the UK on 24 July 2008. Note the following points:

■ There is very little cloud in the centre of the anticyclone. This is because air is sinking, suppressing cloud formation.

■ There is more cloud at the edge of the anticyclone where conditions are slightly less settled. This accounts for the cloudier weather in south-west England and Wales.

■ The lack of cloud in the summer accounts for the high temperatures in much of the UK.

■ Although the easterly winds have brought warm and dry air to much of southern England, note that it is cooler and cloudier over parts of the north-east coast. This is because the wind has travelled over the cooler (and wetter) North Sea.

L   *Satellite image of summer anticyclone, 24 July 2008*

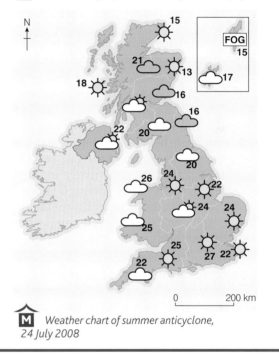

M   *Weather chart of summer anticyclone, 24 July 2008*

4   Study chart **K** and satellite photo **J**.

a   What was the pressure at the centre of the anticyclone and where in the UK was it recorded?

b   What is the evidence on the synoptic chart that explains why the winds were light across the UK?

c   Which parts of the UK were most affected by low cloud and foggy conditions?

d   Where would you have travelled to see some sunshine?

e   Describe some of the typical weather experienced by a winter anticyclone in the UK.

5   Study satellite photo **L** and map **M**.

a   Using an atlas, suggest the location of the centre of the anticyclone.

b   Why is there so little cloud in the centre of the anticyclone?

c   Describe and explain the weather experienced in south-east England.

d   Why did the coast of north-east England experience different weather from that in the south-east?

e   How does the weather around the edge of the anticyclone contrast with the weather experienced at the centre?

Is the weather in the UK becoming more extreme?

Extreme weather events

In recent years there have been a number of **extreme weather** events in the UK:

- 2000 – a series of deep Atlantic depressions brought large quantities of rain accompanied by gale-force winds that lashed parts of southern UK, causing a considerable amount of disruption to people's lives.

- 2003 – Europe suffered from an intense heatwave (photo **A**) and the UK recorded its highest ever temperature of 38.5°C in Kent.

- 2007 – several people were killed and many thousands left homeless following floods that inundated towns and river valleys across the UK. In the north, Hull and Sheffield were badly hit. The village of Toll Bar near Doncaster was particularly badly affected, with some people being unable to return to their homes for over a year. In the south of the UK, several rivers burst their banks. Flooding was particularly serious along the River Severn, affecting Tewkesbury (photo **C**, page 114) and parts of Gloucester. Vast areas of farmland were flooded, destroying crops and drowning livestock, and several major roads including the A38 were impassable.

- 2008 – during a very wet summer there were several flooding incidents caused by torrential rain landing on already saturated ground. In May parts of Somerset were flooded and in September rivers burst their banks in south Wales, Worcestershire and Northumberland, flooding hundreds of properties. The town of Morpeth in north-east England was particularly badly affected.

In this section you will learn

the range and frequency of extreme weather events in the UK

the impacts of extreme weather events on people's lives.

Key terms

Extreme weather: a weather event such as a flash flood or severe snowstorm that is significantly different from the average.

Global warming: an increase in world temperatures as a result of the increase in greenhouse gases (carbon dioxide, methane, CFCs and nitrous oxide) in the atmosphere brought about by the burning of fossil fuels, for example.

A A forest fire caused by the 2003 European heatwave

AQA **Examiner's tip**

Take time to learn a case study of an extreme weather event in some detail so that you can write at length in the exam. Understand the causes and effects of your chosen event.

■ Links to global warming

Some people, particularly in the media, have linked these recent extreme weather events (particularly the European heatwave of 2003) to **global warming**. Although extreme weather events may become more frequent as the atmosphere warms up, no individual event can be blamed on global warming. Evidence would have to be drawn from a much longer period of time – hundreds of years – before any reliable links can be made.

'You can say that due to the Earth getting warmer there will be on average more extreme events,' said Malcolm Haylock of the University of East Anglia's Climate Research Unit, UK, 'but you can't attribute any specific event to climate change.'

Source: BBC

Did you know ??????

An estimated 1,422m litres of water flowed through Boscastle in just two hours during the flood of 2004. This is equal to 21 petrol tanker loads of water every second.

The Boscastle flash flood, 2004

During the afternoon of Monday 16 August, the small picturesque Cornish village of Boscastle (map **B**) was hit by a tremendous thunderstorm. About 200 mm of rain fell in 24 hours, most of it in just five hours. The average rainfall in Boscastle for the whole of August is 75 mm.

This incredible quantity of rainfall poured down the steep-sided valley slopes and swelled the normally small River Valency flowing towards the harbour, causing it to burst its banks. The floodwaters quickly spread across the car park before surging down the narrow streets towards the harbour. Many cars were picked up and carried by the floodwaters (photo **C**). Together with tree trunks and large branches, the cars acted like battering rams smashing into bridges and buildings.

Luckily, no one was killed, but a number of people had to be airlifted to safety.

Case study

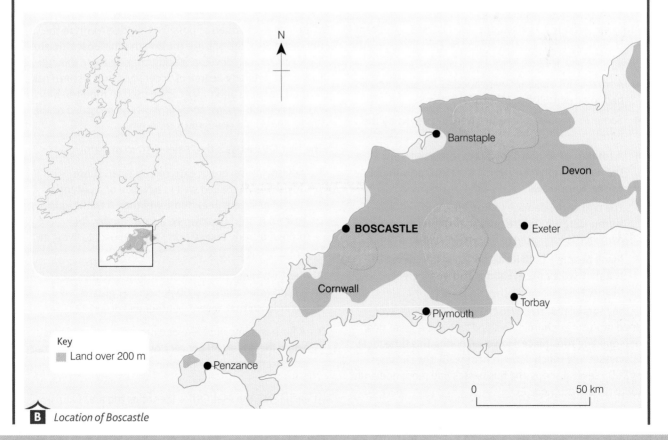

N

Barnstaple

Devon

BOSCASTLE

Exeter

Cornwall

Torbay

Plymouth

Key
■ Land over 200 m

Penzance

0   50 km

B *Location of Boscastle*

C *The flood in Boscastle*

Effects

Although nobody was killed, the flood caused significant short-term and longer-term effects on the village and the local community:

- A total of 58 buildings were flooded, some of which had to be demolished (photo **D**).
- Some 25 business properties were destroyed by the flood.
- A total of 84 wrecked cars were recovered from the harbour and a further 32 were lost, presumably washed out to sea.
- The cost of the damage to Boscastle and the surrounding area where several roads and bridges were damaged is estimated to have been at least £15m.
- Many people – both locals and tourists – were traumatised by the event and suffered increased levels of stress.

In the months that followed the event, buildings were dried out using dehumidifiers and fans and a great deal of rebuilding took place. However, during this time local businesses lost much of their income. By the summer of 2005 several shops and businesses had reopened and tourists had returned.

Planning for the future

Warnings were issued of possible thunderstorms in the Boscastle area. Storms like the one that hit Boscastle are localised and can develop quickly, making them difficult to predict with any degree of accuracy. Once a storm has formed, satellite and radar can be used to monitor and track its development, and warnings can be issued to the public and to the emergency services.

The Met Office has a three-tier system of warnings:

1   Advisory of severe or extreme weather. These advisories are issued well in advance of a potential period of severe or extreme weather. A traffic light system is used (yellow: 'be aware'; amber: 'be prepared'; red: 'take action') to warn people living in areas of the UK deemed to be at risk.

2   Early warnings of severe or extreme weather are issued when there is a 60 per cent or greater confidence that severe weather will occur.

3   A flash warning of severe or extreme weather is issued when there is an 80 per cent or greater confidence that an extreme event will occur in the next few hours.

At each level, the Met Office identifies the areas at risk, the level of risk and the likely duration of the risk.

The Environment Agency is the UK government's body with responsibility for flood management. A dedicated telephone Floodline offers warnings, help and advice to people. It is possible to use the Environment Agency's website to identify 'at-risk' areas based on individual postcodes.

The Boscastle flood event was a 1-in-400-year event. This means that it has a reocurrence rate of once every 400 years.

For an event of such rarity it is simply not cost effective to build expensive flood defences. However, some limited modification to the stream channel and the bridges has been made. People have been allowed to rebuild their homes and their lives with the knowledge that an event of similar magnitude is unlikely to occur again for hundreds of years.

This is the site of a historic building, which was completely destroyed

Youth hostel badly damaged

Fire brigade searching for possible fatalities

Gifts flushed out from nearby shops and deposited here

D　*The immediate effects of the Boscastle flood*

∞ links

There are several extreme weather events that would make excellent research projects. The Met Office has a dedicated website at www.metoffice.gov.uk/education/secondary/students. Look for relevant headings or enter 'Boscastle' or 'extreme weather' into the search box to find out some useful information.

The Environment Agency flood pages are at www.environment-agency.gov.uk/subjects/flood.

Activities

1　Read the case study about Boscastle. Use additional information from the Internet if you wish.

  a　Why was the term 'flash flood' used to describe the weather event in Boscastle?

  b　Boscastle is situated at the confluence of two rivers, both of which flow through steep valleys. Suggest how these physical factors contributed to the flood event.

  c　Despite being full of summer visitors, nobody was killed by the flood. Why do you think this was?

2　Study photo **C**. Describe what is happening in the photo.

3　Use photo **D** to help you describe some of the effects of the flood.

4　Write a few sentences outlining the long-term impacts of the flood on the people of Boscastle.

5　The Boscastle flood has a recurrence rate of 1 in 400 years. Do you think it is right that people have been allowed to rebuild their homes and re-establish their businesses or should the area have been completely cleared to prevent similar flood damage occurring in the future?

The global climate has been changing since the beginning of time and will do so in the future. There have been periods in the geological past when the poles were completely ice-free and other times when ice coverage was much more extensive than it is now. During the **Pleistocene period** (the so-called Ice Age), fluctuating temperatures resulted in cold periods called glacials and warm periods called interglacials (page 124). At the end of the last glacial period about 10,000 years ago, the climate began to warm up. We are living in this post-glacial warm period.

Throughout the last few hundred years, temperatures have continued to fluctuate (graph **A**). During the medieval period, the climate displayed a warming trend. From about 1550 to 1750, there was a period of lower temperatures that became known as the 'Little Ice Age'. These changes took place before there were cars or power stations and can be considered to be completely 'natural'.

Since about 1950, there is evidence of a very steep increase in temperatures. It is this recent dramatic increase in global temperatures that is known as **global warming**.

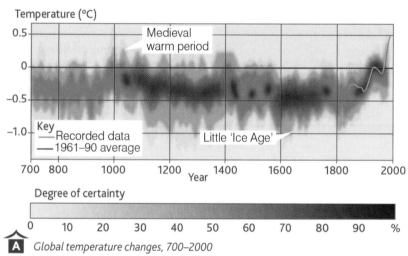

A *Global temperature changes, 700–2000*

In this section you will learn

the principles of climate change and global warming

the debate surrounding the evidence for global warming

the greenhouse effect and its connection with global warming.

Key terms

Pleistocene period: a geological time period lasting from about 2 million years ago until 10,000 years ago. Sometimes this period is referred to as the Ice Age.

Global warming: the recent trend showing an increase in global temperatures.

Assessing the evidence for global warming

Evidence for global warming comes from the direct measurement of temperatures using thermometers (mostly post-1850 as indicated on graph **A**), together with data derived from other sources such as historical records, ice cores and sea-floor sediment samples.

Instrument readings

Measurements recorded directly using thermometers suggest a clear warming trend in the last few decades (graph **B**). Average global temperatures have risen by 0.74°C during the last 100 years and by 0.5°C since 1980. Although scientists have little doubt about the overall trend, there are some factors that might be responsible for causing minor inaccuracies in the data. Site characteristics may change over time, such as the growth of vegetation. This may make comparisons of historical records unreliable.

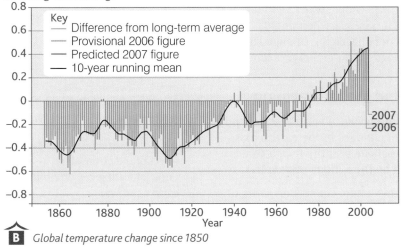

Temperature difference (°C) from long-term average

Key
— Difference from long-term average
— Provisional 2006 figure
— Predicted 2007 figure
— 10-year running mean

B Global temperature change since 1850

Glacier retreat

There is strong photographic evidence that many of the world's glaciers have been retreating over the last 50 to 100 years (photo **C**, page 127). In 2001 the World Glacier Monitoring Service reported a significant retreat of glaciers since 1980. It estimates that up to 25 per cent of global mountain glacier ice could disappear by 2050.

Retreat could, however, be caused as much by a reduction in snowfall as by an increase in melting. Some models of global warming even suggest an increase in snowfall, which would be evidenced by glaciers advancing rather than the retreating.

Arctic ice cover

Over the last 30 years, the Arctic ice has thinned to almost half its earlier thickness. There are concerns that in a few years the Arctic will be completely ice-free during the summer. As the ice continues to thin, less solar radiation will be reflected back to space. Instead, the darker sea will absorb more radiation, increasing temperatures further.

Ice cores

Some of the most compelling evidence for global warming has been the scientific study of deep ice cores extracted from Greenland and Antarctica (photo **C**). When snow falls year on year, it builds up a record going back thousands of years, just like rings in a tree. Water molecules and trapped air can be analysed to detect subtle changes in temperatures and atmospheric gas concentrations at the time when each layer of snow fell. Scientists from the Antarctic Survey in Cambridge have found clear evidence of a rapid increase in temperature in recent decades.

Early spring

In the last 30 years, there have been many signs of seasonal shifts, with spring in particular seeming to arrive earlier than in the past. Birds are nesting earlier and bulbs such as crocuses and daffodils are flowering earlier. Winters also appear to be less severe, with fewer frosts and days of snow cover.

C Ice cores reveal scientific evidence about global warming

The directly recorded data, scientific research of stored data and direct observations all point to the same conclusion – that we are currently witnessing an increase in global temperatures. Global warming does appear to be taking place. This is the conclusion overwhelmingly accepted by the UN Intergovernmental Panel on Climate Change (IPCC), now working towards international solutions to this global problem.

■ Causes of climate change and global warming

Natural climate change is thought to be the result of various shifts and cycles associated, for example, with slight changes in the earth's axis or orbit around the sun. Many scientists believe that the recent trend of global warming is to some extent caused by the actions of people.

One important natural function of the atmosphere is to retain some of the heat lost from the earth. This is known as the **greenhouse effect** (diagram **D**). Without this 'blanketing' effect it would be far too cold for life to exist on earth.

Just like a greenhouse, the atmosphere allows most of the heat from the sun (short-wave radiation) to pass straight through it to warm up the earth's surface. However, when the earth gives off heat in the form of long-wave radiation, some gases such as carbon dioxide and methane are able to absorb it. These gases are called **greenhouse gases** (table **E**). In the same way that panes of glass keep heat inside a greenhouse, the greenhouse effect keeps the earth warm.

Key terms

Greenhouse effect: the blanketing effect of the atmosphere in retaining heat given off from the earth's surface.

Greenhouse gases: gases such as carbon dioxide and methane, which are effective at absorbing heat given off from the earth.

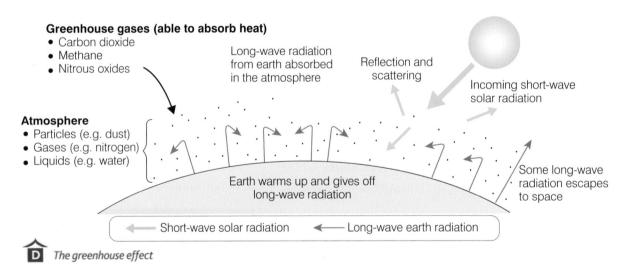

Greenhouse gases (able to absorb heat)
- Carbon dioxide
- Methane
- Nitrous oxides

Long-wave radiation from earth absorbed in the atmosphere

Reflection and scattering

Incoming short-wave solar radiation

Atmosphere
- Particles (e.g. dust)
- Gases (e.g. nitrogen)
- Liquids (e.g. water)

Earth warms up and gives off long-wave radiation

Some long-wave radiation escapes to space

⟵ Short-wave solar radiation   ⟵ Long-wave earth radiation

D *The greenhouse effect*

E *Greenhouse gases*

Greenhouse gas	Sources
Carbon dioxide: accounts for an estimated 60 per cent of the 'enhanced' greenhouse effect. Global concentration of carbon dioxide has increased by 30 per cent since 1850	Burning fossil fuels (e.g. oil, gas, coal) in industry and power stations to produce electricity, car exhausts, deforestation and burning wood
Methane: very effective in absorbing heat. Accounts for 20 per cent of the 'enhanced' greenhouse effect	Decaying organic matter in landfill sites and compost tips, rice farming, farm livestock, burning biomass for energy
Nitrous oxides: very small concentrates in the atmosphere are up to 300 times more effective in capturing heat than carbon dioxide	Car exhausts, power stations producing electricity, agricultural fertilisers, sewage treatment

Human activities and global warming

In recent years, the levels of greenhouse gases (principally carbon dioxide and methane) have increased (graph **F**) and many scientists believe that this is the result of human activities such as the burning of fossil fuels in power stations, the dumping of waste in landfill sites and burning trees during deforestation.

It is the increased effectiveness of the greenhouse effect that scientists believe is causing global warming. For the first time in history, human activities appear to be affecting the atmosphere with potentially dramatic effects on the world's climate. In 2007 the IPCC stated that global climate change is 'very likely' to have a human cause. It made a prediction that by the end of the 21st century temperatures would rise by between 1.8°C and 4°C and that this would lead to a sea-level rise of 28–43 cm.

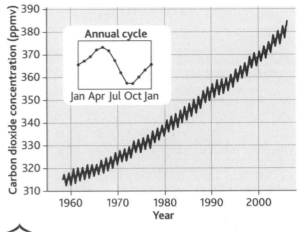

F   Increases in carbon dioxide obtained from direct readings at the Mauna Loa Observatory, Hawaii

Activities

1   Study graph **B**.

a   Describe the changes in temperature from 1850 to 2007, as shown by the 10-year running mean line.

b   With reference to graph **A**, suggest why data obtained before 1850 are of limited use in studying climate change.

c   Are data obtained directly from thermometers totally reliable?

d   How does ice-core analysis provide information about past global temperatures?

e   What additional observations support the recent trend of global warming?

f   Do you think there is sufficient evidence to support global warming? Justify your answer.

2   Study diagram **D**.

a   Make a large copy of the diagram.

b   Use table **E** to add the main sources of greenhouse gases to your diagram. Use simple sketches if you prefer.

c   Why is the greenhouse effect vital in supporting life on earth?

d   Why has the greenhouse effect become more effective in recent years?

3   Study graph **F**.

a   Describe the trend of carbon dioxide concentration in the atmosphere.

b   Suggest why carbon dioxide in the atmosphere increases in the winter but decreases in the summer. (Hint: think about plants.)

c   Does the graph support the suggestion that human activities may be contributing to global warming? Explain your answer.

AQA   *Examiner's tip*

Be prepared to debate the validity of the evidence for global warming. Connections between observations and global warming may appear obvious but they are not absolutely certain. Learn the diagram showing the greenhouse effect.

⚭ links

An excellent summary of global warming can be found on the Met Office's website at **www.metoffice. gov.uk/education/higher/climate_ change.html**.

The Met Office Hadley Centre for Climate Change is at **www. metoffice.gov.uk/research/ hadleycentre/index.html**.

The BBC's Climate Change website is at **www.bbc.co.uk/topics/ climate_change**.

What are the possible effects of global warming?

In this section you will learn
the possible impacts of global warming in the UK and the world.

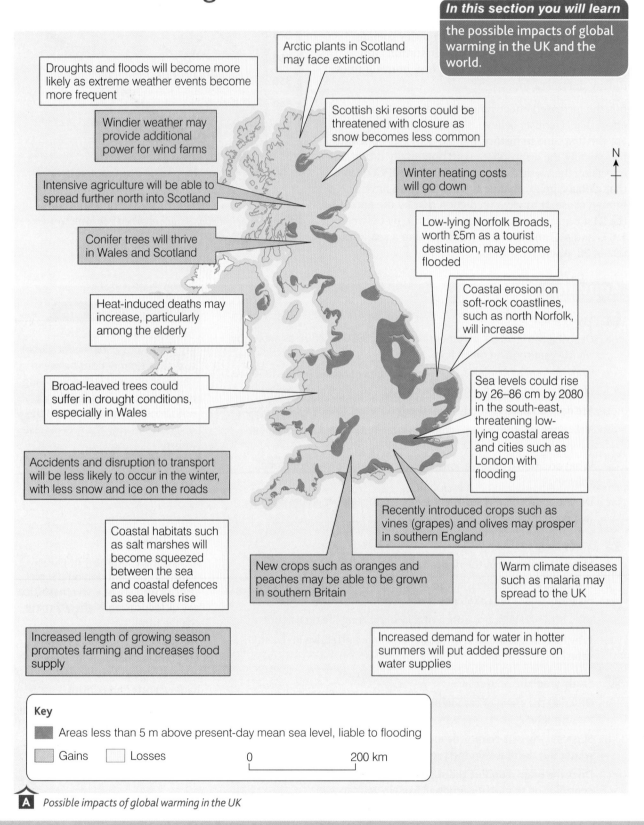

Arctic plants in Scotland may face extinction

Droughts and floods will become more likely as extreme weather events become more frequent

Scottish ski resorts could be threatened with closure as snow becomes less common

Windier weather may provide additional power for wind farms

Winter heating costs will go down

Intensive agriculture will be able to spread further north into Scotland

Low-lying Norfolk Broads, worth £5m as a tourist destination, may become flooded

Conifer trees will thrive in Wales and Scotland

Coastal erosion on soft-rock coastlines, such as north Norfolk, will increase

Heat-induced deaths may increase, particularly among the elderly

Sea levels could rise by 26–86 cm by 2080 in the south-east, threatening low-lying coastal areas and cities such as London with flooding

Broad-leaved trees could suffer in drought conditions, especially in Wales

Accidents and disruption to transport will be less likely to occur in the winter, with less snow and ice on the roads

Recently introduced crops such as vines (grapes) and olives may prosper in southern England

Coastal habitats such as salt marshes will become squeezed between the sea and coastal defences as sea levels rise

New crops such as oranges and peaches may be able to be grown in southern Britain

Warm climate diseases such as malaria may spread to the UK

Increased length of growing season promotes farming and increases food supply

Increased demand for water in hotter summers will put added pressure on water supplies

N

Key

Areas less than 5 m above present-day mean sea level, liable to flooding

Gains   Losses

0   200 km

A *Possible impacts of global warming in the UK*

Impacts of global warming in the UK

The Department for Environment, Food and Rural Affairs (Defra) estimates that annual average temperatures in the UK may rise by between 2°C and 3.5°C by 2080. This could have significant effects on the geography of the UK. Temperatures are expected to rise most in the south and east. High summer temperatures will become more common and cold winters are expected to be increasingly rare. Computer models suggest that winters will become wetter and summers drier. The amount of snow is expected to decline.

Map **A** describes some of the possible impacts of global warming on the UK.

Impacts of global warming in the world

Table **B** lists some of the advantages and disadvantages that global warming may have. It is important to remember that these impacts are highly speculative and based on computer models, which may turn out to be inaccurate. Always consider these impacts to be 'possible' rather than 'probable'. Nonetheless, take note that global warming could have very severe impacts indeed.

> **AQA** *Examiner's tip*
>
> Be prepared to discuss advantages as well as disadvantages of global warming. Remember that the impacts described are far from certain.

> **⦾ links**
>
> More about global warming can be found at www.defra.gov.uk/ environment/climatechange.

B

Advantages of global warming	Disadvantages of global warming	
▪ Frozen regions of the world such as Siberia and northern Canada may be able to grow crops in a milder climate	▪ Higher sea levels may flood low-lying areas such as Bangladesh, Myanmar and the Netherlands, threatening the lives of 80 million people	▪ Tropical storms affecting the Caribbean and the USA may increase in magnitude
▪ Canada's North-west Passage may become ice-free and can be used by shipping	▪ Islands such as the Maldives and Tuvalu may completely disappear as sea levels rise	▪ Loss of glaciers (fresh water) in the Himalayas may threaten agriculture and water supply in India, Nepal and China
▪ Energy consumption may go down as temperatures increase in densely populated parts of the world such as north-west Europe	▪ Parts of Africa may become drier and more prone to droughts, leading to starvation and civil war	▪ Hazards such as landslides, floods and avalanches may become more common in mountainous areas such as the Alps
▪ Fewer deaths or injuries due to cold weather	▪ Cereal yields are expected to decrease in Africa, the Middle East and India	▪ Arctic ice may melt completely
▪ Longer growing season in rich agricultural areas such as Europe and North America will increase food production	▪ An additional 280 million people may be at risk from malaria, particularly in China and central Asia	▪ Some species, whose habitat changes, may become extinct – there is considerable concern about polar bears in the Arctic
		▪ Alpine ski resorts may be forced to close due to lack of snow

Activity

Study table **B**.

a Present the information in the table as labels on a blank outline map of the world, in a similar style to map **A**.

b Conduct an internet search to see if you can find some more advantages of global warming. As you can see, there are not very many in table **B**.

c What do you consider to be the three disadvantages that could have the greatest impact on the geography of the world? Explain your answer.

d Which advantage do you think will bring the greatest benefits to the world? Explain why you think this.

Individual and local responses

Faced with the issue of global warming, individuals often feel helpless. How can the actions of one person possibly help to solve a global crisis? Individuals are, however, important as they set the trends for others to follow. They start campaigns and invent new technology. Individual actions include:

- Conserving energy at home by using low-energy light bulbs, switching off electrical appliances, insulating lofts and wearing an extra sweater rather than turning up the heating.
- Walking or cycling to work or school rather than using cars.
- Reducing waste by reusing materials or **recycling** (page 211).
- Buying organic food to reduce the use of chemical fertilisers.
- Paying a carbon offset when making a journey, particularly a flight. Air passengers can pay a voluntary amount to a recognised organisation to offset the carbon emitted by the plane. Most commonly this money is used to plant trees, which absorb carbon dioxide from the atmosphere.

Local authorities are active in trying to reduce carbon emissions:

- They promote public transport by using park-and-ride schemes in towns and cities.
- Grants are available for people to insulate their homes.
- People are encouraged to recycle waste materials at domestic refuse sites and there are separate waste collections for garden waste and recyclables.
- In some cities, such as London and Edinburgh, **congestion charging** has been introduced to discourage cars from entering the city centre (photo **A**). Bus lanes and car-sharing lanes encourage people to save energy by leaving the car at home.

National responses

The UK government has introduced tougher MOT tests on vehicle exhausts and it has set higher road taxes for 'gas-guzzling' vehicles. It supports transport initiatives such as bus lanes and cycleways, and it encourages recycling and waste reduction initiatives. Power stations in the UK have been fitted with filters to reduce emissions.

The government is committed to reducing carbon dioxide emissions by 60 per cent by 2050. In order to meet this aim, it has set a target of producing 10 per cent of electricity using renewable sources by 2010 (in 2005 it was 4 per cent). To meet these targets, the government is encouraging energy conservation to reduce energy use as well as seeking to develop new sources of energy such as wind and wave power.

> **In this section you will learn**
>
> the individual, local, national and international responses to global warming.

> **Key terms**
>
> **Recycling:** using materials, such as aluminium or glass, time and again.
>
> **Congestion charging:** charging vehicles to enter cities, with the aim of reducing the use of vehicles.
>
> **Kyoto Protocol:** an international agreement aimed at reducing carbon emissions from industrialised countries.
>
> **Carbon credits:** a means of trading carbon between organisations or countries in order to meet an overall target.

A   Congestion charging in London

> **Did you know** ??????
>
> In London, most vehicles are charged £8 to enter the centre of the city. All of the money raised is used to improve London's transport system.

◼ International responses

On 16 February 2005, after seven years of debate between world leaders, politicians and scientists, the **Kyoto Protocol** became international law. It states that:

▪ The 37 industrialised countries that have signed the treaty are legally bound to reduce their carbon emissions by an average of 5.2 per cent below their 1990 levels by 2012.

▪ Of the major greenhouse gas emitters, only the USA and Australia have refused to sign the treaty. Those that have signed it – including Russia – account for over 60 per cent of carbon dioxide emissions. Over 170 countries have signed the agreement.

▪ The USA has refused to sign on the grounds that the costs of reducing carbon emissions would harm its economy. It is also unhappy that the binding agreement only applies to industrialised countries.

Carbon credits can be used nationally or internationally to trade carbon between organisations or countries. Organisations that have not used up their carbon quota, which are set by national governments, can trade their credit on the open market. Organisations that have exceeded their quota may choose to buy carbon credits (photo **B**) rather than installing expensive equipment. The same principle applies to countries. Overall a balance is achieved and the target can be met.

B   *Tree planting buys carbon credits*

C   *Individual, national and international responses to global warming*

Gas	Sources	Individual responses	National responses	International responses
Carbon dioxide				
Methane				
Nitrous oxides				

AQA **Examiner's tip**

Be clear about the different levels of responses to global warming and read the exam question carefully.

Activity

For this activity you should work in pairs or small groups.

a   Make a large copy of table **C**.

b   Look back at table **E** on page 68 to remind yourself of the sources of the three main greenhouse gases. Write these in your table.

c   Discuss the individual, national and international responses that would lead to reductions in the emissions of each of the gases.

d   Complete your table with your suggestions.

∞**links**

An excellent summary of the Kyoto Protocol can be found at www.bbc.co.uk/climate/policies/kyoto.shtml.

What is the hurricane hazard?

How are hurricanes formed?

A **hurricane** or tropical cyclone is a deep area of low pressure formed in the Tropics (satellite photo **A**). In other parts of the world it is known as a cyclone (south-east Asia) or a typhoon (Japan and the Philippines). Hurricanes are immensely powerful storms capable of devastating large areas and causing considerable damage to property and loss of life.

Although scientists are not certain what actually triggers a hurricane, they tend to form:

- over warm water (over 26.5°C), which explains why they are found in the Tropics
- in summer and autumn when sea temperatures are at their highest
- at a latitude greater than 5°N or S – any closer to the Equator there is not sufficient 'spin' resulting from the rotation of the earth
- in tropical regions of severe air instability where air is converging on the surface and rising rapidly.

As air rises over a warm ocean, a huge quantity of water is evaporated quickly. As it rises it cools and condenses to form cloud. The cloud condenses and releases latent heat, which serves to power the hurricane. More and more water is drawn up from the ocean, so the clouds continue to grow and the storm intensifies. Several smaller thunderstorms may coalesce (join together) to form one large storm. The rotation of the earth sets up a spinning motion and the storm assumes its characteristic Catherine-wheel shape (satellite photo **A**). When surface winds reach an average of 120 kph (75 mph), the storm officially becomes a hurricane.

In the centre of the growing storm cooler, denser air sinks down towards the ground. This is the '**eye**' of the hurricane (diagram **C**). Here, conditions are deceptively calm and clear. On either side of the eye is a towering bank of cloud called the **eye wall**. It is here that the winds are usually at their strongest and where the heaviest rainfall occurs. Once formed, the hurricane is carried across the ocean by the prevailing winds (map **B**), generally from east to west. As it moves it continues to pick up moisture from the sea, becoming ever more powerful. On reaching land the supply of water – the hurricane's fuel – is cut off and the storm begins to weaken.

In this section you will learn

how hurricanes are formed

the effects of hurricanes, with particular reference to a rich country and a poor country

how to reduce the hurricane hazard.

Key terms

Hurricane: a powerful tropical storm with sustained winds of over 120 kph (75 mph). Also known as a tropical cyclone, a cyclone or a typhoon.

Eye: the centre of a hurricane where sinking air creates clear conditions.

Eye wall: a high bank of cloud either side of the eye where wind speeds are high and heavy rain falls.

Track: the path or course of a hurricane.

A   The distinctive shape of a hurricane in the Caribbean

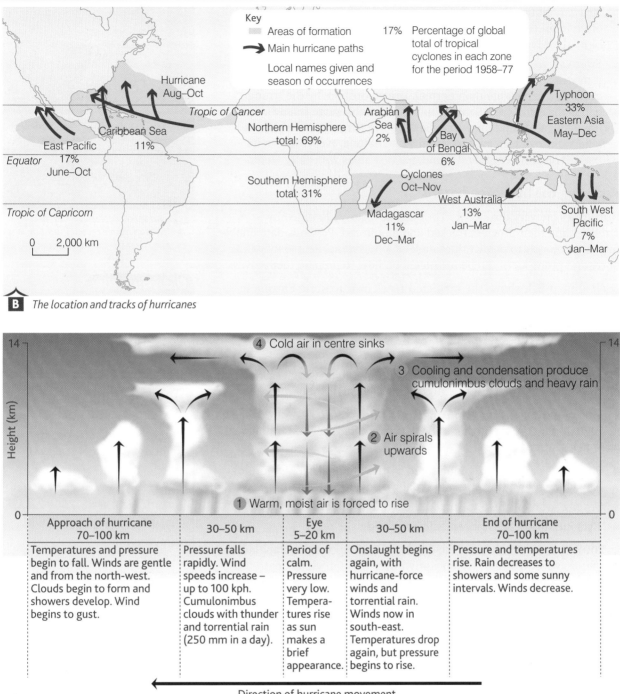

B   *The location and tracks of hurricanes*

C   *The structure of a hurricane (in the northern hemisphere)*

The effects of hurricanes

Hurricanes are capable of causing great destruction. The main causes of destruction are:

- Strong winds – with sustained wind speeds in excess of 120 kph (75 mph) and gusts in excess of 200 kph (125 mph), hurricanes can cause a great deal of damage. Roofs are blown off houses, power lines are torn down. Flimsy houses will be totally destroyed. The damage to crops could have long-lasting economic consequences.

AQA   *Examiner's tip*

Take time to compare and contrast the effects of and responses to hurricanes in rich and poor countries.

- Heavy rainfall – there is the potential for huge amounts of rain to fall, often in excess of 200 mm. This causes widespread flooding as rivers burst their banks. Landslides may be triggered in upland regions.
- Storm surge – a hurricane is an area of intense low pressure. As it moves over the sea, the level of the sea rises as it is under less atmospheric pressure than normal. Driven onshore by the strong winds these high seas, often 3 to 5 m in height, surge inland over low-lying areas and up river valleys. Storm surges are the biggest killers during hurricanes. A storm surge flattens everything in its path. It destroys crops and inundates vast areas with salty water.

Reducing the hurricane hazard

Hurricanes are very distinctive. They can be clearly identified by satellites and their progress tracked and monitored. Previous data enable computer models to predict the course of a hurricane. People living in hurricane-prone areas can be instructed on how to prepare themselves.

Look at map **E**. It shows the predicted **track** of Hurricane Katrina in August 2005. Locate the current position of the hurricane and see how a 'forecast cone' has been drawn ahead of it. Note the two levels of advisory: hurricane watch and the more serious hurricane warning.

Hurricanes tend to be more destructive in poorer parts of the world where communications are less effective. Emergency services are often unable to cope with the aftermath of a hurricane.

Following terrible disasters in Bangladesh (up to 500,000 were killed in 1970 and in 1991 a further 138,000 were killed) several initiatives have been introduced:

- Hurricanes are carefully monitored by the Bangladesh Meteorology Department and warnings are broadcast over the radio.
- Trained wardens help to spread the word in remote villages.
- Cyclone shelters have been built to provide shelter.

In 2007 Cyclone Sidr caused the deaths of 3,000 people in Bangladesh, which, however tragic, is significantly lower than previous death tolls.

D *An aerial view shows the destruction from a recent mudslide, one of many near the town of Panajachel, Guatemala, 2005*

Did you know ??????

The deadliest cyclone was the 1970 Bhola Cyclone, which killed an estimated 500,000 people in Bangladesh.

Did you know ??????

The impact and after-effects of hurricanes are different for countries at different stages of development. Two hurricanes demonstrate this:

- Hurricane Katrina, 2005 (USA)
- Cyclone Nargis, 2008 (Myanmar)

See table **F**.

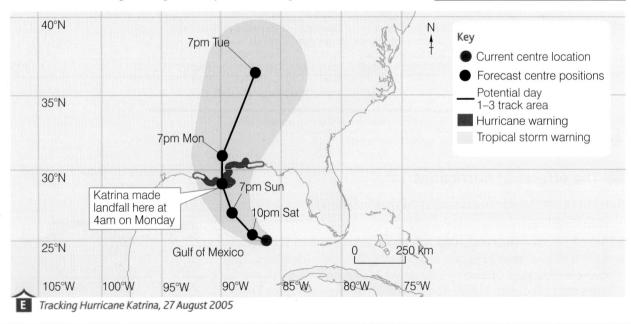

Key
- ● Current centre location
- ● Forecast centre positions
- — Potential day 1–3 track area
- ■ Hurricane warning
- ▨ Tropical storm warning

7pm Tue
7pm Mon
7pm Sun
10pm Sat
Katrina made landfall here at 4am on Monday
Gulf of Mexico
0   250 km

E *Tracking Hurricane Katrina, 27 August 2005*

F *Comparing Hurricane Katrina and Cyclone Nargis*

Impact	Hurricane Katrina, 2005 (USA)	Cyclone Nargis, 2008 (Myanmar)
Event details	• Sixth most powerful North American hurricane ever recorded. Formed over the Bahamas and then passed over Florida. It intensified in the Gulf of Mexico and made landfall close to New Orleans, Louisiana. Strongest winds recorded were 280 kph. Over 200 mm of rain fell. The system of levees (waterway embankments) failed, resulting in widespread flooding.	• Cyclone developed in the Bay of Bengal in April and made landfall in the south-west of Myanmar on 2 May 2008. • Winds of up to 215 kph battered the coastline. • The Irrawaddy delta region was particularly badly affected by a storm surge.
Deaths	• 1,836 confirmed dead.	• At least 140,000 dead but could be a great deal higher.
Homeless	• Hundreds of thousands displaced from their homes; many evacuated.	• Estimated at 2 to 3 million. • In the Irrawaddy delta region an estimated 95 per cent of all homes destroyed.
Number affected	• 3 million people left without electricity. 80 per cent of New Orleans flooded.	• UN estimate 1.5 million but probably many more.
Economic losses	• $89bn (the costliest US hurricane ever).	• $10bn.
Time for help to arrive	• A few hours.	• The government refused to admit foreign aid and relief workers for several weeks.
Services damaged	• All services (electricity, water, sanitation) severely affected in New Orleans.	• Almost total destruction of services (electricity, telephones, water).
Short-term responses	• Rescuing people from floodwaters; treating injured; providing food, water and shelter for those left in New Orleans.	• Identification and burial of the dead. • Treating the injured. • Providing safe water, food and medicines.
Long-term responses	• Massive rebuilding of New Orleans. • Some evacuees have still not returned. • Strengthening of levees.	• Rebuilding homes and workplaces. • Reclaiming farmland from salty water. • Rebuilding services and transport networks.
Environmental effects	• Floodwaters became severely polluted. • Physical changes to the Mississippi delta. • Some coastal habitats lost or damaged.	• Salt water inundation of farmland and water sources. • Burst sewage pipes caused pollution.
Warnings	• Katrina had been monitored and predictions were accurate. • 80 per cent evacuated. • Public transport closed down ahead of the storm.	• Most people had no idea that the cyclone was approaching.

Activities

1 Study map **B**.

a On a blank map of the world, make a copy of the areas of hurricane formation and the main hurricane paths.

b Use an atlas to label some of the countries most at risk from hurricanes.

c What percentage of hurricanes form in the northern hemisphere?

d Which area of the world spawns the greatest percentage of hurricanes?

e What time of year are hurricanes most likely to affect the Caribbean?

f Why do hurricanes form in the Tropics over oceans?

g Why are there no hurricanes at the Equator?

2 Conduct a research project comparing the impacts of hurricanes in rich and poor countries. You could compare Hurricane Katrina and Cyclone Nargis (table **F**) or choose other examples using the Internet. Describe the events and the impacts using maps and photos. Compare the social, economic and environmental impacts of your chosen hurricanes.

4.1 What is an ecosystem?

An **ecosystem** is a natural system that comprises plants (flora) and animals (fauna) and the natural environment in which they live. There are often complex relationships between the living and non-living components in an ecosystem. Non-living components include the climate (primarily temperature and rainfall), soil, water and light.

Ecosystems can be identified at different scales. A local ecosystem can be a pond (diagram **A**) or a hedge. Larger ecosystems can be lakes or woodlands. It is possible to identify ecosystems on a global scale, such as tropical rainforests or deciduous woodland. These global ecosystems are called **biomes**.

In this section you will learn

the concept of an ecosystem and its key components, such as producers, consumers and food chains

how change can have a considerable effect in an ecosystem.

The freshwater pond ecosystem

Case study

Freshwater ponds provide a variety of habitats for plants and animals (diagram **A**). Note that there are considerable variations in the amount of light, water and oxygen available in different parts of a pond. Animals living at the bottom in deep water need different **adaptations** to those living on the margins of the pond. Certain plants such as water lilies tolerate total immersion by sending their flowering stems to the surface of the water. Reeds and other similar plants are better adapted to being right on the edges as they can tolerate drier conditions.

There are a number of important ecological concepts that you need to understand:

- **Producers** and **consumers** – organisms can be either producers or consumers. Producers convert energy from the environment (typically sunlight) into sugars (glucose). The most obvious producers are plants, which convert energy from the sun by the process of photosynthesis. Consumers obtain their energy from the sugars made by the producers. The grasses at the margins of the pond in diagram **A** are good examples of producers. A pond snail is a good example of a consumer because it eats the plants.

- **Food chain** – this shows the links (hence the term 'chain') between producers and consumers. Diagram **B** shows a food chain that might exist in a typical pond. Note that it is a simple linear series of connections.

- **Food web** – this shows the connections between producers and consumers in a rather more detailed way, hence the term 'web' rather than 'chain' (diagram **C**).

- **Scavengers** and **decomposers** – when living elements (plants and animals) of an ecosystem die, scavengers and decomposers break them down and effectively recycle their nutrients. Scavengers eat dead animals and plants. A rat-tailed maggot is a good example of a freshwater pond scavenger. Flies and earthworms are examples of scavengers found on land. Decomposers are usually bacteria and fungi. They break down the remaining plant and animal material, often returning the nutrients to the soil.

Key terms

Ecosystem: the living and non-living components of an environment and the interrelationships that exist between them.

Biomes: global-scale ecosystems.

Adaptations: the ways that plants evolve to cope with certain environmental conditions such as excessive rainfall.

Producers: organisms that obtain their energy from a primary source such as the sun.

Consumer: organisms that obtain their energy by eating other organisms.

Food chain: a line of linkages between producers and consumers.

Food web: a diagram that shows all the linkages between producers and consumers in an ecosystem.

Scavengers: organisms that consume dead animals or plants.

Decomposers: organisms such as bacteria that break down plant and animal material.

Nutrient cycling: the recycling of nutrients between living organisms and the environment.

- **Nutrient cycle** – nutrients are foods that are used by plants or animals to grow, such as nitrogen, potash and potassium. There are two main sources of nutrients: rainwater washes chemicals out of the atmosphere and weathered rock releases nutrients into the soil. When plants or animals die, the scavengers and decomposers recycle the nutrients, making them available once again for the growth of plants or animals.

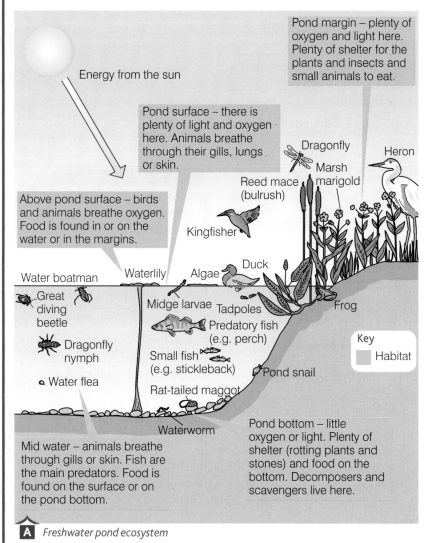

Energy from the sun

Pond margin – plenty of oxygen and light here. Plenty of shelter for the plants and insects and small animals to eat.

Pond surface – there is plenty of light and oxygen here. Animals breathe through their gills, lungs or skin.

Above pond surface – birds and animals breathe oxygen. Food is found in or on the water or in the margins.

Dragonfly

Heron

Marsh marigold

Reed mace (bulrush)

Kingfisher

Duck

Water boatman   Waterlily   Algae

Great diving beetle

Midge larvae   Tadpoles

Dragonfly nymph

Predatory fish (e.g. perch)

Small fish (e.g. stickleback)

Water flea

Pond snail

Rat-tailed maggot

Frog

Key

Habitat

Waterworm

Mid water – animals breathe through gills or skin. Fish are the main predators. Food is found on the surface or on the pond bottom.

Pond bottom – little oxygen or light. Plenty of shelter (rotting plants and stones) and food on the bottom. Decomposers and scavengers live here.

A   *Freshwater pond ecosystem*

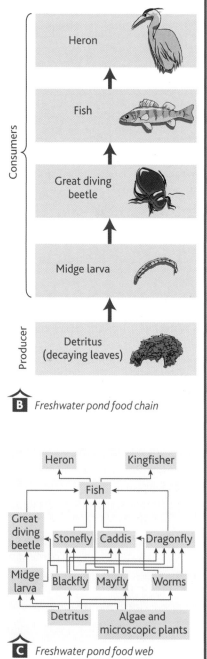

Consumers

Heron

Fish

Great diving beetle

Midge larva

Producer

Detritus (decaying leaves)

B   *Freshwater pond food chain*

The impact of change on the freshwater pond ecosystem

The diversity and relative numbers of the components in an ecosystem can change over time. This can be caused by natural factors such as environmental change (e.g. flood, fire, drought) or human-induced change (e.g. drainage, reclamation, fish stocking). Once a change has occurred it is rarely isolated and often has an impact on other parts of the ecosystem.

If predatory fish are introduced into the pond in diagram **A** they will eat more of the smaller fish and small animals such as frogs. This will affect the numbers of those creatures, which will in turn reduce the amount of food available to creatures further up the food chain. At the same time, with fewer frogs in the pond, numbers of those creatures below frogs in the food chain such as slugs will increase.

Heron   Kingfisher

Fish

Great diving beetle   Stonefly   Caddis   Dragonfly

Midge larva   Blackfly   Mayfly   Worms

Detritus   Algae and microscopic plants

C   *Freshwater pond food web*

kerboodle!

D *Freshwater pond species and energy sources*

Species	Energy source (sunlight or food)
Algae	Sunlight
Dragonfly	Other adult insects
Dragonfly nymph	Tadpoles, young fish, water fleas, beetles
Duck	Water plants, insects, tadpoles, small fish, pond snails
Frog	Insects, water worms, snails
Great diving beetle	Water fleas, midge larvae, pond snails, nymphs, tadpoles, water boatmen
Heron	Fish, frogs and tadpoles, larger insects
Kingfisher	Small fish, tadpoles, small frogs, great diving beetle
Marsh marigold	Sunlight
Midge larvae	Microscopic plants, small particles of dead plants
Perch	Small fish, beetles, water fleas
Pond snail	Large water plants, algae
Rat-tailed maggot	Decaying plants
Reed mace	Sunlight
Sticklebacks	Tadpoles, young fish, water fleas, beetles
Tadpole	Microscopic plants, algae, midge larvae
Water boatmen	Tadpoles, water worms, midge larvae, water fleas
Water flea	Microscopic plants, small particles of dead plants
Water lily	Sunshine
Water worm	Small particles of dead animals

AQA Examiner's tip

It is important that you learn the terminology of ecosystems. Understand the principles of change in an ecosystem with reference to the pond ecosystem. Use the pond ecosystem to give you an example of a food chain.

∞ links

Excellent ecosystems links can be found at **www.geography. btinternet.co.uk/ecosystems.htm**.

Activities

1 Study diagram **A** and table **D**.

a Identify some producers that live in a freshwater pond.

b From where do these producers obtain their energy?

c Bacteria and fungi are decomposers in a pond ecosystem. Can you name some scavengers?

d Which pond species are at the top level in the food chain, i.e. they are not eaten by any other species in the ecosystem?

e Select one of the species listed in **D** and draw a food chain diagram with your chosen species at the top. Include sunlight in your diagram and add sketches to make it look more interesting.

2 Study diagram **A** and table **D**.

a In pairs, draw a food web for the species shown in diagram **A** and listed in table **D**.

b Identify the producers in your food web.

c Add some simple sketches or photos using the Internet if you wish.

3 Study diagram **A** and table **D**.

a Imagine that the landowner cuts down all the vegetation at the side of the pond to create a wooden deck for fishing. How would this affect the ecosystem in the short term and the long term?

b Imagine that disease wipes out all the frogs. How would this affect the ecosystem in the short term and the long term?

c Suggest another change that could happen to this ecosystem and describe the effects that this change might have on the species living in the pond.

4.2 What are the characteristics of global ecosystems?

The distribution of global ecosystems

Global ecosystems are known as biomes. The dominant type of vegetation cover usually defines a biome. Map **A** shows the global distribution of the major world biomes.

The biome in the UK is **temperate deciduous forest**. This is the natural vegetation that would occur in much of the UK in response to climates and soils. It does not mean that the entire country is covered by woodland. However, if no land management took place at all in the UK for 100 years or so, then the landscape would start to revert back to natural deciduous woodland.

> **In this section you will learn**
>
> the distribution of temperate deciduous forests, tropical rainforests and hot deserts
>
> the characteristics and adaptations of vegetation in these three biomes to climate and soils.

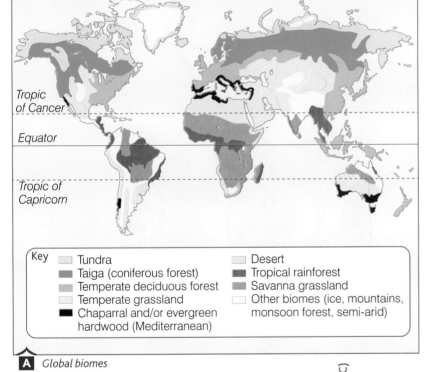

Key
- Tundra
- Taiga (coniferous forest)
- Temperate deciduous forest
- Temperate grassland
- Chaparral and/or evergreen hardwood (Mediterranean)
- Desert
- Tropical rainforest
- Savanna grassland
- Other biomes (ice, mountains, monsoon forest, semi-arid)

A Global biomes

> **Key term**
>
> **Temperate deciduous forest:** forests comprising broad-leaved trees such as oak that drop their leaves in the autumn.

Temperate deciduous forests

Temperate deciduous forests are found across much of north-west Europe, eastern North America and parts of East Asia. They occur in these regions because they are well suited to the moderate climate (graph **B**). Rainfall is distributed evenly throughout the year, summers are warm but not too dry, winters are cool but not too cold. There is a long growing season lasting up to seven months.

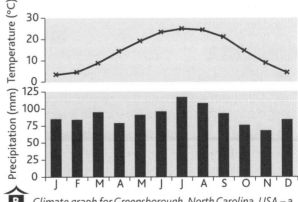

B Climate graph for Greensborough, North Carolina, USA – a temperate deciduous forest biome

Soils that develop under these climatic conditions tend to be rich and fertile. Weathering is active, providing plenty of nutrients, and the annual leaf fall provides organic matter to enrich the soil further. The common soil found in this biome is a brown soil.

Look at photo **C**, which shows oak woodland in southern England. This is typical temperate deciduous woodland. The main feature of the trees in this biome (e.g. oak, beech, birch, ash) is that they shed their leaves in autumn. This is what the term 'deciduous' means. They drop their leaves in response to reductions in light and heat, which enables them to conserve water. Deciduous trees are typically broad-leaved, which means there is a great deal of potential for water loss through the holes (stomata) on the underside of their leaves. Leaf fall comes early in some years if there has been a shortage of water in late spring and summer.

C A typical oak woodland in southern England

Deciduous woodlands are rich in their diversity of vegetation and they provide a great range of habitats for the many plants and animals that live there (diagram **D**). One typical characteristic of deciduous woodland is the layering or **stratification** of the vegetation:

- The top of the fully grown trees provides a canopy, which acts like an umbrella. The main trees forming this layer are oak and ash.
- Beneath this is a sub-canopy of saplings and smaller trees, such as hazel.
- Below this is a herb layer of brambles, bracken, bluebells, wild garlic and ivy.
- Finally there is a ground layer close to the soil surface. Here, it is damp and dark – ideal conditions for moss to grow.

For much of the year it is quite dark in deciduous woodland, which is not ideal for flowering plants. This helps to explain why bluebells, for example, commonly flower in the early spring before the canopy has fully developed.

Rainfall – carries nutrients

Sunlight

Height (m)
40 — Dominant oak Ash sapling in gap   Dominant oak
                        Bramble   Hazel
30
                                    Hazel
20
Hazel
10   Ivy
0

Height (m)
-40
-30
-20
-10

0.1 1.0 10 100
Percentage of full sunlight (log scale)

Roots occupy different soil layers

Moss-litter, with herbs in less dense shade

Bare (leaves)

Bracken, bluebells

Herbs

Bedrock: weathered to release nutrients (red arrows) into the soil

D A temperate deciduous forest ecosystem

Activities

1 Study map **A**.

a On a blank outline map of the world, show the distribution of temperate deciduous forest.

b Use an atlas to identify some of the main regions and countries where this is the natural type of vegetation.

c Why is a map of natural vegetation zones (biomes) slightly misleading? Consider the situation of the UK.

2 Study graph **B**.

a Describe the climate of Greensborough, North Carolina.

b Why is temperate deciduous forest well suited to this climate?

c One of the main characteristics of deciduous trees is that they shed their leaves. Explain why this occurs and why it benefits the trees.

Activity

3 Study diagram **D**.

a What is the evidence from the diagram that the temperate deciduous forest supports a great diversity of wildlife?

b What is meant by stratification and how is this exhibited in a deciduous forest? Draw a diagram to support your answer.

c Why do you think stratification exists?

d How have flowers like bluebells adapted to living in a deciduous forest?

Did you know ??????

The tallest tree in the world is a coast redwood found in Redwood National Park, California, which measures an extraordinary 115.55 m. In 2008 scientists claimed to have discovered the oldest tree in the world in Sweden, a Norway spruce said to be 9,550 years old.

Tropical rainforests

Tropical rainforests are found in a broad belt through the tropics (map **A**), from Central and South America, through central parts of Africa, in South-east Asia and into the northern part of Australia. This biome is characterised by a plentiful supply of rainfall (over 2,000 mm a year) and high temperatures (averaging 27°C) throughout the year. This climate (graph **E**) provides ideal conditions for plant growth.

Tropical rainforests have extremely lush and dense vegetation. If you were to enter a rainforest, you would need a torch and good shoes as it is dark and damp. The trees in a tropical rainforest grow to be extremely tall, often up to 45 m in height. There is a great variety of species, typically up to 100 in a single hectare. This explains why the wood is such a valuable resource.

As with deciduous woodlands, a tropical rainforest has a clear stratification (diagram **F**). It is interesting to note that, unlike deciduous forests where most plants and animals live close to the forest floor, in a tropical rainforest the majority are found in the canopy where there is maximum light. Some tree leaves are specially adapted to twist and turn to face the sun as it arcs across the sky. In contrast, rainforest floors are often too dark to support many plants.

Tropical rainforests support the largest number of plant and animal species of any biome. The constant environmental conditions of its climate promote plant growth and result in a great variety of food sources and natural habitats. Many birds live in the canopy. Some mammals, such as monkeys, are well adapted to living in the trees. Animals such as deer live on the forest floor eating seeds and berries.

Tropical rainforest soils are surprisingly infertile considering the lush growth of the vegetation. Most of the nutrients are found at the surface where dead leaves decompose rapidly in the hot and humid conditions. Many of the trees and plants have shallow roots to absorb these nutrients and fungi growing on the roots transfer nutrients straight from the air. The heavy rainfall quickly dissolves and carries away nutrients. This is called **leaching**. It leaves behind an infertile red-coloured soil called latosol, which is rich in iron (hence the colour) and very acidic (photo **G**).

Key terms

Stratification: layering of forests, particularly evident in temperate deciduous forests and tropical rainforests.

Tropical rainforests: the natural vegetation found in the tropics, well suited to the high temperatures and heavy rainfall associated with these latitudes.

Leaching: the dissolving and removal of nutrients from the soil, typically very effective in tropical rainforests on account of the heavy rainfall.

Arid: dry conditions typically associated with deserts.

Hot deserts: regions of the world with rainfall less than 250 mm per year.

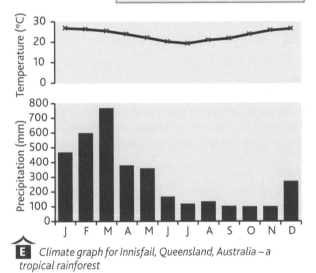

E *Climate graph for Innisfail, Queensland, Australia – a tropical rainforest*

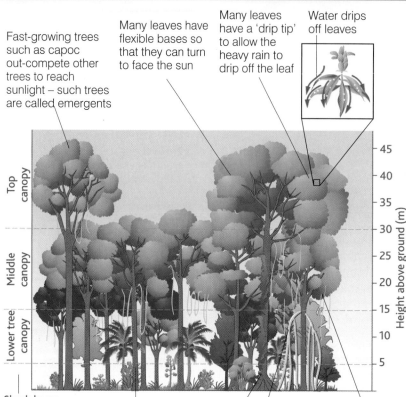

Fast-growing trees such as capoc out-compete other trees to reach sunlight – such trees are called emergents

Many leaves have flexible bases so that they can turn to face the sun

Many leaves have a 'drip tip' to allow the heavy rain to drip off the leaf

Water drips off leaves

Top canopy

Middle canopy

Lower tree canopy

Shrub layer and ground layer

Height above ground (m)

45
40
35
30
25
20
15
10
5

Thin, smooth bark on trees to allow water to flow down easily

Buttresses – massive ridges help support the base of the tall trees and help transport water. May also help oxygen/carbon dioxide exchange by increasing the surface area

Lianas – woody creepers rooted to the ground but carried by trees into the canopy where they have their leaves and flowers

Plants called epiphytes can live on branches high in the canopy to seek sunlight – they obtain nutrients from water and air rather than soil

F *Stratification and vegetation adaptations in a tropical rainforest*

G *Latosol: a typical soil found in the tropical rainforest biome*

Hot deserts

A desert is an area that receives less than 250 mm of rainfall per year. The resulting dryness or **aridity** is the main factor controlling life in the desert. **Hot deserts** are generally found in dry continental interiors in a belt at approximately 30°N and 30°S. It is at these latitudes where air that has risen at the Equator descends, forming a persistent belt of high pressure (anticyclone). This explains the lack of cloud and rain and high daytime temperatures. It also explains why, with the lack of cloud cover, temperatures can plummet to below freezing at night during the winter.

Desert soils tend to be sandy or stony, with little organic matter due to the general lack of dense vegetation. Soils are dry but can soak up water rapidly after rainfall. Evaporation draws salts to the surface, often leaving a white residue on the ground. Desert soils are not particularly fertile.

⚭ **links**

For more information on plant adaptations in arid environments, go to www.cwnp.org/adaptations.html.

AQA **Examiner's tip**

The key aspect in this section is the way that plants respond to climate and soils. Learn some of the ways that plants have adapted to the challenges presented by living in particular environments.

Activities

4 Study graph **E**.

a How does the climate of a tropical rainforest compare with that of a temperate deciduous forest?

b How is the climate ideal for the growth of plants?

c How does the climate provide opportunities as well as problems for the plants and animals living in this environment?

5 Study diagram **F**.

a What name is given to the tall trees that break through the canopy?

b How high can the tallest trees grow?

c On the ground in a tropical rainforest, in which layer of the forest would you be in?

d How have the leaves of the tallest trees adapted to gain maximum sunlight?

e Describe how leaves are designed to shed water quickly during torrential downpours.

f What are lianas and how have they adapted to live successfully in tropical rainforests?

g What is a buttress and what are the possible reasons why some trees have them?

h Select and write about one other plant adaptation in a tropical rainforest.

6 Study photo **G**.

a What name is given to a tropical rainforest soil?

b Describe the characteristics of the soil in the photo.

c Why are most of the nutrients found near the surface of the soil?

d How have plants adapted to this?

e What is leaching and why is it a problem?

7 In the past, some people cut down rainforest trees and replaced them with commercial crops expecting a wonderful harvest. Instead, the new plants grew poorly. From what you have learned, explain why this happened.

8 Use the information in fact file **H**, together with your own Internet research and the text here, to complete a short research project on desert ecosystems. You should include the following information:

■ Draw a map to show the main areas of hot desert. Use an atlas to name the deserts.

■ Using a climate graph, describe the climatic conditions experienced in hot deserts. Why are hot deserts hostile environments?

■ With the aid of labelled photos or sketches, describe the adaptations of plants and animals to hot desert conditions. A good website to get you started is www.cwnp.org/adaptations.html.

Death Valley, California

	J	F	M	A	M	J	J	A	S	O	N	D
Average temperature (°C)	11	15	19	24	29	35	38	37	32	25	17	10
Average precipitation (mm)	0.8	1.2	0.8	0.4	0.2	0.1	0.3	0.2	0.4	0.3	0.6	0.4

Plant adaptations

1 Desert yellow daisy – small linear leaves that are hairy and slightly succulent.

2 Great basin sagebrush – tap roots up to 25 m long and small needle-like leaves to reduce water loss.

3 Giant saguaro cactus – roots very close to the surface so that it can soak up water before it evaporates. Outside skin is pleated so that it can expand when water is soaked up. Grows very slowly.

4 Joshua tree – needle-like leaves coated with a waxy resin.

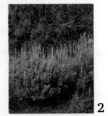
1

2

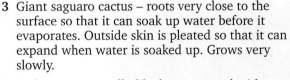
3

4

H *Desert ecosystem fact file*

Epping Forest, Essex

Epping Forest is an ancient deciduous forest that runs north-east of London on a high gravel ridge. It covers an area of about 2,500 ha and is about 19 km long and 4 km wide. It is the largest area of public open space near London.

Although 70 per cent of Epping Forest is deciduous woodland (mostly beech), there are a number of other natural environments including grasslands and marshes. It is home to a rich variety of wildlife including all three native species of woodpeckers and wood-boring stag beetles. Fallow deer still roam the forest.

Early uses and management

Since Norman times, kings and queens of England have used Epping Forest for hunting deer. Local people ('commoners') were able to use the forest to graze their animals and to collect wood for firewood and building.

For many years the practice of **pollarding** was used to manage the woodland. This involves cutting the trees at about shoulder height, above the level of browsing by animals such as deer (photo **A**). Pollarded trees reshoot at this height, thereby producing new wood for future cutting. This is a good example of **sustainable management** as it ensures a supply of wood for future generations. It also accounts for the presence of some ancient trees because, rather than being felled for timber, they were pollarded.

As royal use declined in the 19th century, local landowners made attempts to buy parts of the forest. In response to this threat, in 1878 the Epping Forest Act of Parliament was passed in which it was stated that 'the Conservators shall at all times keep Epping Forest unenclosed and unbuilt on as an open space for the recreation and enjoyment of the people'.

Since 1878 the Forest has been managed by the City of London Corporation.

Recent management

Epping Forest is an excellent example of a natural deciduous forest that is being managed sustainably for the future. Over 1,600 ha of the forest has been designated a Site of Special Scientific Interest and a European Special Area of Conservation. This offers protection under law to its large number of ancient trees, which support a vast variety of flora and fauna.

A *An ancient pollarded tree, Epping Forest*

The overall planning responsibility of Epping Forest lies with the City of London Corporation, which produces management plans to ensure that the forest continues to provide open space for the public while conserving the natural environment. Planning measures adopted include the following:

In this section you will learn

the various uses of a temperate deciduous forest

how deciduous forests can be managed sustainably.

Key terms

Pollarding: cutting off trees at about shoulder height to encourage new growth.

Sustainable management: a management approach that conserves the environment for future generations to enjoy as it is today.

∞ links

Further information about Epping Forest can be found at **www.bbc.co.uk**. Type 'Epping forest' into the search box.

The Epping Forest Information Centre can be found at High Beach, Loughton, Essex IG10 4AF.

Did you know ? ? ? ? ? ?

Epping Forest is a renowned location for the rare stag beetle due to the presence of dead and decaying wood. The stag beetle is one of several species of wildlife that are considered to be rare or endangered.

AQA Examiner's tip

Take time to learn the various measures of sustainable management that have been adopted, both past and present, in Epping Forest.

- managing recreation by providing appropriate car parks, toilets and refreshment facilities and by maintaining footpaths (photo **B**)
- providing three easy-access parks to allow access for people with disabilities
- allowing old trees to die and collapse naturally unless they are dangerous
- controlling some forms of recreation, such as riding and mountain biking, which may damage or affect other forms of recreation
- preserving ancient trees by re-pollarding them to enable new shoots to grow – since 1981, over 1,000 ancient trees have been re-pollarded
- encouraging grazing to maintain the grassland and the flora and fauna associated with it
- preserving ancient earthworks and buildings
- maintaining ponds to prevent them silting up
- preserving the herd of fallow deer.

B   Recreation in Epping Forest

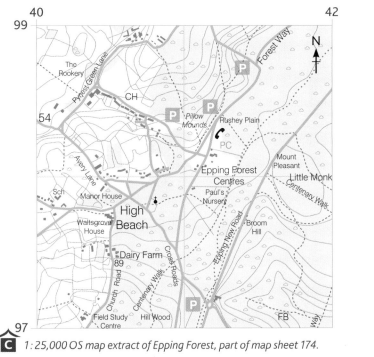

Key
- Non-coniferous trees
- **P** Car park

C   1 : 25,000 OS map extract of Epping Forest, part of map sheet 174.

Activities

1 Study photo **A**.

a What is meant by pollarding? Draw a sketch of the tree in the photo to illustrate your answer.

b Why are trees pollarded?

c Why is pollarding an example of sustainable management?

d How does pollarding lead to the survival of ancient trees?

e Suggest why trees are being re-pollarded today.

2 Study map extract **C**.

a Identify the different types of natural environment.

b What are the attractions and opportunities for recreation?

c Suggest any conflicts that might arise between people visiting Epping Forest.

d Why do you think it is important to have properly designated car parking?

e You may have noticed that there is a field study centre. Why is this a good location for an education centre?

3 Do you think the Forest is being managed sustainably? Explain your answer.

Malaysia's tropical rainforests

Malaysia is a country in south-east Asia. It is made up of Peninsular Malaysia and Eastern Malaysia, which is part of the island of Borneo (maps in **A**). Along with neighbouring countries, the natural vegetation in Malaysia is tropical rainforest. Nearly 63 per cent of Malaysia is forested and commercial tree crops, primarily rubber and oil palm, occupy a further 13 per cent. Trees and forest cover an area equivalent to the whole of the UK.

a World location map

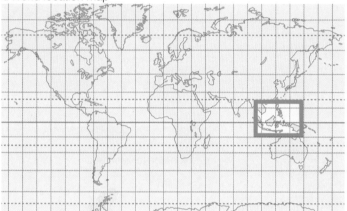

b Regional location map

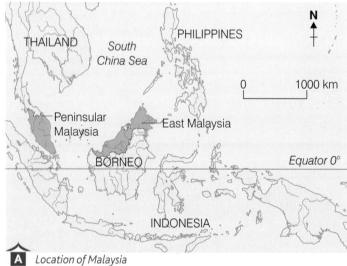

A Location of Malaysia

In the past most of the country was covered by **primary (virgin) rainforest**. In Peninsular Malaysia most of this has now gone and little is left on Borneo. Today an estimated 18 per cent of Malaysia's forest is virgin forest.

Malaysian rainforests support over 5,500 species of flowering plants (the UK has 1,350), 2,600 species of tree (UK 35) and over 1,000 species of butterflies (UK 43). Of the 203 species of mammals, 78 per cent live only in forests. Malaysia's rainforests are clearly special places.

In this section you will learn

the causes of deforestation in tropical rainforests

the effects of deforestation in tropical rainforests.

Key terms

Primary (virgin) rainforest: rainforest that represents the natural vegetation in the region unaffected by the actions of people.

Deforestation: the cutting down and removal of forest.

Clear felling: absolute clearance of all trees from an area.

Selective logging: the cutting down of selected trees, leaving most of the trees intact.

Did you know ??????

The rainforests of South-east Asia are the heartland for the giant dipterocarp trees, of which there are 515 species. These tall, straight trees dominate the tropical timber trade. They are only found in virgin forests and form vital vertical structures in the forests by supporting many species of wildlife.

■ Threats to Malaysia's rainforests

Recent statistics from the United Nations (UN) suggest that the rate of **deforestation** in Malaysia is increasing faster than in any other tropical country in the world, increasing 85 per cent between the periods 1990 to 2000 and 2000 to 2005. Since 2000, some 140,200 ha of forest have been lost on average every year. There are several threats to the rainforests in Malaysia.

Logging

During the 1980s, rampant logging on Borneo led to Malaysia becoming the world's largest exporter of tropical wood. **Clear felling**, where all trees are felled in an area, was common and this led to the total destruction of forest habitats. In recent years the main logging practice has been **selective logging**. Theoretically only fully grown trees are felled and those with important ecological qualities are left unharmed. Although selective logging is far less damaging, it reduces biodiversity. All forms of logging require road construction to bring in machinery and take away the timber (photo **B**).

B *Road construction and logging in Sarawak, East Malaysia*

Malaysia has one of the best rainforest protection policies in the region, but environmental groups claim to have found evidence of illegal logging in Borneo. Here, increasingly marginal slopes have been logged, leading to problems of soil erosion and mudslides.

Not only has logging reduced biodiversity, it has also threatened indigenous tribes. In 2003 a local Penan community in the village of Long Lunyim in Sarawak state protested against the encroachment by a logging company. Some members of the community were imprisoned for their protests, and the company pushed on to the community's forest reserve to exploit the timber.

Energy

The $2bn Bakun Dam project in Sarawak, due for completion in 2009, will result in the flooding of thousands of hectares of forest in order to supply hydroelectric power mainly for industrialised Peninsular Malaysia. An estimated 230 km² of virgin rainforest will have to be cut down for the project. Some 9,000 indigenous Kenyah people have been forced to move from the flooded area. They are traditional subsistence farmers with little money, yet they are being asked to pay to be rehoused. Many now suffer from depression and alcoholism is rife.

Mining

Mining has been widespread in Peninsular Malaysia, with tin mining and smelting dominating. Areas of rainforest have been cleared to make way for mining operations and the construction of roads. In some places, the mining activities have led to pollution of the land and rivers. Drilling for oil and gas has started on Borneo.

Did you know ??????

The orang-utan is a great ape found only in south-east Asia. Able to reason and think, it is one of our closest relatives, sharing 97 per cent of our DNA. It is the largest tree-living mammal in the world.

Commercial plantations

Malaysia is a major producer of oil palm and rubber. In the early 20th century, forest was cleared to make way for the rubber plantations. In recent decades, however, synthetic rubber has led to a steep decline in rubber exports and many plantations have either been abandoned or converted to oil palm (photo **C**).

Today, Malaysia is the largest exporter of palm oil in the world. During the 1970s, large areas of land were converted to palm oil plantations. With plantation owners receiving a 10-year tax break, increasing amounts of land have been converted to plantations. Deforestation for palm oil is taking place on Borneo and threatening the survival of many species of wildlife including the orang-utan.

Resettlement

In the past, poor urban dwellers were encouraged to move into the countryside to relieve pressure on cities. This policy is called transmigration. Between 1956 and the 1980s, an estimated 15,000 ha of rainforest was felled to accommodate the new settlers, many of whom set up plantations.

Fires

Fires are common on Borneo. Some are natural, resulting from lightning strikes, whereas others result from forest clearance or arson. Occasionally, '**slash and burn**' agriculture – where local people clear small areas of land in order to grow food crops – results in wildfires.

C Oil palm plantation in Malaysia

Activities

1 Study this section on the threats to Malaysia's rainforest.

a Briefly outline the main causes of deforestation in Malaysia.

b To what extent do you think deforestation in Malaysia has been driven by economic gain (i.e. making money)?

c What have been the environmental effects of deforestation?

d What have been the social effects (i.e. on the people) of deforestation?

e Despite government policies to preserve Malaysia's rainforests, why do you think deforestation continues to be an issue in Malaysia?

2 Complete a large revision diagram to summarise the threats to Malaysia's rainforests, using an A3 sheet of paper if possible. Include some Internet research information if you can.

a At the centre of your diagram, place a photo or sketch to show the features of Malaysia's rainforest. You could include a map too.

b Around the central feature, create a series of illustrated text boxes describing the main threats to the rainforest. Use arrows to link these boxes to the central image. Use plenty of colour and think carefully about your design to ensure that the final outcome supports your revision.

4.5 Sustainable rainforest management in Malaysia

■ National Forest Policy

Widespread logging in Malaysia started after the Second World War due to improvements in technology (e.g. chainsaws, trucks). The government responded by passing the National Forestry Act in 1977. The Act paved the way for sustainable management of Malaysia's rainforests and had the following aims:

- Develop timber processing to increase the profitability of the exported wood and reduce demand for raw timber. The export of low-value raw logs is now banned in most of Malaysia.
- Encourage alternative timber sources (e.g. from rubber trees).
- Increase public awareness of forests.
- Increase research into forestry.
- Involve local communities in forest projects.

One of the main initiatives of the 1977 Act was to introduce a new approach to forest management known as the **Selective Management System** (diagram **A**). This is recognised as one of the most sustainable approaches to tropical forestry management in the world.

> **In this section you will learn**
>
> the concept of sustainable management of tropical rainforests
>
> the range of national and international options for sustainable management.

> **Key terms**
>
> **Selective Management System:** a form of sustainable forestry management adopted in Malaysia.

Stage	Actions
2 years before felling	Pre-felling study to identify what is there.
1 year before felling	Commercially viable trees marked for felling. Arrows painted on trees to indicate direction of felling to avoid damaging other valuable trees.
Felling	Felling carried out by licence holders.
3–6 months after felling	Survey to check what has been felled. Prosecution may result from illegal felling.
2 years after felling	Treatment plan drawn up to restore forest.
5–10 years after felling	Remedial and regeneration work carried out by state forestry officials. Replacement trees planted.
30–40 years after felling	Cycle begins again.

A *Malaysia's Selective Management System*

Unfortunately, a lack of trained officials to enforce and monitor the system across the country has led to the continuation of abuses and illegal activities. Remedial measures, such as replanting, have not always been carried out satisfactorily. Deforestation is still taking place in Borneo where land is being converted to oil palm plantations.

Permanent Forest Estates and National Parks

Land-use surveys carried out in the 1960s and 1970s have enabled the government to identify Permanent Forest Estates. These areas are protected, with no development or conversion of land use allowed. Large areas of forest are, however, used for commercial logging. Some 10 per cent of the forested land (essentially the primary forest areas) has special conservation status ensuring the survival of the rainforest habitats and species.

Forest Stewardship Council

The Forest Stewardship Council (FSC) is an international organisation that promotes sustainable forestry. Products that have been sourced from sustainably managed forests carry the FSC label. The FSC tries to educate manufacturers and consumers about the need to buy wood from sustainable sources. It also aims to reduce demand for rare and valuable tropical hardwoods.

Developing tourism

In recent years, Malaysia has promoted its forests as destinations for **ecotourism** ('green tourism'). This aims to introduce people to the natural world without causing any environmental damage (photo **B**). The great benefit of ecotourism is that it enables the undisturbed natural environment to create a source of income for local people without it being damaged or destroyed.

B *Ecotourism in Borneo*

Key terms

Ecotourism: nature tourism usually involving small groups, with minimal impact on the environment.

Debt relief: many poorer countries are in debt, having borrowed money from developed countries to support their economic development. There is strong international pressure for the developed countries to clear these debts – this is debt relief.

Carbon sink: forests are carbon sinks because trees absorb carbon dioxide from the atmosphere. They help to address the problem of global carbon emissions.

Non-governmental organisation (NGO): an organisation that collects money and distributes it to needy causes, e.g. Oxfam, ActionAid and WaterAid.

Features of ecotourism

- Usually involves small groups.
- Local guides used.
- Buildings use local materials and are environmentally friendly (sustainable water, energy and waste management). Their construction and maintenance provides employment for local people.
- Mostly nature-based experiences (walks, birdwatching).
- Limited transport involved.

■ Recent worldwide initiatives

Rainforests are valuable resources particularly for
poor countries wishing to expand their economies.
Apart from the timber itself, rainforests occupy land
that could be used for commercial agriculture such as
plantations (e.g. Malaysia) or ranching (e.g. Brazil).
Valuable mineral resources such as bauxite, copper or
iron may be present in the rock beneath the forests.
To expect countries to 'mothball' their rainforests is
naïve, however important they are at the global scale.

Debt relief

One approach is to recognise the international
importance of rainforests by giving them a monetary
value and paying countries to maintain them. This
could take the form of **debt relief**, for example, where
countries are relieved of some of their debt in return for
retaining their rainforests.

C   *Scientists in Sierra Leone's Gola Forest*

Carbon sinks

In 2008 the Gola Forest on Sierra Leone's southern border with Liberia was
protected from further deforestation by becoming a National Park (photo
C). In recognition of the forest's role in reducing global warming by acting
as a **carbon sink**, the 75,000 ha park is supported by money from the
European Commission, the French government and **non-governmental
organisations (NGOs)** such as the Royal Society for the Protection of Birds
(RSPB) and Conservation International.

Activities

1 Study diagram **A**.

  a Make a copy of the diagram showing the Selective Management System.

  b Use the information in the table to add more detail in the boxes in
     your diagram.

  c Why is the system a good example of sustainable management?

  d Can you suggest any modifications that might make it even more
     sustainable?

  e Suggest some possible problems that might exist in trying to
     implement this approach to forest management.

2 Study photo **B**.

  a What is meant by ecotourism?

  b What aspects of ecotourism are evident in the photo?

  c How does ecotourism offer opportunities to protect rainforests from
     deforestation?

  d Ecotourism has become popular and travel companies are keen to
     attach this label to many tours. How might this trend cause problems
     in the future for Borneo's rainforests?

  e Use the internet to look for an example of an ecotourism trip to
     Borneo. Describe the nature of the trip with the aid of photos. Assess
     whether it is genuinely ecotourism.

∞ links

For further information on the
Forest Stewardship Council, visit
www.fsc.org/77.html.

More details about a National Park,
Taman Negara in Malaysia, can be
found at **www.geographia.com/
malaysia/taman.html**.

Ecotourism information can be
found at **www.about-malaysia.
com/adventure/eco-tourism.htm**.

4.6 What are the opportunities for economic developments in hot deserts?

The Thar Desert, Rajasthan, India

Case study

The Thar Desert is one of the major hot deserts of the world. It stretches across north-west India and into Pakistan (map **A**). The desert covers an area of some 200,000 km², mostly in the Indian state of Rajasthan.

Rainfall in the Thar Desert is low – typically between 120 and 240 mm per year – and summer temperatures in July can reach 53°C. Much of the desert is sandy hills with extensive mobile sand dunes and clumps of thorn forest vegetation, a mixture of small trees, shrubs and grasses (photo **B**). The soils are generally sandy and not very fertile, as there is little organic matter to enrich them. They drain quickly so there is little surface water.

> *In this section you will learn*
>
> the economic opportunities of deserts in rich and poor areas of the world
>
> the challenges faced by desert communities and the management responses.

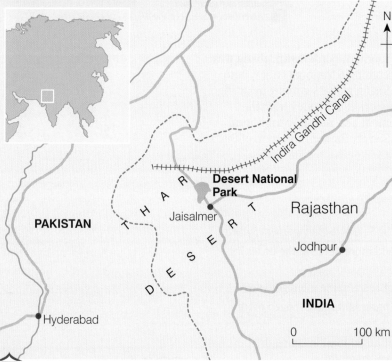

A *Location map of the Thar Desert*

> **Key terms**
>
> **Subsistence farming:** farming to produce food for the farmer and his/her family only.
>
> **Hunter-gatherers:** people who carry out a basic form of subsistence farming involving hunting animals and gathering fruit and nuts.
>
> **Commercial farming:** farming with the intention of making a profit by selling crops and/or livestock.
>
> **Salinisation:** the deposition of solid salts on the ground surface following the evaporation of water.

Economic opportunities in the desert

Subsistence farming

Most of the people living in the desert are involved in farming. The climate presents huge challenges, with unreliable rainfall and frequent droughts. The most successful basic farming systems involve keeping a few animals on the grassy areas and cultivating vegetables and fruit trees. Although a good deal of the farming is **subsistence farming**, some crops are sold at local markets.

Over the border in Pakistan's Thar region, the Kohlis tribe are descendants of **hunter-gatherers** who survived in the desert by hunting animals and gathering fruit and natural products such as honey. This type of subsistence farming is the most basic form of farming and is rarely found in the world today.

Irrigation and commercial farming

Irrigation in parts of the Thar Desert has revolutionised farming in the area. The main form of irrigation in the desert is the Indira Gandhi (Rajasthan) Canal (map **A**). The canal was constructed in 1958 and has a total length of 650 km. Two of the main areas to benefit are centred on the cities of Jodhpur and Jaisalmer, where over 3,500 km² of land is under irrigation. **Commercial farming** in the form of crops such as wheat and cotton now flourishes in an area that used to be scrub desert. The canal also provides drinking water to many people in the desert.

Mining and industry

The state of Rajasthan is rich in minerals. The desert region has valuable reserves of gypsum (used in making plaster for the construction industry and in making cement), feldspar (used to make ceramics), phosphorite (used for making fertiliser) and kaolin (used as a whitener in paper).

There are valuable reserves of stone in the area. At Jaisalmer the Sanu limestone is the main source of limestone for India's steel industry. Limestone is also quarried for making cement. Valuable reserves of the rock marble are quarried near Jodhpur for use in the construction industry. Local hide and wool industries form a ready market for the livestock that are reared in the area.

Tourism

In the last few years, the Thar Desert, with its beautiful landscapes, has become a popular tourist destination. Desert safaris on camels, based at Jaisalmer, have become particularly popular with foreigners as well as wealthy Indians from elsewhere in the country. Local people benefit by acting as guides or by rearing and looking after camels.

B *The natural desert environment near Jaisalmer*

Future challenges

The Thar Desert faces a number of challenges for the future:

- Population pressure – the Thar Desert is the most densely populated desert in the world, with a population density of 83 people per km², and the population is increasing. This is putting extra pressure on the fragile desert ecosystem and leading to overgrazing and overcultivation.

> **Did you know ??????**
>
> The Thar Desert National Park in India is home to the rare great Indian bustard, a large ground-dwelling bird.

- Water management – excessive irrigation in some places has led to waterlogging of the ground. Where this has happened, salts poisonous to plants have been deposited on the ground surface. This is called **salinisation** and is a big problem in deserts (diagram **C**). Elsewhere, excessive demand for water has caused an unsustainable fall in water tables.

- Soil erosion – overcultivation and overgrazing have damaged the vegetation in places, leading to soil erosion by wind and rain. Once eroded away, the soil takes thousands of years to re-form.

- Fuel – reserves of firewood, the main source of fuel, are dwindling with the result that people are using manure as fuel rather than using it to improve the quality of the soil.

- Tourism – although tourists bring benefits such as employment and extra incomes, the environment that they have come to enjoy is fragile and will suffer if tourism becomes overdeveloped.

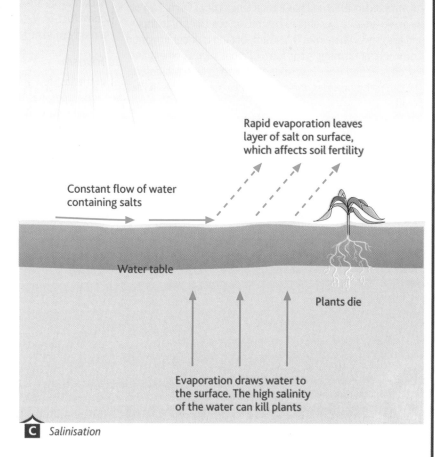

Rapid evaporation leaves layer of salt on surface, which affects soil fertility

Constant flow of water containing salts

Water table

Plants die

Evaporation draws water to the surface. The high salinity of the water can kill plants

C Salinisation

Sustainable management

A number of approaches have been adopted to address the challenges of living in the Thar Desert and to provide its people with a sustainable future. In 1977 the government-funded Desert Development Programme was started. Its main aims are to restore the ecological balance of the region by conserving, developing and harnessing land, water, livestock and human resources. In Rajasthan it has been particularly concerned with developing forestry and addressing the issue of sand dune stabilisation.

Forestry

The most important tree in the Thar Desert is the Prosopis cineraria. It is extremely well suited to the hostile conditions of the desert and has multiple uses (photo **D**). Scientists at the Central Arid Zone Research Institute have developed a hardy breed of plum tree called a Ber tree. It produces large fruits and can survive in low rainfall conditions. The fruits can be sold and there is the potential to make a decent profit.

Stabilising sand dunes

The sand dunes in the Thar Desert are very mobile. In some areas they form a threat to farmland, roads and waterways. Various approaches have been adopted to stabilise the sand dunes, including planting blocks of trees and establishing shelterbelts of trees and fences alongside roads and canals.

Thar Desert National Park

The Thar Desert National Park has been created to protect some 3,000 km² of this arid land and the endangered and rare wildlife that has adapted to its extreme conditions.

- A lot of foliage is produced, which can be used to feed animals, especially in the drier winter.
- The trees can provide good-quality firewood.
- The wood is strong and can be used as a local building material.
- Its pods provide animal fodder.
- Crops can benefit from shade and moist growing conditions if interspersed between the trees.
- This and other tree species are planted in blocks to help stabilise the sand dunes.

D *Sustainable qualities of the Prosopis cineraria tree*

Activities

1 Study photo **B**.

a Describe the environment of the Thar Desert as shown in the photo.

b What are the challenges of this environment for local people?

c What is the difference between subsistence farming and commercial farming?

d Describe the characteristics of these two types of farming in the Thar Desert.

e Apart from farming, what other economic activities take place in the desert?

2 Study diagram **C**.

a With the aid of a diagram, describe the process of salinisation.

b Why does salinisation occur in the Thar Desert?

c Why is salinisation a problem for the future?

d Suggest ways of reducing the problem of salinisation.

e Apart from salinisation, what other challenges face the people of the Thar Desert in the future?

3 Study the information in photo **D**. Apart from stabilising sand dunes, what other benefits do trees provide?

4 a What are the benefits of sand dune stabilisation?

b Is sand dune stabilisation a form of sustainable management? Explain your answer.

The Sonoran Desert, Arizona, USA

The Sonoran Desert is one of North America's largest and hottest deserts. It is also one of the wettest, with over 300 mm of rain falling in some places. It is located in the south-west of the USA, straddling the lower states of Arizona and California and stretching south into Mexico (map **E**). The Sonoran Desert is stunningly beautiful and is home to a great diversity of flora and fauna including the iconic saguaro cactus (fact file **H**, page 85).

The USA is able to respond somewhat differently to the challenges and opportunities of a desert environment compared with poorer countries such as India (Thar Desert) or the African countries bordering the Sahara Desert. Money enables many of the physical difficulties to be overcome.

The physical extremes of the climate can be overcome to some extent by using air conditioning for vehicles, houses, workplaces and shopping centres. With plentiful supplies of relatively cheap energy, this is perfectly possible in the USA.

Water can be relatively easily piped into the area for irrigating crops, to supply drinking water and for filling swimming pools and watering golf courses.

The clear, clean atmosphere and open spaces form an attraction to short-term holidaymakers and long-term migrants. A recent trend in the Sonoran Desert has been **retirement migration**, where people decide to retire to newly built housing complexes with swimming pools and golf courses.

Marana: the tale of one town in the Sonoran Desert

Marana is a town of some 30,000 people located a few kilometres north-west of the city of Tucson in Arizona (map **E**). Over the years it has developed into a thriving business town and leisure resort.

The town began as a mid-19th century ranching and mining community along the Southern Pacific Railroad. In 1920 a new irrigation system enabled it to become an agricultural centre specialising in cotton, a crop that does well in hot conditions provided it is well watered. Families migrated to the town to work in the cotton fields. Agricultural production increased during the 1940s and expanded to include wheat, barley and pecans.

However, since the 1990s farming in the area has declined to be replaced with housing developments. Today only some 15 cotton farms remain. Durum wheat is grown and exported to Italy to make pasta. A heritage park has been opened to celebrate the town's agricultural heritage.

Migration accounts for much of the growth of the town, which is a thriving and wealthy business community.

In 2007 Marana began hosting golf's PGA Matchplay Championship (photo **F**).

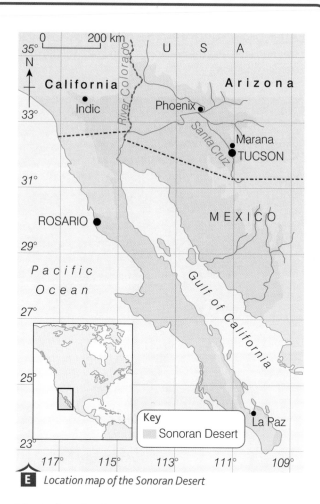

E *Location map of the Sonoran Desert*

Managing the Sonoran Desert

In 1998 the Sonoran Desert Conservation Plan was initiated in Pima County, the administrative region incorporating Tucson in south-west Arizona (map **E**). This is a comprehensive plan to 'conserve the county's most valued natural and cultural resources, whilst accommodating the inevitable population growth and economic expansion of the community'.

The plan resulted from concern about threats to wildlife habitats as housing developments expanded into the desert. An endangered species of pygmy owl was considered to be particularly vulnerable.

Among other initiatives, the plan has led to:

- detailed mapping and inventory of the county's natural and cultural heritage
- development of buffer zones around areas of ecological significance
- native plant protection
- hillside development restrictions
- home design recommendations to conserve energy and water.

F   Golf course in the desert near Marana

Key term

Retirement migration: migration to an area for retirement.

AQA **Examiner's tip**

Be sure to separate the themes of economic opportunities, challenges and management options when studying deserts. Think carefully about the contrasts between poor countries such as India and rich countries such as the USA.

Activities

5   Study photo **F** and the information in the text about Marana.

a   Describe the characteristics and location of Marana using the photo to help you.

b   Draw a timeline to describe the development of Marana since the mid-19th century.

6   What is retirement migration and why is it a significant issue in Arizona?

7   What are the current challenges facing the planners in Marana and how does the Sonoran Desert Conservation Plan address some of these issues?

8   To what extent do you think richer countries such as the USA are better able to exploit the opportunities and address the challenges of desert environments than poorer countries such as India?

⬭⬭ **links**

Some excellent maps are available at www.mapsofindia.com/geography and www.mapsofindia.com/maps/rajasthan.

Information about the Sonoran Desert can be accessed at http://alic.arid.arizona.edu.

Marana Town's website is at www.marana.com.

5.1 How and why do river valleys change downstream?

The river valley is subjected to the three main landscape-shaping processes of erosion, transportation and deposition. It is the extent to which they occur and where they dominate that is critical in shaping the valley.

Processes of erosion

A river near to its source concentrates on erosion, and especially downward erosion. Photo **A** of Golden Clough, Edale, shows one such river **channel** and its valley. There are four ways in which a river erodes. These are **hydraulic action, abrasion, attrition** and **solution**.

- Hydraulic action is the sheer force of the water hitting the bed and the banks. This is most effective when the water is moving fast and there is a lot of it.
- Abrasion occurs when the **load** the river is carrying repeatedly hits the river bed and the banks, causing some of the material to break off.
- Attrition is when the stones and boulders carried by the river knock against each other and over time are weakened, causing bits to fall off and reduce in size.
- Solution occurs only when the river flows on certain types of rock, such as chalk and limestone. These are soluble in rainwater and become part of the water as they are dissolved by it.

Rivers tend to erode in one of two directions: downwards or sideways. The terms for these are vertical and lateral erosion. As a river gets further down its course, vertical erosion becomes less important and lateral erosion takes over.

Processes of transportation

Having been successful in dislodging parts of the bed and banks, the river then moves the load it has in it via transportation. There are four methods by which the river transports its load. These are **traction, saltation, suspension** and **solution** (diagram **B**). Photo **C** shows a river further downstream where the gradient has lessened, becoming much more gentle in photo **D**.

> **In this section you will learn**
>
> the processes of erosion: hydraulic action, abrasion, attrition and solution
>
> the processes of transportation: traction, saltation, suspension and solution
>
> how and why deposition occurs
>
> how and why the long and cross profiles are formed and why cross profiles change downstream.

A Golden Clough, Edale

> AQA **Examiner's tip**
>
> Learn the process terms for erosion and transportation. You must use these terms correctly to access the highest marks.

Traction is the method used for moving the largest material. This is too heavy to lose contact with the bed, so material such as boulders is rolled along.

Saltation moves the small stones and grains of sand by bouncing them along the bed. This lighter load leaves the river bed in a hopping motion.

Suspension is a means of carrying very fine material within the water, so that it floats in the river and is moved as it flows.

Solution is the dissolved load and occurs only with certain rock types that are soluble in rainwater. This is true of chalk and limestone and the load is not visible.

B *Processes of transportation*

Key terms

Channel: the part of the river valley occupied by the water itself.

Hydraulic action: the power of the volume of water moving in the river.

Abrasion: occurs when larger load carried by the river hits the bed and banks, causing bits to break off.

Attrition: load carried by the river knocks into other parts of the load, so bits break off and make the material smaller.

Solution: the dissolving of certain types of rock such as chalk and limestone by rainwater. This is a means of transportation as well as an erosion process.

Load: material of any size carried by the river.

Traction: the rolling along of the largest rocks and boulders.

Saltation: the bouncing movement of small stones and grains of sand along the river bed.

Suspension: small material carried within the river.

C *Grindsbrook Clough, Edale*

D *River Noe, Edale*

Did you know

The Mississippi River is 3,800 km long. It carries on average 42,002 tonnes of sediment each day and 130 million tonnes a year.

Deposition

This is where the river dumps or leaves behind material that it has been carrying. It deposits the largest material first as this is the heaviest to carry. The smaller the load, the further it can be transported, so this is deposited much further downstream than the larger load. Thus, large boulders can be seen in photo **A**, whereas they are absent in photo **D**. The river drops some of its load when there is a fall in the speed of the water or the amount of water is less. This often occurs when the gradient changes at the foot of a mountain or when a river enters a lake or the sea.

Long and changing cross profiles

The **long profile** shows how the river changes in height along its course. Diagram **E** shows a theoretical long profile from the source to the mouth. The steep reduction in height near the source gives way to a more gradual reduction further downstream, giving a typical·concave profile. The river has much potential energy near the source due to the steep drop. Later on, this is replaced by energy from a large volume of water. However, such a perfect long profile is rare. This is due to land being uplifted, sea level changing and bands of hard and soft rock crossing the path of the river. As the river flows downstream, its valley changes shape and the **cross profile** from one side of the valley to the other clearly shows this. Generally, the cross profile shows the valley becoming wider and flatter, with lower valley sides. Map extract **F** shows part of the course of the River Noe and some of its tributaries, where photos **A**, **C** and **D** were taken. This shows how the valley cross profile changes with distance from the source.

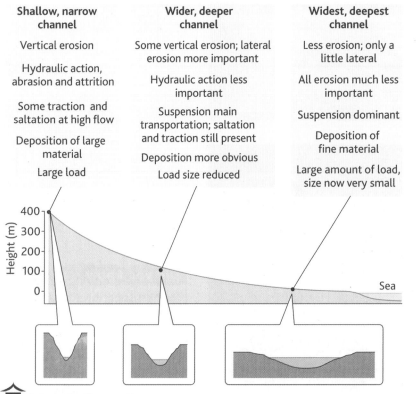

E A theoretical long profile

∞ links

You can find out more facts about large rivers at **http://ga.water. usgs.gov/edu/riversofworld.html** and **http://en.wikipedia.org/wiki/ Categort:Rivers_of_England**. On this website, it would be worth searching for Durham and then the Tees as preparation for the next section.

Key terms

Long profile: a line representing the course of the river from its source (relatively high up) to its mouth where it ends, usually in a lake or the sea, and the changes in height along its course.

Cross profile: a line that represents what it would be like to walk from one side of a valley, across the channel and up the other side.

Activities

1. Study photo **A**. Draw a labelled sketch to show the characteristics of the channel and valley of Golden Clough. Include comments about the following in your labels:
    - width/depth of channel
    - size of load
    - profile of bed and banks
    - what water flow is like
    - valley sides.

2. For the terms relating to erosion, produce an illustrated dictionary to give your own clear definitions, supported by a simple diagram.

3. For the four types of transportation, produce a diagram to illustrate all four processes. Write a definition of your own next to the relevant part of the diagram.

Activities

4   Study photos **A**, **C** and **D** and diagram **B**.

a   How do you think the load shown in photo **A** will be moved? Explain why.

b   What process of transportation is shown in photo **C**? Explain reasons for choosing the process you have selected.

c   Describe how the channel and valleys of the two rivers shown in photos **A** and **C** have changed.

d   Describe further changes that have occurred between photos **C** and **D**.

5   Study diagram **E**.

a   Draw a sketch long profile of a river.

b   Mark with a dot where you think the photos in **A** and **D** could have been taken.

c   Label your long profile to show how and why the size of deposited material changes downstream.

6   Study map extract **F**.

a   Give the approximate height of the rivers where each photo (**A**, **C** and **D**) was taken.

b   Describe the gradient of the long profile at each of the locations where photos **A**, **C** and **D** were taken.

c   Draw a sketch of the cross profile for each of the three locations.

d   Describe how the cross profiles change downstream.

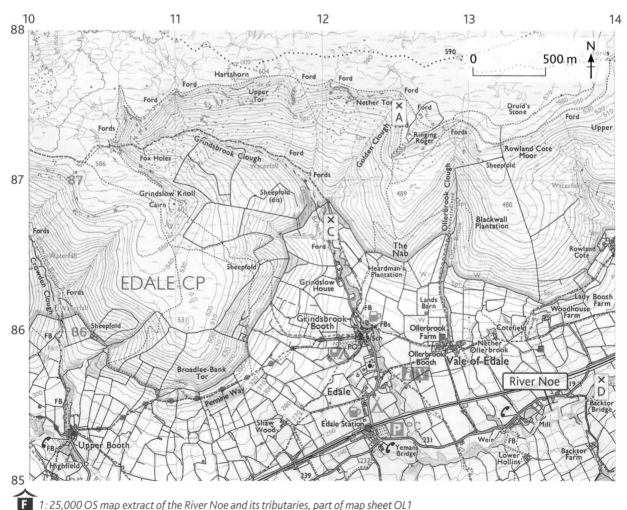

F   *1 : 25,000 OS map extract of the River Noe and its tributaries, part of map sheet OL1*

What distinctive landforms result from the changing river processes?

Different river processes lead to different landforms. Therefore, in areas near the source where vertical erosion is dominant, **waterfalls** and **gorges** are characteristic features. Further down, where lateral erosion and deposition become more important, **meanders** and **oxbow lakes** develop. Nearer to the mouth, where deposition is the most significant process, **floodplains** and **levees** become a key aspect of the landscape. The River Tees will be used here to illustrate different landforms and how they change downstream.

◼ Landforms resulting from erosion: waterfalls and gorges

Waterfalls provide some of the most spectacular scenery in mountainous areas. In their wake, they leave gorges as they retreat back up the valley. Diagram **A** shows the sequence of events that occurs in waterfall formation.

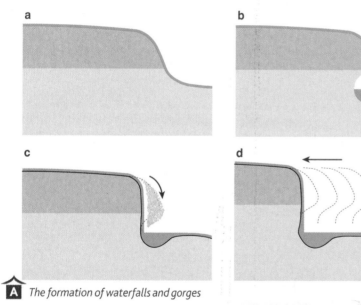

A The formation of waterfalls and gorges

One of the best-known waterfalls in the UK is High Force on the River Tees (photo **B** and map extract **C**). This occurs where whinstone, a resistant igneous rock, overlays softer limestone.

B High Force, River Tees

In this section you will learn

how and why waterfalls and gorges form due to erosion

the formation of meanders and oxbow lakes due to erosion and deposition

the development of levees and floodplains, due to deposition

how the formation of some features is linked to the development of others.

Key terms

Waterfall: the sudden, and often vertical, drop of a river along its course.

Gorge: a narrow, steep-sided valley.

Meander: a bend or curve in the river channel, often becoming sinuous where the loops are exaggerated.

Oxbow lake: a horseshoe or semi-circular area that represents the former course of a meander. Oxbow lakes are cut off from a supply of water and so will eventually become dry.

Floodplain: the flat area adjacent to the river channel, especially in the lower part of the course. This is created as a natural area for water to spill onto when the river reaches the top of its banks.

Levees: raised banks along the course of a river in its lower course. They are formed naturally but can be artificially increased in height.

Did you know ❓❓❓❓❓

Eas a'Chual Aluinn in Scotland is the UK's highest waterfall at 200 m. This is relatively small when compared with the highest waterfall in the world, Angel Falls in Venezuela at a height of 979 m.

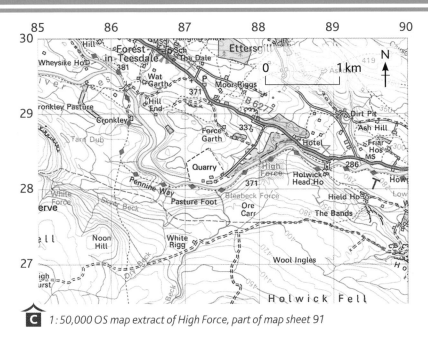

Landforms resulting from erosion and deposition: meanders and oxbow lakes

Meanders and oxbow lakes are characteristic landforms in the middle part of the river. The formation of meanders leads eventually to the development of oxbow lakes, as shown in diagram **D**.

Map extract **E** shows formed meanders further down the course of the Tees. The inside bend of a meander, where deposition is the dominant process, is different from the outside bend of the meander (photos **F** and **G**).

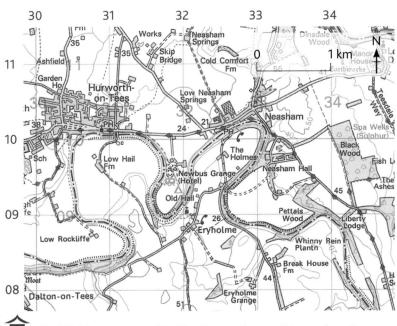

E *1 : 50,000 OS map extract of the River Tees south-east of Huxworth-on-Tees, part of map sheet 93*

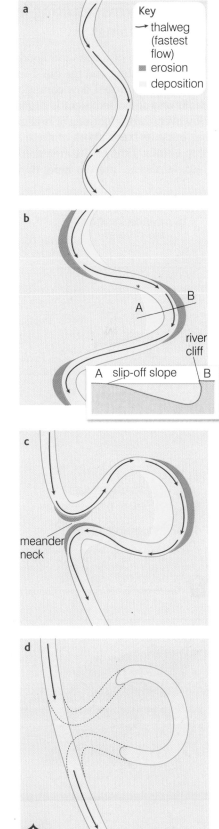

Key
→ thalweg (fastest flow)
■ erosion
░ deposition

D *The formation of meanders and oxbow lakes*

Landforms resulting from deposition: levees and floodplains

The formation of levees and floodplains are linked and involve repeated flooding and the build-up of material during the period of flood. Under normal flow conditions, the river is contained within its banks and so no sediment is available to form levees or the floodplain. However, during periods of high rainfall and discharge when the river has burst its banks, both of these features are formed (diagram **H**). Map extract **J** shows the area much nearer the mouth at Thornaby-on-Tees. Levees are present here; they are far more apparent in photo **K**.

F *Inside bend of a meander*

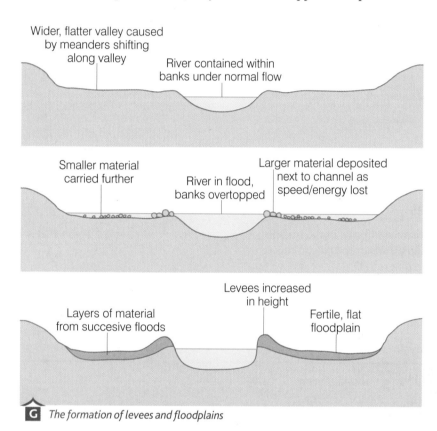

Wider, flatter valley caused by meanders shifting along valley

River contained within banks under normal flow

Smaller material carried further

River in flood, banks overtopped

Larger material deposited next to channel as speed/energy lost

Layers of material from succesive floods

Levees increased in height

Fertile, flat floodplain

G *The formation of levees and floodplains*

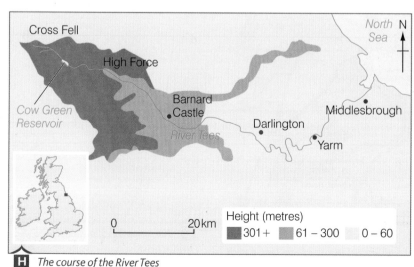

Cross Fell

High Force

Cow Green Reservoir

Barnard Castle

River Tees

Darlington

Middlesbrough

Yarm

North Sea N

Height (metres)
301+   61 – 300   0 – 60

0   20km

H *The course of the River Tees*

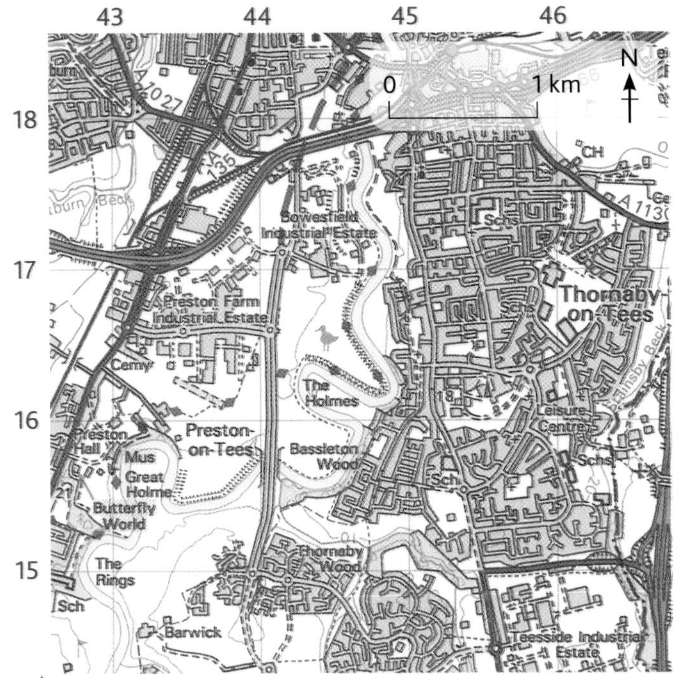

J 1:50,000 OS map extract of the floodplain of the River Tees, part of map sheet 93

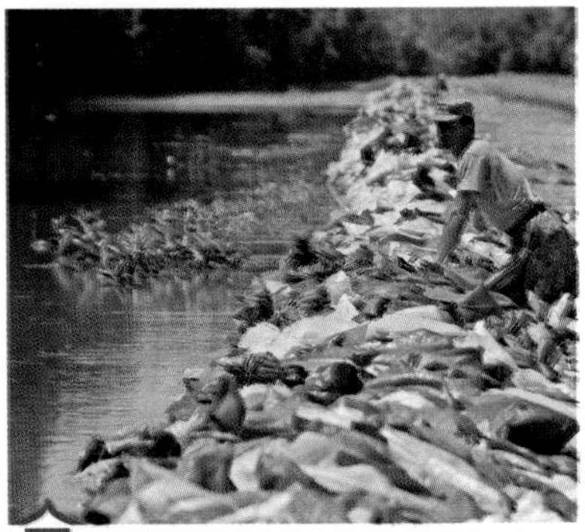

I Artificial levees on the Mississippi

links

You can find out more about the River Tees at www.bookrags.co/wiki/River_Tees.

Activities

1 Study diagram **A**. Write a series of bullet points to describe the sequence of waterfall formation.

2 Study photo **B** and map extract **C**.

a Research High Force using the links suggested to obtain specific facts about the waterfall.

b Produce an information board about High Force. Include on your board a labelled sketch of the waterfall to describe its features and some facts about it that will interest visiting tourists.

c Describe the channel and the valley of the River Tees shown in the map extract.

3 Study diagram **D**.

a Draw simplified copies of the diagrams.

b Label each diagram to show the stages in the formation of meanders and oxbow lakes.

4 Study map extract **E**.

a Draw a sketch map of the meanders showing:

- an inside bend
- an outside bend
- the neck of a meander
- the meander most likely to be cut off first.

b Describe additional information about the meanders that is present on the map.

c Locate Low Hail Farm (309097) and The Holmes (325098). Describe the location of the two farms and suggest why they are located here.

d Comment on the risk of these two farms being flooded by the river.

5 Study photo **F**.

a Would you expect the deepest water in the river to be on the left or the right side? Explain your answer.

b Do you think this photo was taken at low flow or during a flood? Justify your answer.

c What evidence is there that the river rarely flows at a high level?

d Draw a sketch of the meander and add labels to identify the main landforms, the different river depths and the line of fastest flow.

6 Study diagram **G**, map extract **H** and photo **I**.

a Working in pairs, produce a short PowerPoint presentation to include the following:

- the formation of levees
- the formation of floodplains
- the links between the two landforms
- illustrations of the River Tees or another river of your choice.

b Show your presentation to another group in the class.

How and why does the water in a river fluctuate?

The **discharge** of a river shows much variation during a year and in the short term. It can fluctuate a lot in a matter of hours in response to periods of rain. An understanding of the **drainage basin** hydrological cycle (diagram **A**) is useful background in explaining how and why the amount of water in a river is variable.

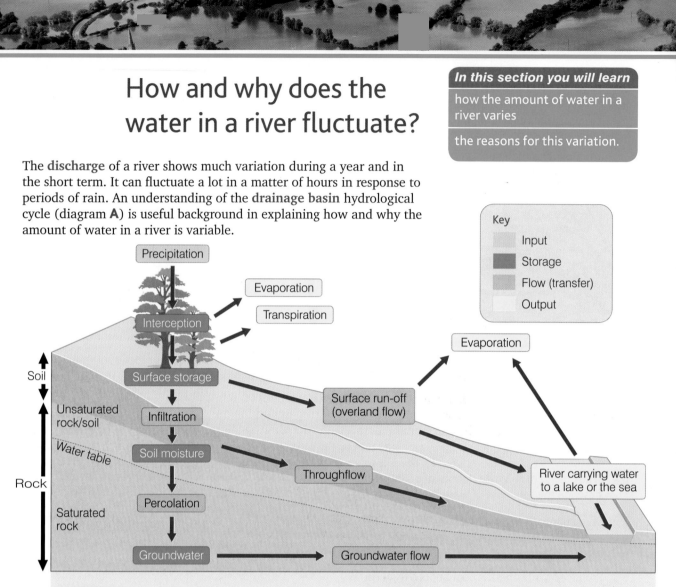

Precipitation: any source of moisture reaching the ground, e.g. rain, snow, frost

Interception: water being prevented from reaching the surface by trees or grass

Surface storage: water held on the ground surface, e.g. puddles

Infiltration: water sinking into soil/rock from the ground surface

Soil moisture: water held in the soil layer

Percolation: water seeping deeper below the surface

Groundwater: water stored in the rock

Transpiration: water lost through pores in vegetation

Evaporation: water lost from ground/vegetation surface

Surface run-off (overland flow): water flowing on top of the ground

Throughflow: water flowing through the soil layer parallel to the surface

Groundwater flow: water flowing through the rock layer parallel to the surface

Water table: current upper level of saturated rock/soil where no more water can be absorbed

A Drainage basin hydrological cycle

The storm hydrograph

The **flood or storm hydrograph** is used to show how a river responds to a period of rainfall (graph **C**). Rivers that respond rapidly to rainfall have a high peak and short lag time and are referred to as **flashy**. A lower peak and long lag time shows a delayed hydrograph. Table **B** gives discharge data for the River Eden in Carlisle between 7 and 9 January 2005.

B *Discharge of River Eden, Carlisle*

Date	Time	Discharge in cubic metres per second (cumecs)
7 January	0000 hours	90
	1200 hours	130
8 January	0000 hours	820
	1200 hours	1,400
	1500 hours	1,520
9 January	0000 hours	1,000
	1200 hours	430

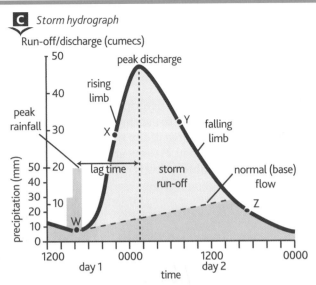

C *Storm hydrograph*

Factors affecting river discharge

River discharge is influenced by a number of factors related to the weather (e.g. rainfall, temperature and previous weather conditions), other physical factors (e.g. **relief** and rock type) and by human land use.

The amount and type of rainfall are important factors influencing river discharge. A lot of rain falling causes high river levels, while less rainfall results in lower river levels. The explanation for this lies in the drainage basin hydrological cycle. High amounts of rain saturate the soil and underlying rock. In the case of drizzle, there is time for water to infiltrate the soil and underlying rock, freeing up space for more rain.

Temperature affects the loss of water from the drainage basin and therefore the level of discharge. When temperatures are higher (table **D**), there is greater water loss via evaporation and transpiration, so river levels go down.

D *Average rainfall and temperature in Sheffield*

Month	Rainfall (mm)	Maximum temperature (°C)
January	87	6.4
February	63	6.7
March	68	9.3
April	63	11.8
May	56	15.7
June	68	18.3
July	51	20.8
August	64	20.6
September	64	17.3
October	74	13.3
November	78	9.2
December	92	7.2

Previous weather conditions also have an impact on river discharge. If it has been dry, it will take longer for the water to reach the river and the amount will be less than if there had been a number of wet days.

Key terms

Discharge: the volume of water passing a given point in a river at any moment in time.

Drainage basin: area from which a river gets its water. The boundary is marked by an imaginary line of highland known as a watershed.

Flood or storm hydrograph: a line graph drawn to show the discharge in a river in the aftermath of a period of rain.

Flashy: a hydrograph that responds quickly to a period of rain so that it characteristically has a high peak and a short lag time.

Relief: height and slope of land.

Impermeable: rock that does not allow water to soak into it.

Porous: rock that has spaces (pores) between particles.

Pervious: rock that allows water to pass through it via vertical joints and horizontal bedding planes.

Deforestation: cutting down trees.

Urbanisation: the increase in the proportion of people living in cities, resulting in their growth.

AQA Examiner's tip

Ensure that you can explain how physical and human features and processes can affect the amount of water flowing in rivers.

Table **E** gives the daily totals for Sheffield from 13 June until flooding occurred on the 26th. Relief affects the rate at which water runs off the land surface and into rivers. Steep slopes encourage fast run-off as the water spills rapidly downwards due to gravity. Gentle slopes allow time for infiltration to occur.

Rock type is important in determining how much water infiltrates and how much stays on the surface. In areas where more water is on the surface, the discharge of the river is higher as it reaches the river fastest. Photos **F**, **G** and **H** show areas of granite, chalk and limestone that are **impermeable**, **porous** and **pervious** respectively.

E	Rainfall in Sheffield, June 2007
Date	Rainfall (mm)
13 June	30.3
14 June	88.2
15 June	16.9
16 June	0.0
17 June	2.2
18 June	0.0
19 June	14.7
20 June	0.1
21 June	9.6
22 June	6.9
23 June	0.2
24 June	36.0
25 June	51.1
26 June	0.0

F Granite rock type

G Chalk rock type

Land use relates to the function of an area. It is a factor that considers the effect of people. **Deforestation** (photo **I**) and **urbanisation** (photo **J**) are the two most important land-use changes with regard to influencing river discharge. If trees are removed, water reaches the surface faster and the trees do not extract water from the ground. Expanding towns create an impermeable surface. This is made even worse by building drains to take the water away from buildings quickly – and equally quickly into rivers!

Activities

1. Study diagram **A**. Work in pairs to produce a series of flash cards for the key terms in the diagram.

   a. On one side of the card, there should be a definition and an illustration of the term.

   b. On the other side, the term should be given so that the answer is clear.

   c. Use your completed flash cards with another pair to see how many terms you got correct and how good your definitions are.

2. Study table **B** and graph **C**.

   a. Draw a line graph to show the discharge of the River Eden in Carlisle.

   b. Using the terms in graph **C**, describe the flood hydrograph you have drawn.

   c. Can the hydrograph be described as 'flashy'?

3. Study tables **D** and **E** and photos **F** to **J**. Work in pairs to produce an A3 spread around the title (placed in the centre) 'The reasons river discharge fluctuates'.

   a. In the central box, give the meaning of 'river discharge'.

   b. Place the factors around the central title.

   c. Add information to each factor to explain how discharge is affected. Each factor should be illustrated by diagrams, sketches or photos. You should use data provided in the tables.

   d. Your finished spread should be informative and accurate but interesting, colourful and original.

H Limestone rock type

I Deforestation in Tayside, Scotland

J Urbanisation in Leicester

5.4 Why do rivers flood?

Floods occur when a river bursts its banks if it is carrying so much water that it cannot be confined in its usual course. Flooding is not a normal condition for the river, but is seen as an extreme situation due to high levels of flow. The extent to which the river exceeds the flow that can be contained in its banks determines the severity of the flood and is sometimes related to how often flooding occurs. Usually, bigger floods occur less often and less severe flood events occur more frequently. Floods are common events. Problems and issues arise when people are affected. Building on floodplains results in property being damaged and lives being lost in what becomes a **hazard**.

Causes of floods

Rivers flood due to a number of physical causes such as prolonged rainfall, heavy rain, snowmelt and steep relief (table **A** and extracts **B** and **C**). People often unintentionally increase the likelihood and severity of flooding. This is mainly the result of deforestation and construction work.

A *Rainfall in Sheffield*

Month	Average rainfall (mm), 1971–2000	Actual rainfall (mm), 2007
March	67.9	44.5
April	62.5	5.8
May	55.5	83.8
June	66.7	285.6

In this section you will learn

why flooding occurs – the natural causes and the ways in which people make it worse

where floods have occurred in the UK

how the frequency of flooding seems to be increasing.

Key terms

Floods: these occur when a river carries so much water that it cannot be contained by its banks and so it overflows on to surrounding land – its floodplain.

Hazard: an event that occurs where people's lives and property are threatened and deaths and/or damage result.

Soil erosion: the removal of the layer of soil above the rock where plants grow.

News

Snowmelt is a main contributing factor to flooding. It was partly responsible for floods in Malton (N. Yorkshire) in November 2000 and is frequently important in Bangladesh. An extract from 'Why Bangladesh floods are so bad' (BBC News website **news.bbc.co.uk**) on the 2004 floods states 'Bangladesh receives enormous amounts of water from four major rivers. All are filled up from melting snow in the Himalayas.' The Himalayas provide some of the steepest relief worldwide and this is also important in ensuring that water reaches the rivers quickly, increasing the flood risk.

Deforestation has an impact on the water cycle in a drainage basin. It is frequently seen as one reason for the increasing severity of flooding in Bangladesh. Chopping trees down in higher-lying areas, including neighbouring countries, such as Nepal, can have unintentional effects elsewhere. As well as increasing rates of surface runoff, **soil erosion** is a consequence. Much of this is washed into rivers, where, when deposited, the amount of water the channel can hold is reduced. Thus, the flood risk is increased.

Building construction can increase flooding. New houses built next to the river in Malton are clearly susceptible to flooding. Building on floodplains is an issue, as outlined in Lincoln.

C *Causes of flooding*

The city experienced the most rain in a single month since records began 125 years ago. A resident said, 'The rain had been almost constant for a week. On the morning the flooding started, the rivers were almost visibly rising. By lunchtime, the city was at a standstill as bridges became impassable and underpasses flooded. We were stranded in the north of the city – it took six hours to travel four miles. By the evening much of the city centre was under water, many roads had collapsed and were impassable and electricity supplies were limited for almost a week. It took well over a year for things to get back to normal.'

Adapted from the Sheffield Star

B *Prolonged rain in Sheffield, 2007*

Frequency and location of flood events

Flooding appears to be becoming an increasingly frequent event. In 1607 a great flood affected Devon, Somerset and South Wales. Major floods, however, were infrequent in the UK. In March 1947 major floods did occur, affecting many areas of southern, central and north-eastern England, including York, Tewkesbury, Shrewsbury, Sheffield, Nottingham and London, following the rapid melting of snow. The combined effects of storm surge and high tides contributed to the floods of January 1953 that hit the east coast, including Suffolk, Essex and Kent, when huge waves washed away sea defences and 307 people died. In 1968 another Great Flood affected counties in south-east England.

After this, there is little reference to major floods until relatively recently, when they have regularly made the headlines. Since 1998 headlines about floods have been an almost annual occurrence. Table **D** summarises some of the most serious floods since 1998.

AQA Examiner's tip

Learn the case study of one recent flood event and its causes.

∞ links

You can find out more about causes of flooding at www.bbc.co.uk.

D *Major flood events, 1998–2007*

Date	Location	Rivers
April 1998	Warwickshire, Gloucestershire, Herefordshire, Worcestershire, Leicestershire, south-east Wales	Avon, Severn, Wye
	Northamptonshire, Bedfordshire and Cambridgeshire	Nene, Great Ouse
October 1998	Course of River Severn from mid-Wales to Gloucester	Severn
March 1999	Malton and Norton flooded by Derwent and its tributaries	Derwent
May 2000	Uckfield, Petworth, Robertsbridge, Horsham (Sussex)	Uck, Rother
June 2000	Calder Valley, Yorkshire and York	Calder, Ouse
August 2004	Boscastle (Cornwall)	Valency
January 2005	Carlisle	Eden
June 2007	Large areas of south and east Yorkshire including Doncaster, Sheffield and Hull	Don, Hull, Witham,
	Parts of Lincolnshire including Lincoln and Louth	Witham, Ludd
July 2007	Large areas of Gloucestershire including Upton-upon-Severn, Tewkesbury, Gloucester	Avon, Severn
	Oxfordshire including Oxford, Banbury and Witney	Thames, Windrush, Cherwell

Activities

1 Study table **A** and extract **B**.

a Draw a comparative bar graph to show the average rainfall between March and June and rainfall that fell in those months in 2007.

b Explain how the figures help to explain the flooding that occurred.

c Consider the comments made by the Sheffield resident in the extract. How do these explain the flooding at the end of June 2007?

2 Study extract **C**.

a Using the example of Bangladesh, explain to a younger student how snowmelt, relief and deforestation are responsible for flooding. You should include one diagram for each reason.

b Explain why building on floodplains might increase the likelihood of flooding.

c To what extent do you agree with the following statement: 'There is not usually one cause of flooding but a combination of reasons.' Support your answer with evidence.

3 Study table **D**.

a On an outline map of the UK, mark and label the places that flooded.

b Add the major rivers to your map.

c Summarise the key points about the locations of flooding shown in your map.

d Present evidence that suggests that the frequency of flooding in the UK is increasing.

5.5   How and why do the effects of flooding and the responses to it vary?

The effects of flooding vary according to their size and location. The impact tends to be more severe in poorer countries. Responses are generally more immediate in countries at further stages of development and the attempts made to reduce the effects come from within the affected area or country. In countries at lesser stages of development, attempts made to reduce the effects may be delayed and require international effort. Long-term responses are likely to show similar differences as a result of variations in wealth and the ability to afford flood protection measures.

In this section you will learn

the effects of flooding and the responses to it in both richer and poorer countries

how and why the effects and responses vary.

Flooding in England

The flooding of many parts of England in June and July 2007 was the most extensive ever experienced. The depth may not have reached the record levels of 1947, but the scale of the areas affected reached a new high. Diagram **A** shows the main areas affected and the damage done when much of central and southern England suffered the effects of record levels of rainfall.

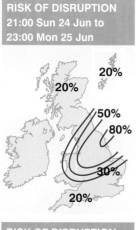

RISK OF DISRUPTION
21:00 Sun 24 Jun to
23:00 Mon 25 Jun

20%
20%
50%
80%
30%
20%

- Surface water flooding in Hull.
- Widespread disruption and damage to more than 7,000 houses and 1,300 businesses in Hull.
- River Don burst its banks, flooding Sheffield and Doncaster.
- Flooding in Derbyshire, Lincolnshire and Worcestershire.
- Highest official rainfall total was 111 mm at Fylingdales (North Yorkshire). Amateur networks recorded similar totals in the Hull area.
- There were fears that the dam wall at the Ulley Reservoir near Rotherham would burst.

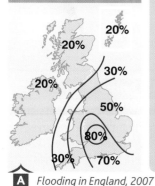

RISK OF DISRUPTION
00:00 Fri 20 Jul to
12:00 Sat 21 Jul

20%
20%
30%
20%
50%
80%
30%
70%

- Widespread disruption to the motorway and rail networks.
- In the following days the River Severn and tributaries in Gloucestershire, Worcestershire, Herefordshire and Shropshire broke banks and flooded surrounding areas.
- River Thames and its tributaries in Wiltshire, Oxfordshire, Berkshire and Surrey flooded.
- Flooding in Telford and Wrekin, Staffordshire, Warwickshire and Birmingham.
- The highest recorded rainfall was 157.4 mm in 48 hours at Pershore College (Worcestershire).

A *Flooding in England, 2007*

B *Victims of the Tewkesbury floods*

Tewkesbury, 2007

Friday 20 July was the day on which 80–90 cm of heavy rain – equivalent to almost two months' rainfall – fell in Tewkesbury. It was to signal the start of five days of flooding and a significantly longer time before things had any chance of getting back to normal. The large extent of the flooding is shown in photo **C**, while map extract **D** covers the flooded area and beyond. The area was in the news for many days and images such as the one in photo **B** were often in the news.

C Flooding at Tewkesbury

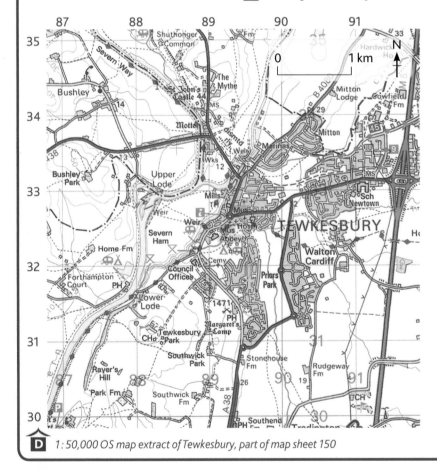

D 1 : 50,000 OS map extract of Tewkesbury, part of map sheet 150

Flooding in Bangladesh

Bangladesh has experienced particularly severe flooding on a regular basis. Annual flooding is expected in this low-lying country that is located largely on the delta of the Ganges. What sets aside years such as 1988, 1998, 2004 and, to a lesser extent 2007, is the scale and severity of the floods.

Did you know ??????

The sale of women's raincoats at John Lewis in July 2007 was over 11 times higher than in July 2006, while the sale of umbrellas was 184 per cent higher.

Bangladesh, 2004

The 2004 floods occurred from July to September, inundating over half of the country at their peak. Map **E** shows the areas of Bangladesh under water and the depth of the water.

At the time of the July 2004 floods, 40 per cent of Dhaka was under water, 60 per cent of the country was submerged, 600 deaths were reported and 30 million left homeless out of a population of 140 million. Some 100,000 people in Dhaka alone were suffering from diarrhoea as the floodwaters left mud and raw sewage in their wake. As the year progressed, things got even worse as Bangladesh experienced its heaviest rain in 50 years with 35 cm falling in one day on 13 September. The death toll rose to over 750, the airport at Dhaka was flooded as were many roads and railways. Bridges were also destroyed – all of which hampered the relief effort that followed. This damage and that to schools and hospitals was put at $7 billion. In many badly hit rural areas, rice – the main food crop – was washed away and other important food supplies such as vegetables were lost along with cash crops such as jute and sugar.

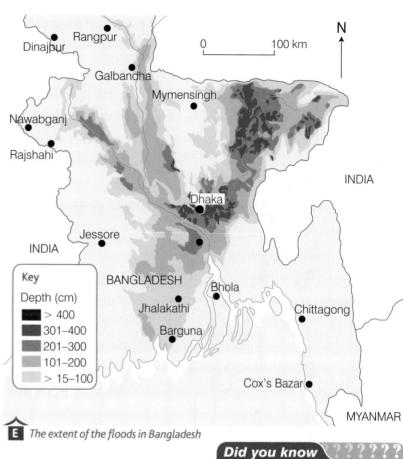

E The extent of the floods in Bangladesh

Key

Depth (cm)
- > 400
- 301–400
- 201–300
- 101–200
- > 15–100

Within Bangladesh, food supplies, medicines, clothing and blankets were distributed. The effects of the floods on the transport system made this difficult (photo **F**). Local communities began to rebuild their homes. Disease from contaminated water remained a major threat. International help began to arrive. The United Nations launched an appeal for $74 million, but had received only 20 per cent of this by September. An appeal by WaterAid sought to supply water purification tablets and posters highlighting the hygiene risks in flood water. All of these represent immediate responses to the floods.

Longer-term approaches included embankments built along the river, which have not really achieved their goal, whereas flood warning and the provision of flood shelters have been more successful. These are areas of raised land where people can move to temporarily with their cattle and have access to items such as dried food and obtain water before supplies are contaminated.

Did you know

Seventy per cent of Bangladesh is less than 1 m above sea level. In 2006 almost half the population lived below the poverty line.

F The effects of flooding in rural areas of Bangladesh

1 Study diagram **A**. Imagine you are a newscaster presenting a special report on 'The 2007 floods in England'. Write your script, summarising the location of areas hit and the effects. Make sure that you stress the scale of the flooding and refer to evidence to back up what you are saying. You should have a map to refer to in your broadcast.

2 Study photo **C** and map extract **D**.

a Tewkesbury Abbey can be seen in the map extract. Give the six-figure grid reference for it.

b A caravan site is located at 888323. Find this on the photo. Is it under water?

c Notice that several campsites and caravan parks are sited close to rivers. Do you think this is an appropriate use of this land? Explain your answer.

d Describe the extent of flooding in Tewkesbury and suggest how it affected people's lives in the town.

3 Study photos **B** and **C** and carry out some Internet research on the Tewkesbury floods. Imagine you keep a daily diary where you describe what happens to you and in the town of Tewkesbury where you live. Write your diary entries for 20 to 25 July 2007. Include facts and figures about the flooding; the effects and responses (you must distinguish between these); yours and your family's feelings about events during those days. You should try to capture what it would have been like in the town.

4 Study map **E** and photo **F**. Imagine you are working as a volunteer for an aid agency in Bangladesh during the floods of 2004. Write a letter home, describing the situation in the floods in Bangladesh, its effects and responses to it. Try to make your account as real as possible so that the reader can imagine the experience.

5 Consider the Tewkesbury flood (2007) and the Bangladesh flood (2004).

a Compare the effects and responses to the floods.

b To what extent do you think the effects and responses reflect the different levels of development of the two countries?

	Tewkesbury	Bangladesh
Location		
Date		
Causes		
Effects		
Responses		

AQA *Examiner's tip*

When aiming to compare two case studies, it can be effective to record information in a simple table like the one below and then to use this as a writing frame to produce a more detailed answer.

links

You can find out more about flooding in the UK and Bangladesh at **www.bbc.co.uk**.

The website **www.sin.org.uk** has information about Bangladesh.

5.6 Hard and soft engineering: which is the better option?

Hard engineering strategies involve the use of technology in order to control rivers, while soft engineering adopts a less intrusive form of management, seeking to work alongside natural processes. Hard engineering approaches tend to give immediate results and control the river, but are expensive. However, in the future, they may make problems worse or create unforeseen ones. Soft engineering is much cheaper and offers a more sustainable option as it does not interfere directly with the river's flow.

Hard engineering

Dams and reservoirs exert a huge degree of control over a river. The natural flow of water is prevented by a dam (often a concrete barrier across the valley), water fills the area behind it and is released or held depending on circumstances such as current and expected rainfall. Dams and reservoirs are normally constructed as part of a **multi-purpose project** rather than with just a single aim in mind.

In this section you will learn

how hard and soft engineering are used to try to manage rivers and flooding

why there is debate about the two options

how to evaluate the two strategies and come to a supported view about them.

Key terms

Hard engineering: this strategy involves the use of technology in order to try to control rivers.

Soft engineering: this option tries to work within the constraints of the natural river system and involves avoiding building on areas especially likely to flood, warning people of an impending flood and planting trees to increase lag time.

Multi-purpose project: a large-scale venture with more than one aim. Many water projects relate to flood control, water supply, irrigation and navigation.

Case study

The Three Gorges Dam, China

The Three Gorges dam was constructed at Yichang on the River Yangtse (map **A**, photo **B** and table **C**). The capacity of the reservoir should reduce the risk of flooding downstream from a 1-in-10-year event to a 1-in-100-year event. Not only will this benefit over 15 million people living in high-risk flood areas, it will also protect over 25,000 ha of farmland.

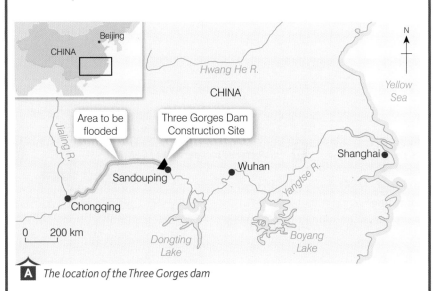

A *The location of the Three Gorges dam*

The dam is already having a positive impact on flood control, navigation and power generation, but it has caused problems. The Yangtse used to carry over 500 million tonnes of silt every year. Up to 50 per cent of this is now deposited behind the dam, which could quickly reduce the storage capacity of the reservoir.

Did you know ???????

The Hoover Dam over the Colorado river near Las Vegas was the biggest concrete structure when it was completed in 1935. There would have been enough concrete to make a two-lane road between San Francisco and New York.

AQA Examiner's tip

Learn examples of hard and soft engineering strategies used to manage river flow and be prepared to argue for and against each strategy.

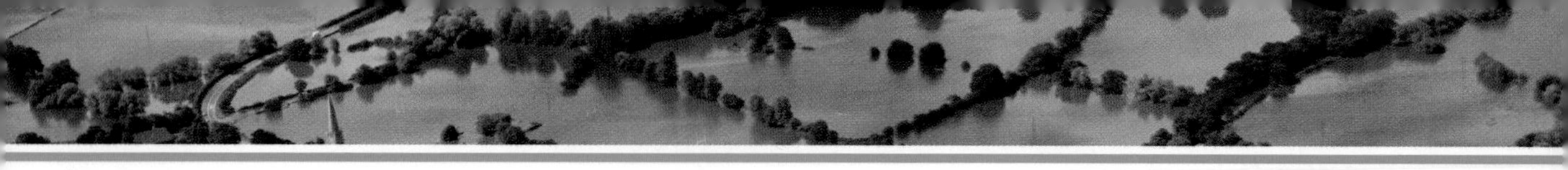

The water in the reservoir is becoming heavily polluted from shipping and waste discharged from cities. For example, Chongqing pumps in over 1 billion tonnes of untreated waste per year. Toxic substances from factories, mines and waste tips submerged by the reservoir are also being released into the reservoir.

Most controversially, at least 1.4 million people were forcibly moved from their homes to accommodate the dam, reservoir and power stations. These displaced people were promised compensation for their losses, plus new homes and jobs. Many have not yet received this, and newspaper articles in China have admitted that so far over $30 million of the funds set aside for this has been taken by corrupt local officials.

C *The Three Gorges Dam Project fact file*

Dimensions	181 m high and 2.3 km wide
Area flooded	632 km^2
Cost	$25.5 billion
Built	Started 1994; finished 2006
Increased depth	110 m (reduced to 80 m when flood risk downstream)

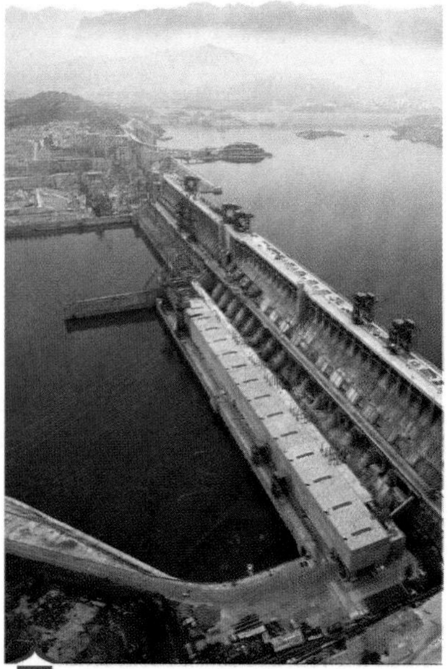

B *The Three Gorges Dam*

Straightening meanders represents a smaller-scale approach to managing rivers. Meanders are circuitous courses. Like following a route in a car, a semicircular way is longer and slower than a straight one. Therefore, water in a meander takes longer to clear an area than water in a straight section of a river. A possible solution to flooding in areas where there are many meanders on a river's course is to straighten them artificially. In this way, the river is made to follow a new shorter, straight section and abandon its natural meandering course (diagram **D**).

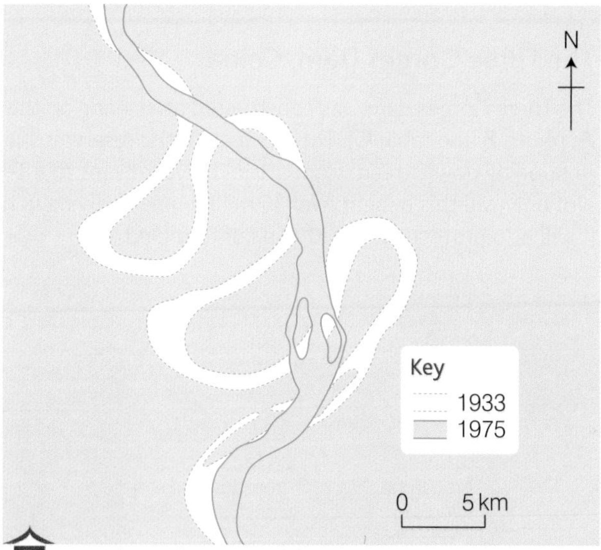
N
Key
1933
1975
0
5 km

D *Straightened sections of the Mississippi*

Soft engineering

Soft engineering is a strategy that accepts the natural processes of the river and seeks to work with it to reduce the effects of flooding, rather than attempting to gain control of it. A conscious decision can be made to 'do nothing' but simply to allow natural events to happen, even if this involves the risk of flooding. In some poorer areas of the world, this is a necessary approach. In richer areas, it could mean money is set aside in years when flooding does not occur to provide relief after the event. However, there are many more positive approaches that can be adopted to reduce the risk of flooding without exerting a major force over the river and its processes.

Flood warnings and preparation are complementary approaches. Telling people in advance of a flood gives them valuable time to prepare for it. The Environment Agency identifies areas at risk of flooding and issues warnings. For example, information was given on local radio, television and the Internet during the Tewkesbury flooding of 2007. Floodline Warnings Direct sends messages to registered users. A flood watch was issued on 20 July at 18.32, followed by a flood warning approximately 20 minutes later. This allows people time to take possessions upstairs; turn off gas, water and electricity; gather together important papers; and take some basic precautions against flooding. The Environment Agency's website contains general information on how to prepare for a flood and what to do during and afterwards.

Floodplain zoning occurs where the flood risk across different parts of the floodplain is assessed and resulting land use takes this into account. Diagram **E** summarises the land uses and changes across the floodplain. It takes into account the frequency and severity of flooding.

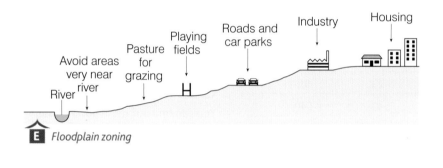

E *Floodplain zoning*

links

You can find out more about the Three Gorges Dam at **www.ctgpc.com**.

The Environment Agency gives extensive advice about flooding, which you can investigate at **www.environment-agency.gov.uk**.

Key terms

Straightening meanders: these occur when the natural curve in a river's course is left as the river follows an artificially more direct course that has been created for it, speeding up its flow out of an area.

Floodplain zoning: controlling what is built on the floodplain so that areas that are at risk of flooding have low-value land uses.

Economic: this relates to costs and finances at a variety of scales, from individuals up to government.

Social: this category refers to people's health, their lifestyle, community, etc.

Environmental: this is the impact on our surroundings, including the land, water and air as well as features of the built-up areas.

Activities

1 Study map **A**, photo **B** and table **C**.

a Produce a fact file giving six key items of background information on the Three Gorges Dam. Include such things as its location, when it was completed, etc.

b Create a table to give one **economic**, one **social** and one **environmental** cost and benefit of the project.

2 Study diagram **D**.

a Use the scale to estimate the reduction in the course of the river shown in the diagram after it had been straightened.

b Describe in your own words why river straightening reduces flood risk.

c Suggest some environmental effects of river straightening.

3 Access the Environment Agency website from the link above. Working in pairs, produce a leaflet or poster on 'What to do in the event of a flood'. Include information on:

■ warnings available

■ how people can get general and specific information

■ what to do before, during and after a flood in their home.

Make your leaflet or poster clear, eye-catching, informative and colourful.

4 Study diagram **E**. Vista Homes has submitted a planning application to build on the area labelled 'Playing fields'. Explain why, as the planner considering the application, you have rejected the scheme.

How are rivers in the UK managed to provide our water supply?

People in the UK use between 124 and 177 litres of water per day. This is an average of 151 litres per person per day. Table **A** shows actual water use, while the maps in **B** show the current and projected demands in England and Wales. The amount of rainfall over England and Wales varies, as does **water stress** (map **C**). Areas with high water stress can be seen as **areas of deficit** while those with low stress are often **areas of surplus**.

A *Water use for a selection of activities*

Activity	Average weekly use	Litres used per activity	Total number of litres
Bath	2	80	160
Flushing the toilet	35	8	280
Power shower	7	80	560
Washing machine	3	65	195
Dishwasher	4	25	100
Watering the garden	1	540	540
Washing car with bucket	1	32	32
Washing car with hose pipe	1	450	450

∞ links

You can find out more about water management at **www.environment-agency.gov.uk**. Click on Business and industry, then Water.

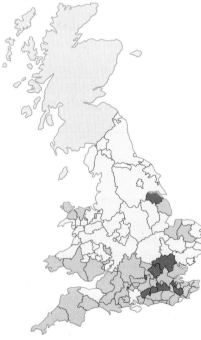

a Actual household use, 2005–06

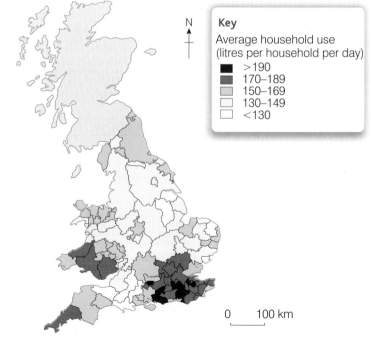

b Forecast household use, 2029–30

N

Key
Average household use
(litres per household per day)
■ >190
■ 170–189
□ 150–169
□ 130–149
□ <130

0   100 km

B *Household water use in England and Wales*

Key terms

Water stress: This occurs when the amount of water available does not meet that required. This may be due to an inadequate supply at a particular time or it may relate to water quality.

Areas of deficit: locations where the rain that falls does not provide enough water on a permanent basis. Shortages may occur under certain conditions, e.g. long periods without rain.

Areas of surplus: areas that have more water than is needed – often such areas receive a high rainfall total, but have a relatively small population.

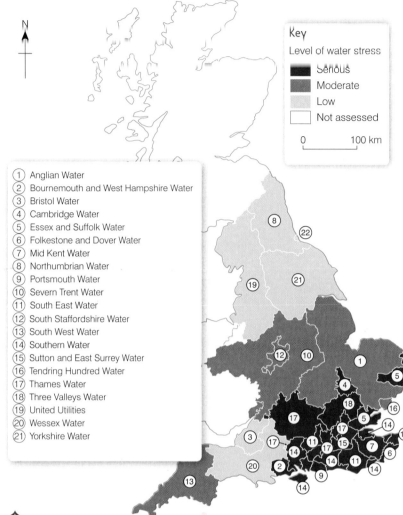

C *Water stress in England and Wales*

Key

Level of water stress

■ Serious
■ Moderate
□ Low
□ Not assessed

0   100 km

① Anglian Water
② Bournemouth and West Hampshire Water
③ Bristol Water
④ Cambridge Water
⑤ Essex and Suffolk Water
⑥ Folkestone and Dover Water
⑦ Mid Kent Water
⑧ Northumbrian Water
⑨ Portsmouth Water
⑩ Severn Trent Water
⑪ South East Water
⑫ South Staffordshire Water
⑬ South West Water
⑭ Southern Water
⑮ Sutton and East Surrey Water
⑯ Tendring Hundred Water
⑰ Thames Water
⑱ Three Valleys Water
⑲ United Utilities
⑳ Wessex Water
㉑ Yorkshire Water

It is expected that demand for water will increase due to an increased number of households and population in certain areas. In 2000, the UK's population was 59.4 million. By 2005 this had risen to over 60 million for the first time. Estimates predict a level of about 66 million by 2031. A more affluent lifestyle increases the demand for water as we buy more time-saving goods. We also demand foodstuffs out of season, which contributes to an increase in overall use of water.

There is a need to ensure that demand can be met in a **sustainable** way. A focus on local schemes is one way of ensuring this, rather than large-scale transfers. Generally, households that have a water meter use less water than those that do not – on average 19 litres per person per day less. When supplies are limited during drought, people use less water. Encouraging **conservation** is another strategy:

- Houses are being designed with better water efficiency.
- Devices are fitted to toilet cisterns to reduce water use.
- Rainwater can be collected.
- Bath water can be recycled to flush toilets.
- More people are taking showers than baths.

A significant amount of water is also lost in leakage.

AQA Examiner's tip

Learn a case study of one dam or reservoir and the positive and negative features associated with it.

Did you know ??????

Around 3 per cent of water in UK is used for drinking, although all water is of drinking quality. Many people in the world survive on less than 10 litres of water a day – a single flush of a toilet uses four-fifths of this.

reservoir. The new village, which retained the name Llanwddyn, was built 5 km from the original village.

Since the dam was built, new transfer schemes have been proposed. In 1973 such an approach was favourably received and three new reservoirs were constructed (Brenig in Wales, Kielder in Northumberland and Carsington in Derbyshire), but the remaining plans never became reality. Predicted demand was never realised and there was over-capacity.

In 1994 the idea of transfer returned, with grand schemes to transfer water from the Severn to the Thames and Trent. However, costs were high and in 2004 it was concluded that local schemes, including small reservoirs, could meet the demand for water in areas such as the south-east. If, in the future, this is not possible, transfers may become a possibility.

E *The location of Lake Vyrnwy*

Activities

1 Study table **A**.

a List the number of times you have used water for the activities in the table during the last 24 hours.

b Multiply the number of times by the litres shown.

c Add this up to give the total number of litres.

2 Study the maps in **B**.

a Which parts of England and Wales are forecast to have increased water demands by 2029–30?

b Suggest reasons why demand in these areas is likely to rise.

3 Study map **E** and photo **D**.

Working in pairs, describe orally one economic, one social and one environmental issue that resulted from the dam over the River Vyrnwy.

4 With the aid of Internet research, produce a colourful poster encouraging water conservation in the home.

Key terms

Sustainable: ensuring that the provision of water is long term and that supplies can be maintained without harming the environment.

Conservation: the thoughtful use of resources; managing the landscape in order to protect existing ecosystems and cultural features.

6   Ice on the land

6.1   How has climate change affected the global distribution of ice?

Temperature fluctuations

We are aware of global warming and the possible impacts it may have on our weather and on sea levels. However, concern about climate change is nothing new. As recently as the 1970s, some scientists suggested that the climate was actually cooling and there was talk of another ice age.

Although the current warming trend may be linked to human activities, climate change in the past has been entirely natural. The most likely causes are to do with subtle variations in the earth's orbit, slight changes in its tilt towards the sun or variations in the global pattern of ocean currents, which move heat around the world. Currently, slight changes in the ocean currents off the west coast of South America result in the El Niño effect. This has a significant impact on patterns of rainfall and on the development of tropical storms in some parts of the world.

Climate change during the Pleistocene period

Geological time is divided into periods. The most recent (apart from the present warm Holocene period) is called the Pleistocene period. Sometimes known as the Ice Age, the Pleistocene was a period of some 2 million years when global temperatures fluctuated considerably (graph **A**).

> **In this section you will learn**
>
> the fluctuations of temperature in the recent geological past
>
> changes in the extent and distribution of ice in the world resulting from changes in global temperature.

> **Key terms**
>
> **El Niño effect:** a periodic 'blip' in the usual global climatic characteristics caused by a short-term reduction in the intensity of the cold ocean current that normally exists off the west coast of South America. It results in unusual patterns of temperature and rainfall and can lead to droughts and floods in certain parts of the world.

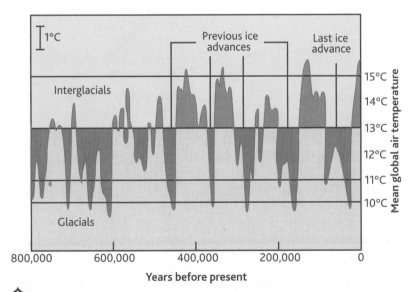

A Fluctuating temperatures during the Pleistocene period

Evidence from ice cores and deep-sea sediments suggests that there may have been as many as 20 cold periods or glacials during the Pleistocene period. During these **glacial periods** ice advanced south in the northern hemisphere to cover large parts of Europe and North America. Just 18,000 years ago ice reached its maximum extent during the last glacial period (map **B**). Map **C** shows how ice spread over the British Isles as far south as the Severn estuary at this time. Try to imagine ice covering almost all of the British Isles to a thickness of hundreds of metres in places. Even southern England would have been completely frozen, rather like parts of northern Canada today.

Between the glacial periods were warmer **interglacials**, which were at least as warm as today's climate if not warmer. There is evidence, for example, that hippopotamuses lived as far north as Leeds. These fluctuations in temperature had nothing to do with human activities – there were no power stations or cars. These changes occurred over thousands of years. You can see why we should treat yearly trends and one-off weather events with extreme scepticism when linking them to climate change.

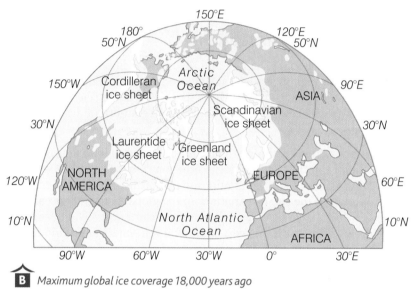

B Maximum global ice coverage 18,000 years ago

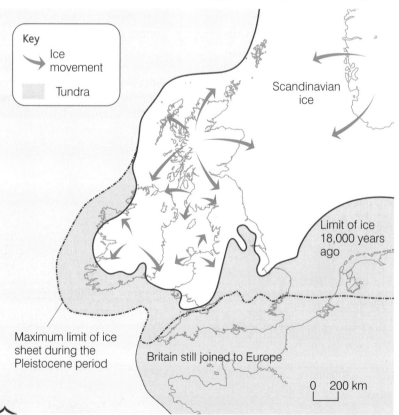

Key
→ Ice movement
▨ Tundra

Key terms

Glacial period: a period of ice advance associated with falling temperatures.

Interglacial: a period of ice retreat associated with rising temperatures.

Ice sheet: a large body of ice over 50,000 km² in extent.

Ice cap: a smaller body of ice (less than 50,000 km²) usually found in mountainous regions.

Glacier: a finger of ice usually extending downhill from an ice cap and occupying a valley.

C Ice coverage over the British Isles 18,000 years ago

Present-day global ice coverage

Currently there are two large areas of ice in the world called **ice sheets** (map **D**). The largest ice sheet is in Antarctica. It covers an area of 14 million km² and holds 90 per cent of all fresh water on the earth's surface. In places it is several kilometres thick. The Greenland ice sheet covers an area of 1.7 million km² (over 80 per cent of Greenland) and is currently showing increasing evidence of melting due to global warming.

AQA *Examiner's tip*

Take time to learn how far ice extended over the British Isles. You must be careful not to refer to glacial activity in parts of the country that were not actually covered by ice.

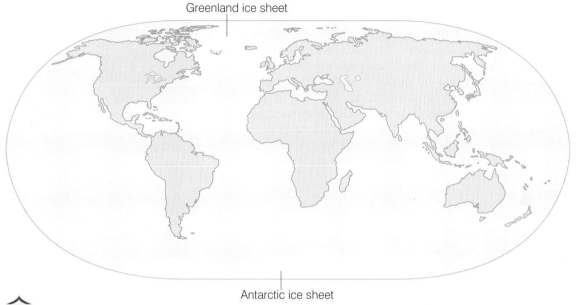

Greenland ice sheet

Antarctic ice sheet

D *Present-day distribution of ice sheets and ice caps*

Smaller bodies of ice covering an area of less than 50,000 km² are called **ice caps** or ice fields. They are usually found in mountainous areas where the temperatures are lower, such as in Iceland and the European Alps. Spreading out from ice caps are individual 'fingers' of ice called **glaciers**. They often follow former river valleys and extend down to an altitude where melting converts the ice to running water. Glaciers are found in every continent in the world and in some 47 individual countries.

∞ links

You can find out more about ice sheets, ice caps and glaciers on the Wikipedia website at **en.wikipedia. org/wiki/Ice_sheet**.

Activities

1   Study map **B**.

a   How many ice sheets existed in the northern hemisphere 18,000 years ago?

b   What were the names of the ice sheets in North America?

c   Using an atlas, describe the extent of the ice coverage in North America and Europe 18,000 years ago. Refer to lines of latitude and longitude. Which countries and present-day major cities would have been covered by ice?

d   Suggest why the glacial period 18,000 years ago had little impact on land masses in the southern hemisphere.

2   Study map **C**.

a   On a blank outline map, make a copy of the maximum extent of ice covering the British Isles. Include the arrows to show the direction of ice flow.

b   Using an atlas, label the four upland areas in the British Isles from where ice spread onto lower ground.

c   Why do you think these areas were source areas for ice during this glacial period?

d   Locate your nearest town or city on your map.

e   During this period it would have been possible to walk from the UK to the Netherlands, as there was no North Sea. Explain why.

Glaciers

A glacier acts as a system with inputs (**accumulation**) and outputs (**ablation**) (diagram **A**). The main input to the system is snow. When snow falls it becomes compacted as more snow settles on top. Air is expelled and the individual snowflakes turn into granular ice crystals. In much the same way that fluffy snow is squeezed into a hard snowball, the ice becomes denser and eventually turns into clear glacier ice. Avalanches of snow and ice also provide inputs into the system. Ablation mostly involves melting. This is likely to occur near the **snout** of the glacier where the air temperature is higher, particularly in summer. It can also occur on the surface of the glacier during summer. Occasionally chunks of ice break away at the snout. This is called calving and is another output from the system. A final output is the loss to the air by evaporation (water liquid to water vapour) and sublimation (water solid to water vapour).

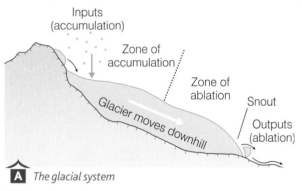

A *The glacial system*

The glacier budget

The **glacier budget** is the balance between the inputs and the outputs. If accumulation exceeds ablation over several years, the glacier will advance. If ablation exceeds accumulation over several years, the glacier will retreat. Note on diagram **A** that accumulation tends to dominate near the top of the glacier whereas ablation dominates at the snout.

The glacier budget varies between the seasons (Graph **B**). In the winter there will be a lot of accumulation with little ablation. In the summer, when it is warmer, ablation will tend to dominate over accumulation.

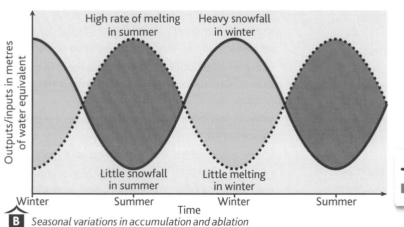

B *Seasonal variations in accumulation and ablation*

Key terms

Accumulation: inputs to the glacier budget, such as snowfall and avalanches.

Ablation: outputs from the glacier budget, such as melting.

Snout: the front of a glacier.

Glacier budget: the balance between the inputs (accumulation) and the outputs (ablation) of a glacier.

AQA Examiner's tip

Make sure you learn the terminology associated with the glacier budget and understand how it fluctuates both seasonally and yearly.

Activity

1   Study diagram **A**.

a   Make a copy of the diagram. Add the following labels:

■   snowfall

■   avalanches

■   melting

■   evaporation/sublimation

■   calving.

b   Why does accumulation dominate in the upper part?

c   Why is ablation at its greatest near to the snout?

The retreat of South Cascade Glacier, USA

The South Cascade Glacier in Washington State, USA is one of three 'benchmark' glaciers in the USA that have been monitored by the United States Geological Survey (USGS) for over 40 years. Data have been collected on stream run-off, air temperature and mass balance (glacier budget). All three glaciers have shown a marked retreat, which may well be caused by global warming. Worldwide, this trend is mirrored by other studies.

Look at the photos in **C**. They show how the South Cascade Glacier changed between 1900 and 2008. Table **D** contains statistics illustrating the changes that have taken place in the glacier budget since 1985.

C *Changes in the South Cascade Glacier*

D *Glacier budget data for South Cascade Glacier, 1985–2005*

Year	Winter (metres of water equivalent)	Summer (metres of water equivalent)	Net glacier budget (metres of water equivalent)	Year	Winter (metres of water equivalent)	Summer (metres of water equivalent)	Net glacier budget (metres of water equivalent)
1985	2.18	−3.38	−1.20	1996	2.94	−2.84	
1986	2.45	−3.06	−0.61	1997	3.71	−3.08	
1987	2.04	−4.10		1998	2.76	−4.62	
1988	2.44	−3.78		1999	3.59	−2.57	
1989	2.43	−3.34		2000	3.32	−2.94	
1990	2.60	−2.71		2001	1.90	−3.47	
1991	3.54	−3.47		2002	4.02	−3.47	
1992	1.91	−3.92		2003	2.66	−4.76	
1993	1.98	−3.21		2004	2.08	−3.73	
1994	2.39	−3.99		2005	1.97	−4.42	
1995	2.86	−3.55					

Activities

2 Study the photos in **C**.

a Describe the changes that took place in the South Cascade Glacier between 1900 and 2008.

b Why do you think the lower part of the glacier has melted between 1900 and 2008?

c Do you think these photos support the concept of global warming? Explain your answer.

d Assuming the melting continues at much the same rate, when do you think the glacier will disappear completely?

3 Study table **D**.

a Copy and complete the table by calculating the figures for the net glacier budget column. The first two have been calculated for you.

b Present the data from the table in the form of a line graph.

c Describe the trends shown by your graph.

d To what extent does your graph support the evidence of glacier retreat indicated by the photos in **C**?

Which processes operate in glacial environments?

Freeze–thaw weathering

If you visit a glacial area you will notice huge piles of jagged rocks lining the sides of the valleys (photo **A**). These are piles of scree and they result from freeze–thaw weathering (page 37).

Glacial erosion

The angular rock fragments produced by freeze–thaw weathering are vital tools for glacial erosion. They work their way under the ice, acting like sand on a sheet of sandpaper enabling the ice to grind away at the valley floor and sides. The scree fragments themselves become shattered and pulverised by the weight of the ice, turning them into tiny pieces. This material is called rock flour and it causes glacial rivers to look milky.

Although glaciers move slowly – usually at only a few centimetres a year – they are capable of carrying out a tremendous amount of erosion. There are two main types of glacial erosion:

- **Abrasion** – this has a sandpaper effect caused by the weight of the ice scouring the valley floor and sides using the angular rock material trapped underneath (diagram **B**). In the same way that sandpaper smoothes wood, abrasion results in a smooth and often shiny rock surface. Scratches caused by large rocks beneath the ice can often be seen. These are called striations.

A Glacier movement in the French Alps and its valley

- **Plucking** – when meltwater beneath a glacier freezes, it bonds the glacier base to the rocky surface below rather like glue. As the glacier moves, any loose fragments of rock are plucked away, like extracting loose teeth. This process leaves behind a jagged rocky surface as opposed to the smooth surface resulting from abrasion.

Glacial movement

In areas such as the Alps in Europe, melting ice in the summer produces a great deal of meltwater. This water helps to lubricate the glacier, enabling it to slide downhill. This type of movement is called basal slip and it can sometimes cause sudden movements of a glacier. In hollows high up on the valley sides, this movement may be more curved, in which case it is called rotational slip (photo **A**).

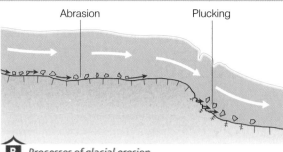

B *Processes of glacial erosion*

In winter, when the glacier is frozen to the rocky surface, the sheer weight of the ice and the influence of gravity cause individual ice crystals to deform in a plastic-like way. This process, called internal deformation, also causes the glacier to slowly move downhill.

Glacial transportation

Rock fragments resulting from freeze–thaw weathering and those that have been eroded by the ice are transported by the glacier, which acts like a giant conveyor belt. This sediment, called **moraine**, can be transported on the ice, in the ice (having been buried by snowfall) and below the ice.

Deposition

Most deposition occurs when the ice melts. As most melting occurs at the snout of a glacier, this is where the bulk of the material carried by the ice ends up. Renewed forward movement of the glacier pushes this debris further downhill rather like a bulldozer, which is why this process is called **bulldozing**.

As a glacier slowly retreats, it leaves behind a sheet of broken rock fragments and pulverised rock flour. Although some of this sediment (moraine) may remain to form **hummocks**, a good deal of it is washed away by meltwater rivers.

Key terms

Abrasion: a process of erosion involving the wearing away of the valley floor and sides (glaciers) and the shoreline (coastal zones).

Plucking: a process of glacial erosion where individual rocks are plucked from the valley floor or sides as water freezes them to the glacier.

Rotational slip: slippage of ice along a curved surface.

Moraine: sediment carried and deposited by the ice.

Bulldozing: the pushing of deposited sediment at the snout by the glacier as it advances.

Hummock: a small area of raised ground, rather like a large molehill.

AQA *Examiner's tip*

Make sure that you learn the terminology associated with the glacial processes described. Refer to them by name in exam answers.

Activities

1. Study photo **A** and diagram **B**.
   a. What is the evidence that freeze–thaw weathering is active in this landscape?
   b. Three conditions are required for freeze–thaw weathering to operate. The availability of water is one. Name the other two.
   c. Draw a series of simple diagrams to show the process of freeze–thaw weathering. Add annotations to describe what is happening.
   d. Why is freeze–thaw weathering absolutely vital if a glacier is going to be able to erode its valley?

2. Study diagram **B**.
   a. Make a copy of the diagram.
   b. Add labels and annotations to your diagram to describe how abrasion and plucking operate. Use colours to improve the clarity of the diagram.

3. Imagine you are in a glaciated area standing on a rocky surface that has been eroded by a glacier. What evidence would you look for to determine which of the two processes had eroded the rock on which you were standing?

4. Why does most of the material eroded by a glacier end up being deposited at the snout?

What are the distinctive landforms resulting from glacial processes?

Landforms of glacial erosion

Ice is an incredibly powerful agent of erosion and it can form spectacular landforms in mountainous areas. Look at photo **A**, which shows a glaciated area in the European Alps. Look closely at the labelled landforms. They are commonly found in areas of past and present glaciation.

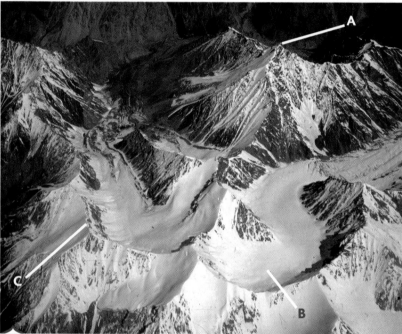

A *Glaciers with major features of erosion*

Corries

Corries, also known as cirques and cwms, are large, hollowed-out depressions found on the upper slopes of glaciated valleys. They are characterised by having a steep back wall and a raised lip at the front. A corrie may contain a lake called a tarn.

Study diagram **B**, which describes the formation of a corrie. Snow accumulates in a sheltered hollow on a hillside. Gradually the snow turns to ice and a small corrie glacier is formed. Through the process of rotational slip, the glacier scoops out an over-deepened hollow in an action similar to that of an ice cream scoop. Reduced erosion at the front of the corrie due to thinner ice forms a raised lip, allowing a tarn to form behind it.

In this section you will learn

the characteristics and formation of glacial landforms.

Key terms

Corrie: a deep depression on a hillside with a steep back wall, often containing a lake.

Arête: a knife-edged ridge, often formed between two corries.

Pyramidal peak: a sharp-edged mountain peak.

Glacial trough: a wide, steep-sided valley eroded by a glacier.

Truncated spur: an eroded interlocking spur characterised by having a very steep cliff.

Hanging valley: a tributary glacial trough perched up on the side of a main valley, often marked by a waterfall.

Ribbon lake: a long narrow lake in the bottom of a glacial trough.

Long profile: this shows the changes in height and shape along the length of a glacier, from its source high in the mountains to its snout.

Lateral moraine: a ridge of frost-shattered sediment running along the edge of a glacier where it meets the valley side.

Medial moraine: a ridge of sediment running down the centre of a glacier formed when two lateral moraines merge.

Terminal moraine: a high ridge running across the valley representing the maximum advance of a glacier.

Drumlin: an egg-shaped hill found on the floor of a glacial trough.

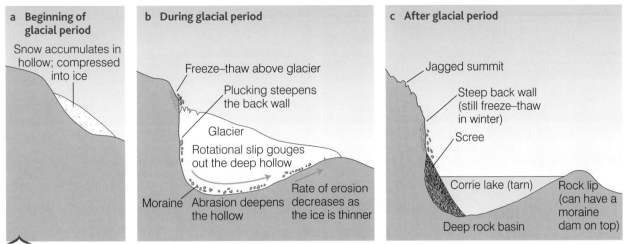

a **Beginning of glacial period**

Snow accumulates in hollow; compressed into ice

b **During glacial period**

Freeze–thaw above glacier

Plucking steepens the back wall

Glacier

Rotational slip gouges out the deep hollow

Moraine   Abrasion deepens the hollow

Rate of erosion decreases as the ice is thinner

c **After glacial period**

Jagged summit

Steep back wall (still freeze–thaw in winter)

Scree

Corrie lake (tarn)

Rock lip (can have a moraine dam on top)

Deep rock basin

B   *The formation of a corrie*

Arêtes and pyramidal peaks

An **arête** is a knife-edged ridge often found at the back of a corrie or separating two glaciated valleys (diagram **C**). Arêtes are often extremely narrow features and, although popular with hill walkers, strong crosswinds can make walking along them hazardous.

A typical arête forms when erosion in two back-to-back corries causes the land in between to become ever narrower. If three or more corries have formed on a mountain, erosion may lead to the formation of a single peak rather than a ridge. This feature is called a **pyramidal peak**.

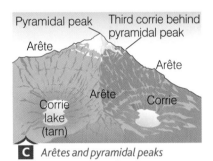

Pyramidal peak   Third corrie behind pyramidal peak

Arête

Arête

Arête

Corrie lake (tarn)

Corrie

C   *Arêtes and pyramidal peaks*

Glacial valley landforms

Glaciers most commonly flow along pre-existing river valleys. Unable to flow around obstacles, they tend to carve relatively straight courses (diagram **D**). Their incredible strength enables them to form dramatic features including deep **glacial troughs**, **truncated spurs**, **hanging valleys** and **ribbon lakes**.

A glacial trough is a steep-sided, wide and relatively flat-bottomed valley. The process of abrasion is mostly responsible for the formation of glacial troughs – the moving glacier grinds into the base and sides of the valley over a period of many hundreds of years. Unable to flow around previously existing interlocking spurs, the glacier simply cuts straight through them, forming steep-edged truncated spurs. Former tributary valleys, occupied with small glaciers unable to erode down to the same level as the main glacier, are left perched up on the valley side as hanging valleys. Spectacular waterfalls often mark these features.

Erosion of the valley floor is usually erratic and results in an uneven **long profile**. Certain stretches of the valley are prone to increased downcutting, for example where the ice becomes thicker after a tributary glacier has joined or where a weaker band of rock is encountered. At the end of a glacial period, water may occupy this deepened section to form a long, narrow ribbon lake often several tens of metres deep. Ribbon lakes are common in Scotland, with one of the most famous being Loch Ness. Occasionally, a much more shallow ribbon lake forms in a glacial trough behind a dam of deposited moraine.

Activities

1   Study photo **A**.

Name the glacial features labelled **A** to **C**.

2   Study diagrams **B** and **C**.

a   Draw a series of simple diagrams to show how the erosion of two back-to-back corries can result in the formation of an arête. Add detailed labels to describe what is happening. Use diagram **B** to help you.

b   Why are arêtes dangerous features to walk along?

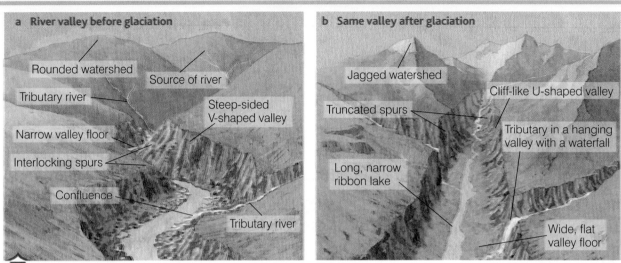

a River valley before glaciation

Rounded watershed

Source of river

Tributary river

Steep-sided
V-shaped valley

Narrow valley floor

Interlocking spurs

Confluence

Tributary river

b Same valley after glaciation

Jagged watershed

Cliff-like U-shaped valley

Truncated spurs

Tributary in a hanging
valley with a waterfall

Long, narrow
ribbon lake

Wide, flat
valley floor

D *Features and landforms in a glacial trough*

Landforms of glacial transportation and deposition

Moraine

Moraine is the general term given to the largely angular rock material transported and then deposited by the ice. It can sometimes be so extensive that it forms a thick blanket of material. Geologists refer to this glacially deposited material as till or boulder clay, on account of the range of sizes of sediment present. Till is found along much of the east coast of northern England. Although it forms fertile soils, it is weak and easily eroded when exposed at the coast.

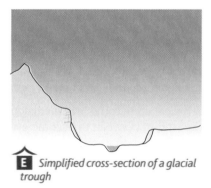

E *Simplified cross-section of a glacial trough*

It is possible to identify a number of types of moraine (diagram **F**):

- **Ground moraine** – material that was dragged underneath the glacier which is simply left behind when the ice melts. It often forms uneven hummocky ground.
- **Lateral moraine** – forms at the edges of the glacier. It is mostly scree that has fallen off the valley sides following freeze–thaw weathering. When the ice melts, it forms a slight ridge on the valley side.

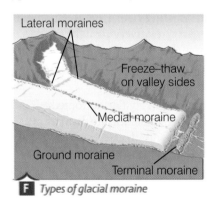

Lateral moraines

Freeze–thaw
on valley sides

Medial moraine

Ground moraine

Terminal moraine

F *Types of glacial moraine*

- **Medial moraine** – when a tributary glacier joins the main glacier, two lateral moraines merge to produce a single line of sediment that runs down the centre of the main glacier. On melting, the medial moraine forms a ridge down the centre of the valley.
- **Terminal moraine** – huge amounts of material pile up at the snout of a glacier to form a high ridge, often tens of metres high across the valley. This is a terminal moraine. It represents the furthest extent of the glacier's advance, hence the term 'terminal'.

Meltwater often erodes away moraine features, leaving only fragments.

Activities

3 Study diagram **E**.

a Draw a large copy of the diagram.

b Use diagram **D** to help you label the main features shown on the cross-section.

4 Standing in a glacial trough, you notice a steep rock cliff on the valley side. What evidence would you look for to identify whether it was a truncated spur or a hanging valley?

5 How do ribbon lakes form? Use simple diagrams to help answer this question.

OS map study: glacial features in the Nant Ffrancon valley, Wales

During the last ice advance, North Wales was buried beneath a thick sheet of ice. Glaciers scoured the valleys to create spectacular landforms – none more so than in the Nant Ffrancon valley (map extract **G** and photo **H**). Note that the Welsh term *cwm* denotes a corrie and *llyn* a lake. Take a few moments to orientate the photo on the map and then attempt the following activities.

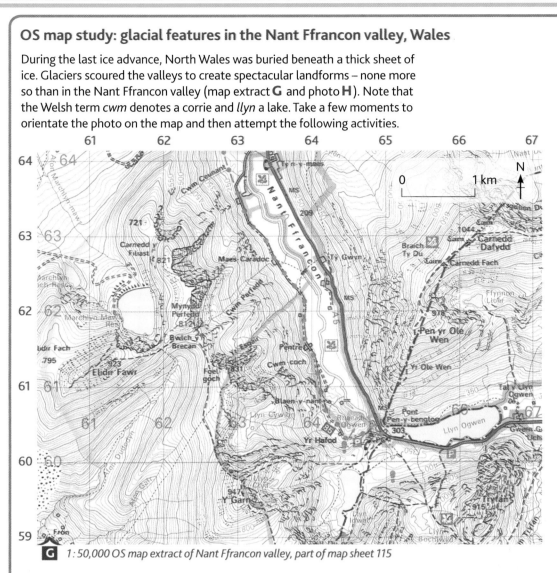

G   *1 : 50,000 OS map extract of Nant Ffrancon valley, part of map sheet 115*

H   *Aerial photo of Nant Ffrancon valley*

AQA   *Examiner's tip*

Practise drawing glacial features and adding labels, including process terms to describe their characteristics and formation.

Did you know ??????

In 1967, in his boat Bluebird K7, Donald Campbell used a ribbon lake called Coniston Water in the Lake District to try to beat his existing world water speed record of 444 kph. Tragically, he was killed as Bluebird flipped and somersaulted at 515 kph.

Drumlins

Drumlins are smooth, egg-shaped hills commonly about 10 m in height and up to a few hundred metres in length, which often occur in clusters on the floor of a glacial trough (photo **I**). They are made of morainic material, but instead of simply being dumped they were actually moulded and shaped by the moving ice. Drumlins usually have a blunt end, which faces up-valley, and a more pointed end facing down-valley. This makes them useful indicators of the direction of glacial movement.

I *Drumlins in Cumbria*

∞ links

A useful glossary with photos can be found at www.uwsp.edu/geo/faculty/Lemke/alpine_glacial_glossary/glossary.html.

Activities

6 Study diagram **F**.

a What is the difference between a lateral moraine and a medial moraine?

b What sort of deposited material would you expect to find making up a terminal moraine?

7 Study photo **I**. Draw a sketch of the photograph. Add labels to describe its main features. Add an arrow to show the likely direction of movement.

8 Imagine you have discovered two small hills in the centre of a glacial trough. What evidence would you look for to identify whether the feature was a fragment of a medial moraine or a drumlin?

9 Why are moraines often fragmentary?

10 Study map extract **G** and photo **H**.

a What are the names of the landforms labelled **A** to **D**?

b What is the name of the corrie lake (tarn) at **E**?

c What is the name of the peak at **F** and what is its six-figure grid reference?

d In what direction do you think the glacier flowed along the valley? Support your answer with evidence from the map.

e Why do you think some corries have lakes and others do not?

f What attractions are there for tourists to the area? Include references to the map in your answer.

11 Study map extract **G** and photo **H**.

a Draw a cross-section across the Nant Ffrancon valley from grid reference 630620 to the 978 m spot height at grid reference 657620. (If you feel confident about drawing cross-sections, you may choose to take the section further west.)

b Use the same horizontal scale as the map (2 cm = 1 km) and use a vertical scale of 1 cm = 500 m.

c Locate and label the following features:

- frost-shattered crags on the east side of the valley
- glacial trough
- the river (the Afon Ogwen)
- the A5 road.

d Complete your cross-section by adding a title and labelling the axes.

12 Study map extract **G**. Complete a summary table to help with your revision. For each of the following features, select an example from the map and draw a simple sketch of the contour patterns. Include scales, contour values and place names. Keep it simple but make it accurate.

- corrie (cwm)
- arête
- ribbon lake
- glacial trough
- peak (not ideal pyramidal peaks, but not bad examples!)

6.5   What opportunities do glacial areas offer for tourism?

Tourism in the French Alps: Chamonix

Chamonix is situated in the north-westerly part of the Alps, just 15 km from the Swiss border and 15 km from Italy via the Mont Blanc tunnel (map **A**). Chamonix and its valley are dominated by the rounded summit of Mont Blanc, Europe's highest mountain at 4,808 m (photo **B**).

Chamonix has been a centre for tourism for over 250 years. Its stunning landscape has a huge amount to offer outdoor enthusiasts. The resident population of 10,000 is swollen by up to 100,000 visitors a day in summer and about 60,000 a day in winter.

> **In this section you will learn**
>
> the attractions and opportunities available for tourism in the Alps
>
> the need for responsible tourism and sustainable management.

Case study

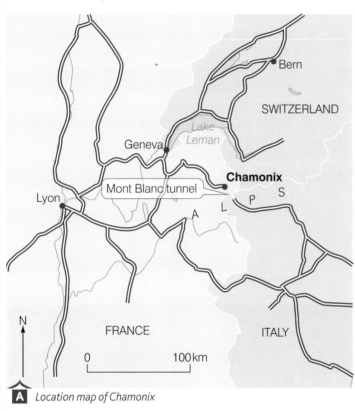

A *Location map of Chamonix*

B *Chamonix and Mont Blanc*

Winter attractions

Chamonix provides a huge range of options for skiers and snowboarders of all abilities (map **C** and photo **D**). Cable cars and cog railways provide easy access to the pistes. Cross-country skiing has become popular and two local courses have been established nearby. There are opportunities for ice climbing, free riding and paragliding. Snowshoe trails offer hikers the opportunity to walk in the area. Chamonix itself caters for winter tourists by providing hotels, restaurants, heated swimming pools and spas. With its museums, shops and historical buildings, there is much to do away from the slopes.

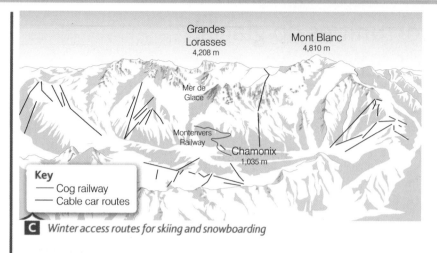

Grandes
Lorasses
4,208 m

Mont Blanc
4,810 m

Mer de
Glace

Montenvers
Railway

Chamonix
1,035 m

Key
— Cog railway
— Cable car routes

C *Winter access routes for skiing and snowboarding*

Summer attractions

In summer, the mountain landscapes offer tremendous potential for outdoor activities. The famous Montenvers railway takes visitors to the Mer de Glace, where visitors can witness the awesome scale of the glacier and the valley that it has carved (photo **A**, page 128). An ice cave allows visitors to step inside the glacier (photo **E**).

The Chamonix area boasts 350 km of marked hiking trails, 40 km of mountain bike tracks, rock climbing, mountaineering, paragliding, rafting, canyoning, pony trekking and summer luging. The town comes alive in the summer with live music, outdoor cafés and colourful flowers.

Impacts of tourism in the French Alps

Tourism to Chamonix and the surrounding area brings both benefits and problems.

Benefits of tourism:

- Tourists bring huge economic benefits, providing employment for local people in hotels and restaurants, in sports facilities and as guides and instructors. Construction and maintenance provides jobs for local people.

- The extra income supports local services such as shops. Local people benefit from improvements in public transport and health care.

- Chamonix is maintained as an attractive town. Pedestrian streets give people safe access to shops and the town is clean and well lit.

Problems of tourism:

- The town can become noisy and congested at peak times. Access to Chamonix via motorway is good, but in Chamonix itself the roads are narrow and become jammed easily.

- Mountain footpaths have become eroded due to the sheer volume of visitors, both walking and using mountain bikes.

D *Winter sports in Chamonix*

E *Tourists explore the ice grotto in the Mer de Glace*

- The shops, cafés and restaurants are tourist-orientated and expensive. Local people often have to pay more for everyday items. Houses are expensive and many are second homes for wealthy visitors.

- Conflicts can arise between different groups of people. Mass tourism activities can create unwelcome noise and damage to the environment, which can detract from the enjoyment of those seeking more peaceful activities such as walking or bird watching. Tourism can conflict with local people. Farm animals can be harmed by thoughtless actions of tourists, such as leaving gates open or dropping litter.

Managing tourism in Chamonix

- Chamonix is keen to promote **responsible tourism** as a means of balancing the demands of tourism with the need to conserve and protect the environment.

- The Chamonix municipality (local authority) provides an environmentally friendly transport service with clean energy buses and free public transport.

- An initiative called Espace Mont-Blanc involves cooperation between France, Italy and Switzerland on issues of international transport, nature conservation, forests and water resources.

- A further initiative called Tomorrow's Valley brings together representatives from the local community and tourist groups to plan for **sustainable management**. Current projects include:
  - burying service networks such as electricity lines underground
  - renovating and preserving historic buildings and monuments
  - preserving natural wetlands and peat bogs
  - minimising the impact of skiing on the landscape by planting trees and using local building materials that blend in with the natural environment
  - maintaining and way-marking footpaths and cleaning rivers – this provides seasonal employment for local people
  - supporting local traditional employment sectors, particularly farming.

Activity

1 Study the photos, maps and text.

a In the form of a table, draw up two lists of the activities available to visitors in winter and summer. (You could add illustrations to your table using pictures from the Internet and present your information electronically if you wish.)

b Why do you think Chamonix promotes the area as both a winter and a summer destination?

c Suggest some possible conflicts that might arise between different groups of tourists in winter and summer.

d How might conflicts arise between tourists and local people?

e In what ways does tourism have a negative impact on the environment?

f What management measures have been adopted to protect the environment?

g Why is it important to involve local people in designing management plans for the local area?

Key terms

Responsible tourism: the idea of encouraging a balance between the demands of tourism and the need to protect the environment.

Sustainable management: a form of management that ensures that developments are long lasting and non-harmful to the environment.

links

The main Chamonix tourism site, which has plenty of excellent information, is at **www.chamonix.com**.

The Espace Mont-Blanc website is at **www.espace-mont-blanc.com**.

A detailed map of Chamonix can be found at **www.skifrance.com/chamonix/resortmap.html**.

6.6   What is the impact of climate change on Alpine communities?

Evidence for recent climate change

Many of the Alps' most popular resorts, such as Morzine and Megève, lie at relatively low levels in France – about 1,000 m. They are all in danger of running out of snow as the world warms up.

The UN estimates that in 30 years' time the snowline will have risen by 300 m and that up to half of all resorts in Europe will be forced to close by 2050. This will cause a great deal of hardship as hotels, restaurants and shops are forced to close due to lack of business. Switzerland could lose up to £1bn a year if its resorts close down.

Responses in lower-level resorts

Lower-level resorts have responded in a number of ways in order to cope with the problem:

- Tourists have been transported by bus to higher-level resorts for skiing.
- Artificial snow is cannoned onto the slopes (photo **A**). In low-level parts of Austria and Italy, up to 40 per cent of resorts now have to make their own snow. This is expensive and can have a serious effect on vegetation, which may take 30 years to recover.
- Resorts have had to re-invent themselves. Some have started to promote themselves as centres for cross-country skiing, hiking, climbing, sledding or snowshoeing.
- There are plans to build new ski lifts to link resorts, but this could cause considerable damage to the environment.

A A snow cannon in action on the Alps

Abondance, France

Abondance is a typical Alpine ski resort in the Haute-Savoie region of France (photo **B**). It is one of several traditional ski resorts in the area that depends on income from winter skiers for the survival of its many hotels, restaurants and shops.

In 2007, following 15 years of unreliable snowfall, the ski lifts in Abondance closed for the last time. The local council is considering two options in an attempt to secure its future and provide employment for its local population:

- To develop other forms of winter sports, such as ski touring, snowshoeing and snow-mobiling that are less dependent on deep snow than traditional skiing.

- To develop its summer programme of activities to include hiking, water sports and mountain biking. This would enable the town to become more of an all-year round resort rather than just a winter one.

New developments

The High Alps form a pristine landscape rich in wildlife and free from pollution. With low-level resorts in decline, there is considerable pressure from developers to develop these wilderness areas. However, tourism developments can harm **fragile environments** in a number of ways:

- Road construction and the building of ski lifts, houses and hotels can have a big impact on natural ecosystems and habitats.

- Trees are often cut down to make way for developments. They have an important function in binding the soil on steep slopes and breaking up avalanches.

- Overuse of slopes for skiing can strip a hillside of its natural vegetation, which can take many years to re-establish. Mountain biking can lead to gullies, which can be enlarged following heavy rain or snowmelt to form scars on the landscape.

- With increased levels of pollution (e.g. noise, visual), the natural landscape loses some of its appeal and this may reduce the attractiveness of the area for tourism.

B   *The Abondance valley in the French Alps*

In Austria new ski lifts and cable cars are being constructed to open up the Gepatsch Glacier to skiers above 3,500 m. However, with almost all Alpine glaciers retreating, there is no guarantee that glacier skiing will last for long. At Lisenser Fernerkogel new cable cars and pistes are being developed (photo **C**). In other areas there are plans to construct cable cars to link resorts, which would enable lower-level resorts to remain viable even if tourists have to travel by cable car to another higher-level resort to ski.

Should the environment be protected at all costs or should some development be allowed in order to provide employment for local people and secure the region's economic future?

C *Off piste group in the Alps*

Activities

1 Study photo **B**.

a Why are low-level ski resorts like Abondance suffering from a decrease in winter tourism?

b How has this affected the lives of local people living in these resorts?

c What strategies can low-level resorts adopt to respond to the problem?

d Imagine that you are the mayor of Abondance. Which of the two options do you favour and why? Perhaps you think that both options should be adopted.

2 Study photo **C**.

a What is meant by 'off piste' skiing?

b Why are many people attracted to this type of skiing?

c Imagine that the area in the photograph was to be devloped as a permanent ski resort. What developments would you expect to take place and how would they impact on the environment?

d Make a list of the advantages and disadvantages of developing new areas such as photo **C** as ski resorts.

e Do you think the area in photo **C** should be developed as a ski resort? Explain your answer.

AQA *Examiner's tip*

Be aware of the arguments for and against developments in the Alps. Be prepared to express your own opinion.

⊂⊃ links

The BBC has information at **www.bbc.co.uk**. Type 'Alpine ski future' into the search box.

6.7   What is the avalanche hazard?

What are avalanches?

Avalanches are masses of snow, ice and rocks that move downhill at speeds of up to 300 kph. They occur naturally in mountain environments and only pose a hazard when they impact on people or human activity, such as transport routes or houses.

There are two main types of avalanche:

- **Loose snow avalanche** – this type of avalanche usually starts from a single point on the hillside and involves loose, powdery snow (photo and diagram **A**).
- **Slab avalanche** – this tends to be a more deadly type of avalanche. It involves a large slab of ice and snow shearing away from a hillside and moving rapidly downhill, carrying rocks and trees as it does so (photo and diagram **B**). It has immense power and can cause a great deal of damage.

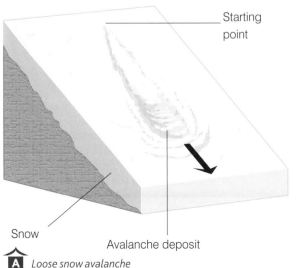

Starting point

Snow

Avalanche deposit

A *Loose snow avalanche*

In this section you will learn

the risk posed by avalanches

the factors contributing to the avalanche hazard and ways of reducing the hazard.

Key terms

Avalanche: a rapid downhill movement of a mass of snow, ice and rocks, usually in a mountainous environment.

Loose snow avalanche: a powdery avalanche usually originating from a single point.

Slab avalanche: a large-scale avalanche formed when a slab of ice and snow breaks away from the main ice pack.

Did you know ??????

If you are buried by an avalanche, you have a 90 per cent survival rate if you are found in the first 15 minutes. This drops to 30 per cent after 35 minutes. The speed of rescue is critical if you are to survive.

AQA *Examiner's tip*

Be aware that avalanches usually result from a combination of factors, although there is usually a single trigger such as heavy snowfall or an earthquake. Use the Internet to update your case studies.

Causes of avalanches

There are a number of factors that contribute to the risk of avalanches:

- Heavy snowfall – this adds weight to earlier snowfalls. Uneven rates of freezing, together with occasional melting, can create distinct layers within the snow and ice making slab avalanches more likely to occur as one layer slips over another.
- Steep slopes – avalanches are more likely to occur on steep slopes in excess of 30°.
- Tree removal – the removal of trees for ski developments enables avalanches to move downhill unimpeded. When present on a hillside, trees can break up an avalanche and prevent it becoming too large.
- Temperature rise – sudden rises in temperatures and associated melting often lead to avalanches in the spring.
- Heavy rainfall – this can lubricate a slope and trigger an avalanche.
- Human factors – almost all deaths from avalanches kill the people who actually triggered them. Off-piste skiing is a major cause of avalanches because it often involves skiing in areas of fresh snow that have not been assessed for the avalanche risk.

Avalanches as hazards

Deaths, injuries and property damage due to avalanches have increased in the last 50 years. This is mainly due to the growth of winter sports and the expansion of ski resorts to cater for increasing numbers of visitors. In Switzerland an average of 40 people die each year from avalanches, over 80 per cent of whom are involved in winter sports. In 2007–08, avalanches in France killed 15 people. Four were climbing and the rest were skiing or snowboarding.

Increasing numbers of people are putting themselves at risk from avalanches. If avalanches become more frequent as the climate warms, the death toll and damage to property may well increase.

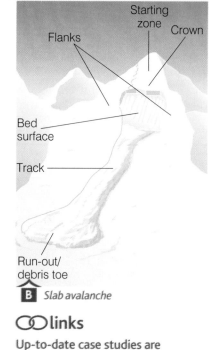

B *Slab avalanche*

⚭ links

Up-to-date case studies are important, so make use of online resources and the Internet. The main media sites, such as the BBC (www.bbc.co.uk) are excellent sources of information.

The Avalanche Center is an excellent source of information at www. avalanche-center.org.

Recent avalanches

Kashmir, February 2008

Avalanches in the Himalayan foothills killed at least 30 people in February 2008 (photo **C**). The avalanches swept through villages, burying residents and destroying buildings. The cause of the avalanches was unusually heavy snowfall. Avalanches and freezing winter weather in the same region killed over 300 people in February 2005.

Europe, 2006

The winter of 2005–06 was a record year for fatalities in the French Alps, with 49 people being killed in off-piste avalanches alone. Irregular weather patterns, with fluctuating temperatures and variable amounts of rain and snow, led to a greater number of avalanches than usual. The fashion for extreme sports was also thought to put more people at risk from avalanches.

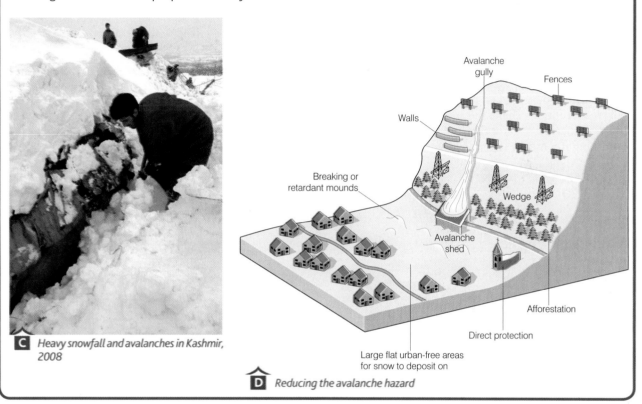

C *Heavy snowfall and avalanches in Kashmir, 2008*

D *Reducing the avalanche hazard*

1 Study figures **A** and **B**, and diagram **D**.

a With the aid of simple diagrams and sketches, describe the differences between a loose snow avalanche and a slab avalanche.

b Why do you think slab avalanches pose a greater threat than loose snow avalanches?

c Present the information describing the factors contributing to the risk of avalanches in the form of a large diagram of a mountain slope similar to that in diagram **D**. Use labels to identify the various factors on your drawing.

d For each scenario in the case study, which factors seemed to have been important in triggering the avalanches that caused the death and destruction?

2 Study diagram **D**.

a Draw a simplified version of the diagram.

b For each of the measures identified, add an annotated label to describe how it reduces the avalanche threat.

c Which of the measures in your diagram do you think are most appropriate for poor mountainous parts of the world such as Kashmir? Give reasons for your answer.

7.1 How do waves shape our coastline?

How waves form

Waves are usually formed by the wind blowing over the sea. Friction with the surface of the water causes ripples to form and these develop into waves. The stretch of open water over which the wind blows is called the **fetch**. The longer the fetch, the more powerful a wave can become.

Waves can also be formed more dramatically when earthquakes or volcanic eruptions shake the seabed. These waves are called tsunamis. In December 2004 giant tsunami waves devastated the countries bordering the Indian Ocean and 240,000 people were killed (photo **A**).

 A *The Indian Ocean tsunami, Sri Lanka*

When waves reach the coast

In the open sea, despite the wavy motion of the water surface, there is little horizontal transfer of water. It is only when the waves approach the shore that there is forward movement of water as waves break and wash up the **beach**. Diagram **B** shows what happens as waves approach the shore. Note how the seabed interrupts the circular orbital movement of the water. As the water becomes shallower, the circular motion becomes more elliptical. This causes the **crest** of the wave to rise up and then eventually to topple onto the beach. The water that rushes up the beach is called the **swash**. The water that flows back towards the sea is called the **backwash**.

> **In this section you will learn**
>
> how waves are formed
>
> why waves break at the coast
>
> the characteristics of constructive and destructive waves.

> **Did you know** ?????
>
> When the volcano Krakatoa erupted in 1883, a tsunami said to be 35 m high killed 36,000 people on the Indonesian island of Sumatra.

C *Surfers on a beach in Newquay*

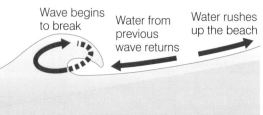

Top of wave moves faster

Wave begins to break

Water from previous wave returns

Water rushes up the beach

Circular orbit in open water

Friction with the seabed distorts the circular orbital motion

Increasingly elliptical orbit

Shelving seabed (beach)

 B *Waves approaching the coast*

Types of wave found at the coast

It is possible to identify two types of wave at the coast: **constructive waves** and **destructive waves**.

Constructive waves are waves that surge up the beach with a powerful swash. They carry large amounts of sediment and 'construct' the beach, making it more extensive. These are the waves that are loved by surfers (photo **C**). They are formed by distant storms, which can be hundreds of kilometres away. The waves are well spaced apart and are powerful when they reach the coast. Diagram **D** shows the typical characteristics of a constructive wave.

Destructive waves are formed by local storms close to the coast. They are so named because they 'destroy' the beach. Destructive waves are closely spaced and often interfere with each other, producing a chaotic, swirling mass of water. They rear up to form towering waves before crashing down onto the beach (diagram **E**). There is little forward motion (swash) when a destructive wave breaks, but a powerful backwash. This explains the removal of sediment and the destruction of the beach.

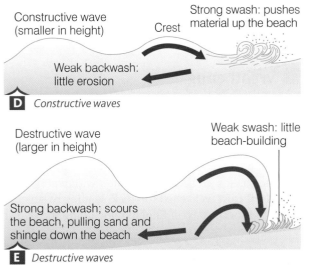

Constructive wave (smaller in height)   Crest   Strong swash: pushes material up the beach

Weak backwash: little erosion

D   *Constructive waves*

Destructive wave (larger in height)   Weak swash: little beach-building

Strong backwash; scours the beach, pulling sand and shingle down the beach

E   *Destructive waves*

Activities

1   Study photo **A**.

a   What evidence is there in the photo that tsunamis are extremely powerful?

b   Are there any safe places in the photo?

c   Many of the people who died in 2004 lived in small coastal communities around the Indian Ocean. Why do people choose to live close to the sea?

2   Study diagram **B** make a copy of it.

a   Draw an arrow to show the waves approaching the coast.

b   Write the labels 'swash' and 'backwash' in the correct places.

c   What causes the waves to rise up and break on the beach?

d   When waves break on a sandy or pebbly beach, the amount of backwash is often less than the amount of swash. Why do you think this is?

e   On a pebble beach, larger pebbles are often found near the top of the beach with smaller ones near the bottom. Use your answer to d to suggest why this happens.

3   Study photo **C**, which shows a constructive wave. They are sometimes known as a surging waves because they 'surge' up the beach.

a   Why do surfers prefer constructive waves to destructive waves?

b   Are constructive waves generated by storms close to the coast or a long way off?

c   Wave power is being considered as a source of **renewable energy** off the north Cornwall coast. Do you think this is a good idea? Explain your answer.

∞ links

You can find out more about the Indian Ocean tsunami at **www.guardian.co.uk/world/tsunami2004**.

Watch the surfers in Cornwall by webcam at **www.fistralsurfcam.com**.

Which land processes shape our coastline?

Weathering

Processes of weathering affect rocks exposed at the coast. Look at pages 37 to 39 to find out how common weathering processes operate. At the coast, freeze–thaw weathering is particularly effective if the rock exposed is porous (contains holes) and permeable (allows water to pass through it). Look at photo **A**. The autumn of 2000 was exceptionally wet and the chalk rock became saturated with water. During late winter, long periods of frost weakened the rock, leading to several dramatic **rockfalls** along the south coast of England, including the one shown.

In this section you will learn

how weathering affects rocks at the coast

how processes of mass movement operate at the coast.

Key terms

Rockfall: the collapse of a cliff face or the fall of individual rocks from a cliff.

Did you know ??????

The lighthouse at Beachy Head has had to be moved inland to avoid collapsing into the sea.

A Rockfall at Beachy Head, Sussex

Mass movement

The rockfall shown in photo **A** is one example of mass movement. Mass movement is the downhill movement of material under the influence of gravity. In 1993, 60 m of cliff slid onto the beach near Scarborough in North Yorkshire, taking with it part of the Holbeck Hall Hotel (photo **B**). Teetering on the brink, the hotel had to be demolished. Diagram **C** describes some of the common types of mass movement found at the coast.

B Landslip at Holbeck Hall, Scarborough

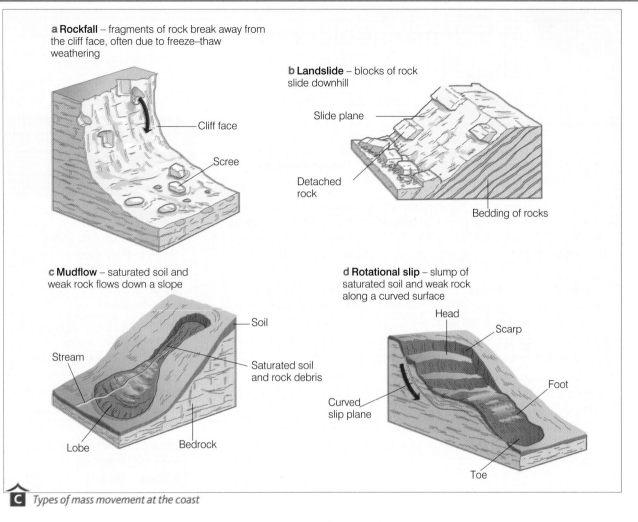

a Rockfall – fragments of rock break away from the cliff face, often due to freeze–thaw weathering

Cliff face

Scree

b Landslide – blocks of rock slide downhill

Slide plane

Detached rock

Bedding of rocks

c Mudflow – saturated soil and weak rock flows down a slope

Stream

Soil

Saturated soil and rock debris

Lobe

Bedrock

d Rotational slip – slump of saturated soil and weak rock along a curved surface

Head

Scarp

Foot

Curved slip plane

Toe

C *Types of mass movement at the coast*

Both mass movement and weathering provide an input of material to the coastal system. Much of this material is carried away by the waves to be deposited elsewhere along the coast.

Activity

Study photo **A** and the section about chalk on pages 42–43.

a   Why is chalk vulnerable to freeze–thaw weathering?

b   How does freeze–thaw weathering operate?

c   What other weathering processes affect chalk? Describe how these processes work.

d   How might rockfalls such as that shown in the photo be a hazard to people?

e   What do you think will happen to the pile of rocks at the base of the cliff in the photo?

f   Imagine the local council has decided to place an information board at the top of the cliff to warn people of the dangers of cliff collapse. It wants to inform people why the cliff is dangerous. Design an information board explaining why the cliff is vulnerable to rockfalls. Use diagrams to illustrate your board and do not forget to warn people to keep well away from the cliff edge!

AQA   *Examiner's tip*

Weathering processes take place over long periods of time and often require many repetitions before changes occur. Always refer to the correct geographical terms when writing about weathering and mass movement.

∞ links

You can find out more about a Beachy Head rockfall in 1999 at **www.bbc.co.uk**. Enter 'Beachy Head rockfall' into the search box.

Some excellent information about the processes of mass movement can be found at **www.intute.ac.uk/ sciences/hazards/mass.html**.

Which marine processes shape our coastline?

Coastal erosion

When a wave crashes down on a beach or smashes against a cliff, it carries out the process of erosion (photo **A**). There are several processes of coastal erosion:

- **Hydraulic power** – this involves the sheer power of the waves as they smash onto a cliff (photo **A**). Trapped air is blasted into holes and cracks in the rock, eventually causing the rock to break apart. The explosive force of trapped air operating in a crack is called cavitation.
- **Corrasion** – this involves fragments of rock being picked up and hurled by the sea at a cliff. The rocks act like erosive tools by scraping and gouging the rock.
- **Abrasion** – this is the 'sandpapering' effect of pebbles grinding over a rocky platform, often causing it to become smooth.
- **Solution** – some rocks are vulnerable to being dissolved by seawater. This is particularly true of limestone and chalk, which form cliffs in many parts of the UK.
- **Attrition** – this is where rock fragments carried by the sea knock against one another, causing them to become smaller and more rounded.

Coastal transportation

Four main types of sediment transportation can be identified (diagram **B**). The size and quantity of sediment transported by the sea depends on the strength of the waves and tidal currents. During storms, quite large pebbles can be flung up on to seawalls and promenades where they can cause damage to buildings and cars.

> **In this section you will learn**
>
> the processes of coastal erosion, transportation and deposition.

> **Key terms**
>
> **Hydraulic power:** the sheer power of the waves.
>
> **Corrasion:** the effect of rocks being flung at the cliff by powerful waves.
>
> **Solution:** the dissolving of rocks, such as limestone and chalk.
>
> **Attrition:** the knocking together of pebbles, making them gradually smaller and smoother.
>
> **Traction:** heavy particles rolled along the seabed.
>
> **Solution:** the transport of dissolved chemicals.
>
> **Saltation:** a hopping movement of pebbles along the seabed.
>
> **Suspension:** lighter particles carried (suspended) within the water.
>
> **Longshore drift:** the transport of sediment along a stretch of coastline caused by waves approaching the beach at an angle.

A *Waves crashing onto cliffs in Iceland*

Solution: dissolved chemicals often derived from limestone or chalk

Suspension: particles carried (suspended) within the water

Traction: large pebbles rolled along the seabed

Saltation: a 'hopping' or 'bouncing' motion of particles too heavy to be suspended

B *Types of coastal transportation*

The movement of sediment on a beach is largely determined by the direction of wave approach. Look at diagram **C**. Note that where the waves approach 'head on', sediment is moved up and down the beach. However, if the waves approach at an angle, sediment moves along the beach in a zig-zag pattern. This is called **longshore drift**.

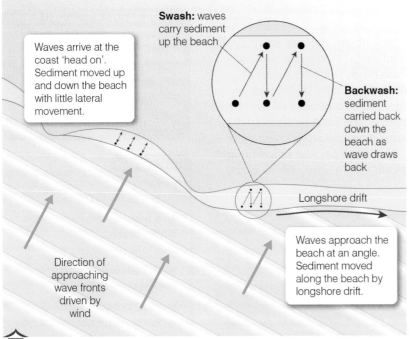

Swash: waves carry sediment up the beach

Waves arrive at the coast 'head on'. Sediment moved up and down the beach with little lateral movement.

Backwash: sediment carried back down the beach as wave draws back

Longshore drift

Direction of approaching wave fronts driven by wind

Waves approach the beach at an angle. Sediment moved along the beach by longshore drift.

C *Longshore drift*

■ Coastal deposition

Coastal deposition takes place in areas where the flow of water slows down. Sediment can no longer be carried or rolled along and has to be deposited. Coastal deposition most commonly occurs in bays, where the energy of the waves is reduced on entering the bay. This explains the presence of beaches in bays and accounts for the lack of beaches at headlands, where wave energy is much greater.

Activities

1 Study diagram **B**. Draw an annotated diagram similar to this diagram to describe the processes of coastal erosion. Draw a simple diagram of a wave breaking against the foot of a cliff. Add detailed labels (annotations) in their correct places to describe the five processes of erosion.

2 Study diagram **C**.

a What is meant by the term 'longshore drift'?

b Why does longshore drift occur on some beaches but not on others?

c Draw your own diagram or a series of diagrams to show how the process of longshore drift operates. Add labels to describe what is happening.

d Imagine that you are going to carry out a fieldwork investigation to look for evidence of longshore drift along a stretch of coastline. What evidence would you look for? Explain your reasons.

◯◯ links

A good summary of coastal processes and landforms can be found at www.georesources.co.uk/leld.htm.

A fun animation using a rubber duck to show the process of longshore drift can be found at www.geography-site.co.uk/pages/physical/coastal/longshore.html.

7.4   What are the distinctive landforms resulting from erosion?

Headlands and bays

Cliffs rarely erode at an even pace. Sections of cliff that are particularly resistant to erosion stick out to form **headlands**. Weaker sections of coastline that are more easily eroded form **bays**. Diagram **A** shows the changes that take place to a coastline where rocks of different resistances meet the coast. Headlands are most vulnerable to the power of the waves, which explains the presence of erosional features such as cliffs and **wave-cut platforms.** In contrast, bays are often much more sheltered from the full fury of the sea. The waves are less powerful and deposition tends to dominate. This explains why a sandy beach is the most common feature found in bays.

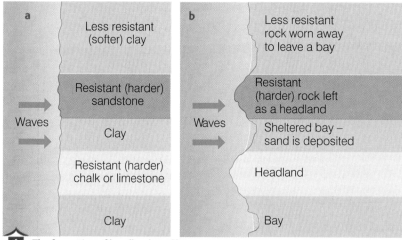

a
- Less resistant (softer) clay
- Resistant (harder) sandstone
- Clay
- Resistant (harder) chalk or limestone
- Clay

Waves

b
- Less resistant rock worn away to leave a bay
- Resistant (harder) rock left as a headland
- Sheltered bay – sand is deposited
- Headland
- Bay

Waves

A   *The formation of headlands and bays*

Cliffs and wave-cut platforms

When waves break against a cliff, erosion close to the high-tide line takes a 'bite' out of the cliff to form a feature called a **wave-cut notch**. Over a long period of time – usually hundreds of years – the notch gets deeper until the overlying cliff can no longer support its own weight and it collapses. Through a continual sequence of wave-cut notch formation and cliff collapse, the cliff line gradually retreats. In its place will be a gently sloping rocky platform called a wave-cut platform (photo **B**).

A wave-cut platform is typically quite smooth due to the process of abrasion, but in some places it may be pockmarked with rock pools. During long periods of constructive waves, the wave-cut platform may become covered by sand or shingle. Destructive waves associated with local winter storms remove the beach once again, exposing the wave-cut platform.

In this section you will learn

the formation and characteristics of features of coastal erosion (headlands and bays; wave-cut platforms and caves; caves, arches and stacks).

Key terms

Headland: a promontory of land jutting out into the sea.

Bay: a broad coastal inlet often with a beach.

Wave-cut platform: a wide, gently sloping rocky surface at the foot of a cliff.

Wave-cut notch: a small indentation (or notch) cut into a cliff roughly at the level of high tide caused by concentrated marine erosion at this level.

Cave: a hollowed-out feature at the base of an eroding cliff.

Arch: a headland that has been partly broken through by the sea to form a thin-roofed arch.

Stack: an isolated pinnacle of rock sticking out of the sea.

B   *Wave-cut platform and cliff near Beachy Head*

Caves, arches and stacks

Lines of weakness in a headland, such as joints or faults, are particularly vulnerable to erosion. The energy of the waves gouges out the rock along a line of weakness to form a **cave** (diagram **C**). Over time, erosion may lead to two back-to-back caves breaking through a headland to form an **arch**. Gradually, the arch is enlarged by erosion at the base and sides and by weathering processes acting on the roof. The roof collapses eventually to form an isolated pillar of rock known as a **stack**.

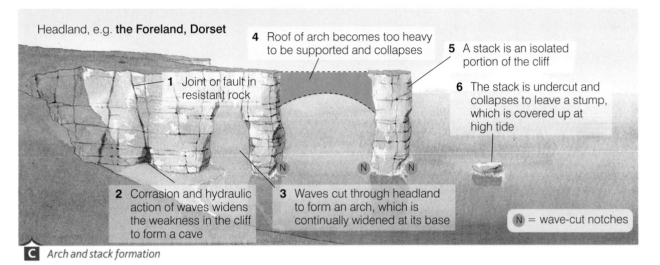

Headland, e.g. **the Foreland, Dorset**

4 Roof of arch becomes too heavy to be supported and collapses

5 A stack is an isolated portion of the cliff

1 Joint or fault in resistant rock

6 The stack is undercut and collapses to leave a stump, which is covered up at high tide

2 Corrasion and hydraulic action of waves widens the weakness in the cliff to form a cave

3 Waves cut through headland to form an arch, which is continually widened at its base

N = wave-cut notches

C *Arch and stack formation*

Study photo **B**.

a   Draw a simple sketch of the cliff and wave-cut platform in the photo.

b   Add labels to identify the cliff and the wave-cut platform. Suggest where you think the wave-cut notch is located at the foot of the cliff.

c   Add labels to indicate where the processes of erosion operate at high tide. The rock in the photo is chalk.

d   Use a series of simple annotated diagrams to show how a cliff (such as the one in the photo) is undercut by the sea and then collapses to form a wave-cut platform. Include reference to the processes and landforms in your annotations.

e   Under what conditions might a wave-cut platform become covered by sand or shingle?

f   Is it a good time to go rock-pooling after a period of stormy weather? Explain your answer.

⚭ **links**

Good diagrams of coastal features can be found at **www.georesources. co.uk/leld.htm**.

Check out Geography at the Movies at **www.geographyatthemovies. co.uk/Coasts.html**.

Coastal erosion at Swanage, Dorset

OS map study: coastal erosion at Swanage, Dorset

The coast at Swanage in Dorset is a classic stretch of coastline in the UK. It exhibits several of the key features of erosion that you have been studying, including headlands, bays, cliffs, arches and stacks (photo **B**).

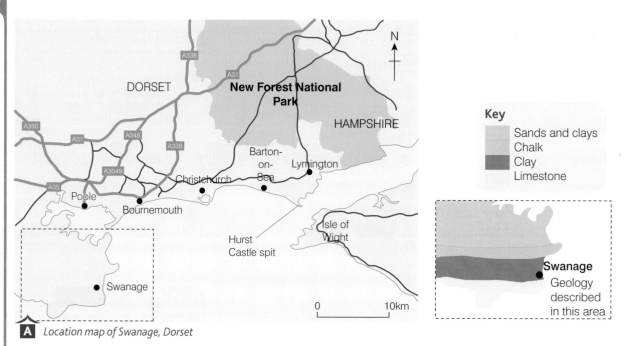

A *Location map of Swanage, Dorset*

Key
- Sands and clays
- Chalk
- Clay
- Limestone

B *Old Harry Stack and the Foreland, Dorset*

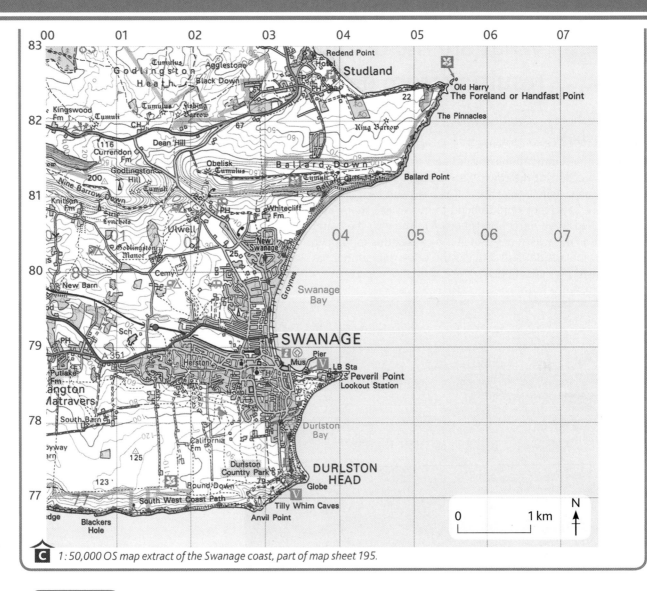

C 1 : 50,000 OS map extract of the Swanage coast, part of map sheet 195.

Activities

1 Study map extract **C**.

a The Foreland, Peveril Point and Durlston Head are all examples of which coastal landform?

b In which grid square is the Foreland?

c Locate Swanage Bay. Approximately how wide is the bay from Ballard Point to Peveril Point?

2 Study map **A** and map extract **C**.

a Use map **A** to explain the formation of Swanage Bay. Include simple diagrams to support your explanation.

b What map evidence is there that deposition is occurring in Swanage Bay?

c How does this deposition help to explain the growth of Swanage as a tourist resort?

3 Study photo **B** and map extract **C**.

a What are the features labelled **A**, **B** and **C** on the photo?

b Why do you think feature **A** has been formed in this position?

c What processes of erosion have led to the formation of feature **A**?

d Use map extract **C** to give the local name of feature **C**.

4 Design an information board to be located at the Foreland to explain the formation of feature **C**. Draw a series of annotated diagrams to describe its formation. Refer to the processes of erosion in your annotations. Your information board is aimed at the general public, so make sure it is clear and colourful.

What are the distinctive landforms resulting from deposition?

Beaches

Beaches are accumulations of sand and shingle (pebbles) found where deposition occurs at the coast. Sandy beaches are often found in sheltered bays, where they are called bay head beaches. When waves enter these bays, they tend to bend to mirror the shape of the coast. This is called wave refraction (photo **A**). The way the water gets shallower as the waves enter the bays causes this to happen. Wave refraction spreads out and reduces the wave energy in a bay, which is why deposition occurs here.

In this section you will learn

the formation and characteristics of features of coastal deposition (beaches, spits and bars).

Key terms

Spit: a finger of new land made of sand or shingle, jutting out into the sea from the coast.

Salt marsh: low-lying coastal wetland mostly extending between high and low tide.

Bar: a spit that has grown across a bay.

Sea

Bending (refraction) of waves

Land

A Wave refraction at the head of a bay

Elsewhere along the coast, pebble beaches may form. These are most commonly found in areas where cliffs are being eroded and where there are higher-energy waves, such as along the south coast of the UK.

Ridges or berms are common characteristics of a beach (photo **B**). They are small ridges coinciding with high-tide lines and storm tides. Some beaches may have several berms, each one representing a different high-tide level.

Spits

A **spit** is a long, narrow finger of sand or shingle jutting out into the sea from the land (diagram **C**). Spits are common features across the world. As sediment is transported along the coast by longshore drift, it becomes deposited at a point where the coastline changes direction or where a river mouth occurs. Gradually, as more and more sediment is deposited, the feature extends into the sea. Away from the coast, the tip is affected by waves approaching from different directions and the spit often becomes curved as a result.

B Berms on a beach at Deal on the Kent coast

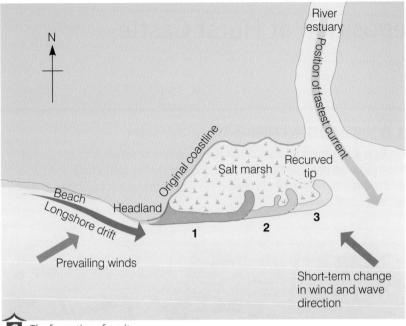

 *The formation of a spit*

D   *Coastal bar at Slapton Ley, Devon*

Over time, the sediment breaks the surface to form new land and a spit is formed. It soon becomes colonised by grass and bushes, and eventually trees will grow. On the landward sheltered side of a spit where the water is a calm, mudflats and **salt marshes** form. These are important habitats for plants and birds. Being close to sea level, spits are vulnerable to erosion, especially during storms.

Bars

Occasionally, longshore drift may cause a spit to grow right across a bay, trapping a freshwater lake or lagoon behind it. This feature is called a **bar** (photo **D**). In the UK some offshore bars have been driven onshore by rising sea levels following ice melt at the end of the last glacial period some 10,000 years ago. This type of feature is called a barrier beach. Chesil Beach in Dorset is one of the best examples of this feature.

AQA Examiner's tip

A spit is a land feature, so it does not get covered at high tide. On a map, it will be clearly bordered by the high tide line. Be careful not to confuse it with sediment exposed only at low tide.

⚭ links

Further information on Dungeness spit can be found at **www.reefnews. com/reefnews/oceangeo/ washngtn/dnwr.html**.

Activities

1   Study photo **A**.

a   What is wave refraction? Draw a simple sketch based on the photo to support your answer.

b   How does wave refraction explain the formation of a bay head beach?

c   A sandy beach is more likely to be found in a sheltered bay, whereas a shingle (pebble) beach is more likely to be found on an open and exposed stretch of coastline. Explain why.

2   Study photo **B**. What are berms and how do they form on a beach? Use a simple diagram to help your explanation.

3   Study diagram **C** and photo **D**.

a   What is a spit?

b   Draw a series of three simple diagrams to show the formation of a spit. Add detailed labels to describe what is happening using the text and the diagram to help you.

c   What is the difference between a spit and a bar?

d   What is the difference between a bar and a barrier beach?

Coastal deposition at Hurst Castle, Hampshire

OS map study: coastal deposition at Hurst Castle, Hampshire

Hurst Castle spit (map extract **A** and photo **B**) is a shingle spit located on the Hampshire coast close to the city of Southampton. Its formation is complex, but longshore drift has been actively shaping the landform for hundreds of years. Henry VIII built a castle near the tip of the spit to help defend England from possible invasions. Today, English Heritage manages the castle and it is a popular tourist destination.

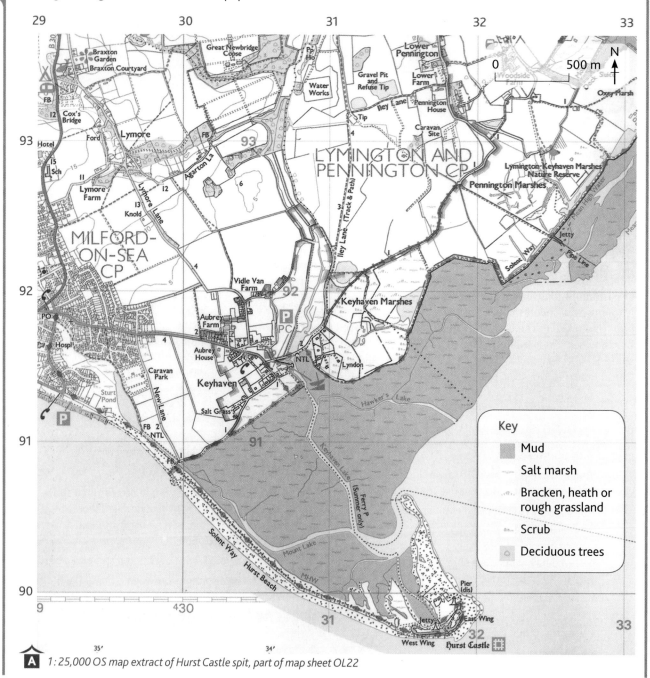

A *1:25,000 OS map extract of Hurst Castle spit, part of map sheet OL22*

B *Hurst Castle spit, Hampshire*

Activities

1 Study map extract **A** and photo **B**.

a In what direction is the photo looking?

b How long is the spit? Is it roughly 1 km, 2 km or 3 km in length?

c Use a ruler to measure the width of the spit in grid square 3090. Give your answer to the nearest 50 m. Remember that the spit is the area of land above the high-tide line, shown by the bold blue line on the map.

d Which symbol is used to indicate that the spit is made of shingle (pebbles)?

e Use the key to describe the nature of the land behind the spit in grid square 3090.

f What is the name of the spit feature at **A** on the photo?

g How has this feature been formed?

h In what direction do you think longshore drift is operating? Explain your answer.

i Summarise the answers you have given by writing a paragraph describing the spit. Use facts and figures in your description.

2 Study map extract **A**.

a Draw a sketch map to show the spit and its main features. Use a pencil for the sketch and a pen for adding labels.

b Begin by drawing a grid of squares to represent the grid squares on the map. This will enable you to copy the spit accurately. Use the same scale: 4 cm = 1 km or enlarge it if you wish.

c Carefully draw the spit. Remember to follow the bold blue high-tide line. This marks the outline of the spit.

d Locate Hurst Castle on your map and label it.

e Add the following additional labels to your sketch.

   ▪ spit

   ▪ recurved tip

   ▪ salt marsh

   ▪ direction of longshore drift.

f Complete your sketch by adding a north point, scale and title.

How will rising sea levels affect the coastal zone?

■ The causes of rising sea levels

One of the effects of global warming is sea-level change. Over the last 15 years, global average sea levels have risen by 3 mm a year. The latest estimates from the Intergovernmental Panel on Climate Change (IPCC) suggest a rise in global sea levels of between 28 and 43 cm by the end of the century.

In this section you will learn

the causes and possible consequences of rising sea levels on the coastal zone.

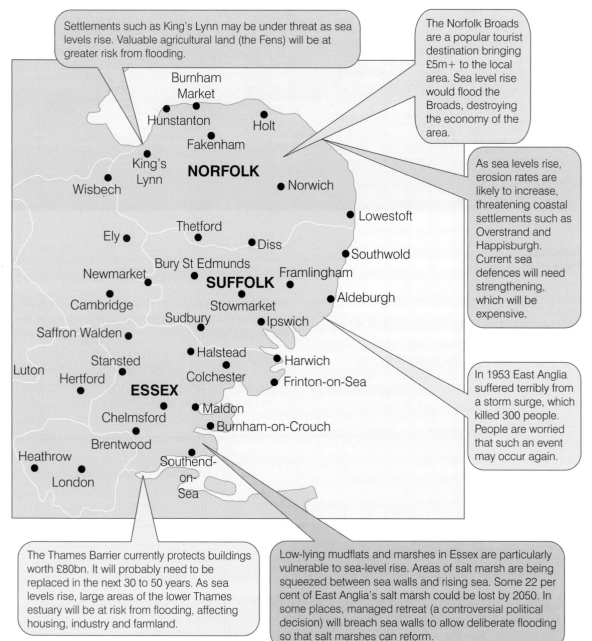

Settlements such as King's Lynn may be under threat as sea levels rise. Valuable agricultural land (the Fens) will be at greater risk from flooding.

The Norfolk Broads are a popular tourist destination bringing £5m+ to the local area. Sea level rise would flood the Broads, destroying the economy of the area.

As sea levels rise, erosion rates are likely to increase, threatening coastal settlements such as Overstrand and Happisburgh. Current sea defences will need strengthening, which will be expensive.

In 1953 East Anglia suffered terribly from a storm surge, which killed 300 people. People are worried that such an event may occur again.

The Thames Barrier currently protects buildings worth £80bn. It will probably need to be replaced in the next 30 to 50 years. As sea levels rise, large areas of the lower Thames estuary will be at risk from flooding, affecting housing, industry and farmland.

Low-lying mudflats and marshes in Essex are particularly vulnerable to sea-level rise. Areas of salt marsh are being squeezed between sea walls and rising sea. Some 22 per cent of East Anglia's salt marsh could be lost by 2050. In some places, managed retreat (a controversial political decision) will breach sea walls to allow deliberate flooding so that salt marshes can reform.

A *Possible impact of sea-level rise in East Anglia*

The main cause of sea-level rise is thermal expansion of the seawater as it absorbs more heat from the atmosphere. The melting of ice on the land, for example from glaciers on Greenland, will increase the amount of water in the oceans but scientists do not believe that this will significantly affect sea levels. Melting sea ice, such as from the Arctic, will have no direct effect on sea levels.

The actual amount of sea-level rise will vary from place to place as it is complicated by relative rises and falls in the level of the land as well as variations in the amount of deposition occurring at the coast. The amount of sea-level rise will also depend on the rate of global warming. However uncertain the future may be, most scientists agree that sea levels will rise and that coastal zones will be at an increasing risk from wave attack and flooding.

In the UK, East Anglia (diagram **A**) is likely to be hardest hit as rising sea levels threaten coastal defences and natural ecosystems. Elsewhere in the world, vast areas of low-lying coastal plains such as Bangladesh and whole chains of islands such as the Maldives and Tuvalu could disappear.

Already, two Pacific islands have been submerged, and many others have been flooded. Since more than 70 per cent of the world's population live on coastal plains, the effects of rising sea levels are likely to be devastating.

Did you know ??????

If sea levels rise by 1 m, several major cities of the world will be affected by flooding including London, Tokyo and New York.

AQA Examiner's tip

Be sure to understand the meaning of economic (money), social (people), environmental (natural world) and political (decision-making) when describing the possible impacts of sea-level rise.

Activities

1   Study diagram **A**.

a   Describe some of the likely economic impacts of sea-level rise in East Anglia.

b   Suggest some likely social and political (decision-making, often by the government) impacts of sea-level rise. Some of these will be linked to the economic factors identified in a.

c   With the aid of a diagram, describe how salt marshes could be lost as sea levels rise.

d   Why is there concern about loss of salt marsh environments in East Anglia?

2   This activity will involve internet research.

a   Select a coastal zone under threat from sea-level rise, such as Bangladesh, the Mississippi delta in the USA or the Maldives.

b   Conduct an internet search ('sea level rise + Maldives', for example) to find out about the economic, social, environmental and political impacts of sea-level rise on your chosen locality.

c   Draw a diagram similar to diagram **A**, using a photo or map of your locality in the centre.

∞ links

Flood London is an excellent source of information at www. floodlondon.com/floodtb.htm.

How can cliff collapse cause problems for people at the coast?

Various factors can contribute to cliff collapse. These include weathering processes such as heavy rainfall that can saturate the land and make it unstable, mass movement such as sliding and slumping, which is more likely if the land is made of soft weak rock types, and of course the power of the waves continually crashing against the cliffs and undercutting them from below. Various stretches of UK coastline are vulnerable to cliff collapse, including Christchurch Bay in Hampshire, where Barton-on-Sea is sited.

> **In this section you will learn**
>
> the causes and consequences of cliff collapse for people living in the coastal zone.

Case study

Barton-on-Sea, Hampshire

Locate the small settlement of Barton-on-Sea on map **B**. This stretch of coastline in Christchurch Bay has long been affected by coastal erosion and cliff collapse. Over the years a number of buildings, and most recently a café, have been lost to the sea.

Extensive coastal defences have been built to try to prevent coastal erosion. However, in 2008 a fresh landslip occurred (photo **A**). This has once again raised concerns among local residents about the vulnerability of this part of the coast to cliff collapse. An older development of houses in Barton Court is just 20 m from the cliff edge. The local authority predicts that the houses will be lost in the next 10 to 20 years.

The cliffs at Barton-on-Sea are prone to collapse due to a number of factors:

- The rocks are weak sands and clays. They are easily eroded by the sea and have little strength to resist collapse.

- The arrangement of the rocks (permeable sands on top of impermeable clay) causes water to 'pond-up' within the cliffs. This increases the weight of the cliffs. The increase in water pressure within the cliffs (called pore water pressure) encourages collapse.

A *Cliff collapse at Barton-on-Sea, 2008*

- This stretch of coastline faces the direct force of the prevailing south-westerly winds. With a long fetch, the waves approaching Barton-on-Sea are powerful and can carry out a great deal of erosion. Rates of erosion have been as much as 2 m a year in places.

- Several small streams (with the local name 'Bunny') flow towards the coast but disappear into the permeable sands before they reach the sea. This adds to the amount of water in the cliffs.

- Buildings on the cliff top have increased the weight on the cliffs, making them more vulnerable to collapse. They can also interfere with drainage.

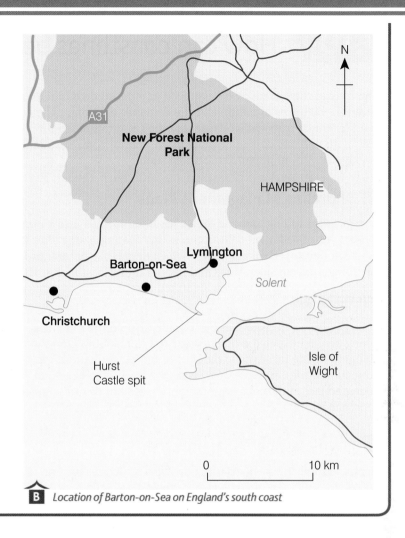

B *Location of Barton-on-Sea on England's south coast*

Activity

Study photo **A**.

a What evidence is there that there has been a cliff collapse?

b Estimate the drop in the grass surface due to the collapse.

c Look closely at the rock material forming the cliffs. How strong do you think it is? Explain your answer.

d What action has been taken to keep people at a safe distance from the cliff?

e Using evidence from the photo, suggest how this and future cliff collapses might affect people living in the area.

f How does the cliff collapse affect the local environment?

g Draw a simplified sketch of the cliffs at Barton-on-Sea and add annotations to identify the factors that have led to the cliff collapse.

AQA Examiner's tip

Cliff collapse usually results from the interaction of several factors. Make this clear in your answers.

∞ links

A website with fantastic photos is www.soton.ac.uk/~imw/barteros.htm.

The coastal zone needs to be managed in order to maintain a balance between the forces of nature and the demands of people. The coastline is under threat from cliff collapse, flooding and future sea-level rise. With millions of people living in the coastal zone, sustainable management is an important consideration.

Shoreline Management Plans

The coastline of England and Wales has been divided into a number of self-contained sediment cells. A **Shoreline Management Plan (SMP)** has been developed for each area, which details the natural processes, environmental considerations and human uses. Coastlines at risk from erosion or flooding have been identified and plans put in place to cope with the issues.

In most cases, a decision has been made to 'hold the line'. This means taking action to keep the line of the coast as it is now. Occasionally, planners decide to 'advance the line' of the coast to afford greater protection by, for example, increasing the size of the beach. In taking action, authorities have the option of hard or soft engineering approaches.

Hard engineering approaches

Hard engineering involves using artificial structures to control the forces of nature. For many decades people have used hard engineering structures such as sea walls (photo **A**) and groynes (photo **B**) to try to control the actions of the sea and protect property from flooding and erosion. Table **C** outlines the common hard engineering approaches used in coastal management.

In this section you will learn

the options of coastal management including hard and soft engineering and managed retreat.

Key terms

Shoreline Management Plan (SMP): an integrated coastal management plan for a stretch of coastline in England and Wales.

Hard engineering: building artificial structures such as sea walls aimed at controlling natural processes.

Soft engineering: a sustainable approach to managing the coast without using artificial structures.

Did you know ??????

In 1902 a sea wall was constructed at Galveston, USA following the devastating hurricane of 1900. The wall is 16 km long, 5.2 m high and 4.9 m thick at its base. So far, it has never been overtopped.

A *The sea wall at Dawlish, Devon*

B *Groynes at Eastbourne*

C *Hard engineering schemes*

Hard engineering	Description	Cost	Advantages	Disadvantages
Sea wall	Concrete or rock barrier to the sea placed at the foot of cliffs or at the top of a beach. Has a curved face to reflect the waves back into the sea, usually 3–5m high.	Up to £6 million per km (south sea zones).	• Effective at stopping the sea. • Often has a walkway or promenade for people to walk along.	• Can be obtrusive and unnatural to look at. • Very expensive and has high maintenance costs.
Groynes Beach sand	Timber or rock structures built out to sea from the coast. They trap sediment being moved by longshore drift, thereby enlarging the beach. The longer beach acts as a buffer to the incoming waves, reducing wave attack at the coast.	£10,000 each (at 200 m intervals).	• Result in a bigger beach, which can enhance the tourist potential of the coast. • Provide useful structures for people interested in fishing. • Not too expensive.	• In interrupting longshore drift, they starve beaches downdrift, often leading to increased rates of erosion elsewhere. The problem is not so much solved as shifted. • Groynes are unnatural and rock groynes in particular can be unattractive.
Rock armour	Piles of large boulders dumped at the foot of a cliff. The rocks force waves to break, absorbing their energy and protecting the cliffs. The rocks are usually brought in by barge to the coast.	Approximately £1,000 –£4,000 per metre.	• Relatively cheap and easy to maintain. • Can provide interest to the coast. Often used for fishing.	• Rocks are usually from other parts of the coastline or even from abroad. Can be expensive to transport. • They do not fit in with the local geology. • Can be very obtrusive.

Hard engineering approaches are less commonly used today. Not only are they expensive and involve high maintenance costs, they are also obtrusive and unnatural. They tend to interfere with natural coastal processes and can cause destructive knock-on effects elsewhere. In altering wave patterns, for example, erosion can become concentrated further down the coast leading to new problems of cliff collapse.

Soft engineering approaches

Soft engineering approaches (table **D**) try to fit in and work with the natural coastal processes. They do not involve large artificial structures. Often more 'low key' and with low maintenance costs – both economically and environmentally – soft engineering approaches such as beach nourishment (photo **E**) are more sustainable. They are usually the preferred option of coastal management.

Soft engineering	Description	Cost	Advantages	Disadvantages
Beach nourishment	The addition of sand or shingle to an existing beach to make it higher or broader. The sediment is usually obtained locally so that it blends in with the existing beach material. Usually brought onshore by barge.	Approximately £3,000 per metre	• Relatively cheap and easy to maintain. • Blends in with existing beach. • Increases tourist potential by creating a bigger beach.	• Needs constant maintenance unless structures are built to retain the beach.
Dune regeneration	Sand dunes are effective buffers to the sea yet they are easily damaged and destroyed, especially by trampling. Marram grass can be planted to stabilise the dunes and help them to develop. Areas can be fenced to keep people off newly planted dunes.	Approximately £2,000 per 100 m	• Maintains a natural coastal environment that is popular with people and wildlife. • Relatively cheap.	• Time-consuming to plant the marram grass and fence off areas. • People do not always respond well to being prohibited from accessing certain areas. • Can be damaged by storms.
Marsh creation (managed retreat)	This involves allowing low-lying coastal areas to be flooded by the sea to become salt marshes. This is an example of managed retreat. Salt marshes are effective barriers to the sea.	Depends on the value of the land. Arable land costs somewhere in the region of £5,000 to £10,000 per hectare	• A cheap option compared with maintaining expensive sea defences that might be protecting relatively low-value land. • Creates a much-needed habitat for wildlife.	• Land will be lost as it is flooded by sea water. • Farmers or landowners will need to be compensated.

Managed retreat

A further option for coastal management is to allow for some retreat of the coastline. This is called **managed retreat** or coastal realignment. It is a real option if there is a high risk of flooding or cliff collapse and where the land is relatively low value. Poor quality grazing land, for example, is seldom worth protecting if the costs of defences outweigh the benefits of protection (see Marsh creation in table **D**). This appraisal is known as a cost–benefit analysis. With sea levels forecast to rise, this approach is likely to become an increasingly popular option.

E *Beach nourishment in operation at Poole in Dorset*

Case study

Activity

Use photos **A**, **B** and **E**, and tables **C** and **D**, to help you answer the following questions.

a   Why is a sea wall an example of hard engineering?

b   What is the purpose of a sea wall?

c   What are the advantages and disadvantages of a sea wall?

d   What are groynes and what is their purpose? Draw a simple diagram to support your answer.

e   How effective do you think the groynes are in photo **B**?

f   Beach nourishment is a good example of soft engineering. What is soft engineering?

g   With reference to photo **E**, outline some of the advantages of beach nourishment.

h   Imagine a local council wishes to defend a 1 km stretch of coastline. Use tables **C** and **D** to calculate comparative costs for a sea wall, groynes, rock armour and beach nourishment.

i   Why is the economic cost only one of several considerations when deciding which coastal defence measure to adopt?

Coastal defences at Minehead

Minehead on the north coast of Somerset is one of the region's premier tourist resorts. It is home to a large Butlin's resort and every year it is visited by many thousands of tourists.

By the early 1990s it became clear that the current sea defences were going to be inadequate in the future. Storm damage was estimated to be £21m if nothing were done. The Environment Agency developed a plan to defend the town and improve the amenity value. Work started in 1997 and the sea defences were officially opened in 2001. The total cost was £12.3m, which represents a considerable saving on the potential losses due to storm damage.

The main features of the scheme (diagram **F**) are:

- A 0.6 m high sea wall with a curved front to deflect the waves. It has a curved top to deter people from walking on it and its landward side is faced with attractive local red sandstone.

- Rock armour at the base of the wall to dissipate some of the wave energy.

- Beach nourishment (sand) to build up the beach by 2 m in height. This forces the waves to break further out to sea and provides an excellent sandy beach for tourists.

- Four rock groynes to help retain the beach and stop longshore drift moving sand to the east.

- A wide walkway with seating areas alongside the sea wall. This is popular with tourists and local people.

The scheme has been extremely successful. Not only does it protect the town from storms and high tides, but it has also enhanced the seafront by creating an attractive beach environment (photo **G**).

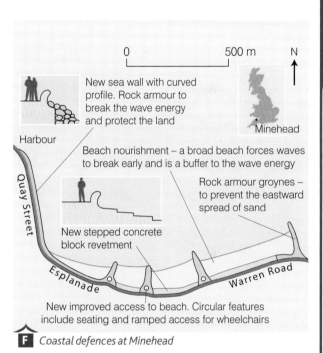

F *Coastal defences at Minehead*

G *The beach at Minehead*

Coastal defences at Wallasea Island

Wallasea Island is a low-lying coastal island formed at the confluence of the River Crouch and River Roach in Essex (map extract **H**). Protected from the sea by a ring of embankments, until recently it has been mostly used for growing wheat.

With the north coast defences falling into disrepair, the government decided to realign the northern part of the island by constructing a new embankment inland and allowing the old sea defences to be breached (map extract **H**). The new mudflats and salt marsh will help to protect the new sea wall and offer additional protection to land and property to the south.

The £7.5m scheme aims to replace bird habitats lost to development, improve flood defences on the island and create new leisure opportunities. In July 2006 the old sea wall was breached in three places to allow 115 ha of land to be flooded (photo **I**). Eventually the flooded area will contain islands and a number of salty lakes. Much of it will revert to salt marsh, providing a much-needed breeding ground for birds. Since 2007 the site has been managed by the RSPB.

I Wallasea Island looking east

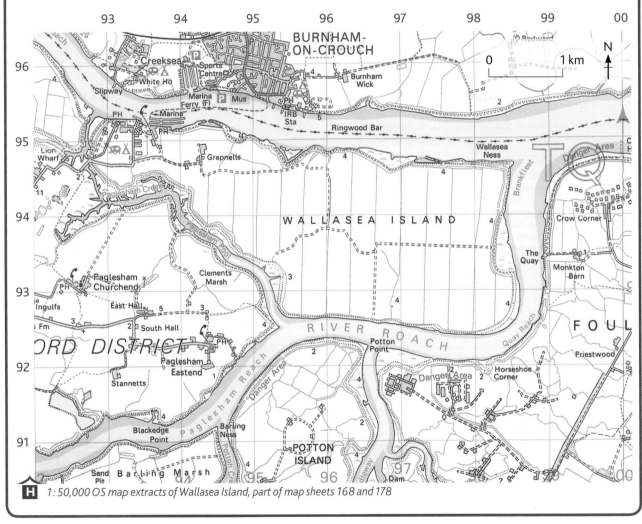

H 1 : 50,000 OS map extracts of Wallasea Island, part of map sheets 168 and 178

∞links

The Geography Site has information at www.geography-site.co.uk/pages/physical/coastal/defences.html.

7.11 What are salt marshes and why are they special?

Salt marshes are areas of periodically flooded low-lying coastal wetlands. They are often rich in plants, birds and animals (photo **A**).

A *Keyhaven Marshes*

A salt marsh begins life as an accumulation of mud and silt in a sheltered part of the coastline, for example in the lee of a spit or bar. As more deposition takes place, the mud begins to break the surface to form mudflats. Salt-tolerant plants such as cordgrass soon start to colonise the mudflats. These early colonisers are called **pioneer plants**. Cordgrass is tolerant of the saltwater and its long roots prevent it from being swept away by the waves and the tides. Its tangle of roots also helps to trap sediment and stabilise the mud.

As the level of the mud rises, it is less frequently covered by water. The conditions become less harsh as rainwater begins to wash out some of the salt and decomposing plant matter improves the fertility of the newly forming soil. New plant species such as sea asters start to colonise the area and gradually, over hundreds of years, a succession of plants develops. This is known as a **vegetation succession** (diagram **B**).

In this section you will learn

the characteristics of a salt marsh environment

sustainable approaches to the management of salt marshes.

Key terms

Pioneer plant: the first plant species to colonise an area that is well adapted to living in a harsh environment.

Vegetation succession: a sequence of vegetation species colonising an environment.

Did you know ??????

In France, salt is produced commercially from salt marshes.

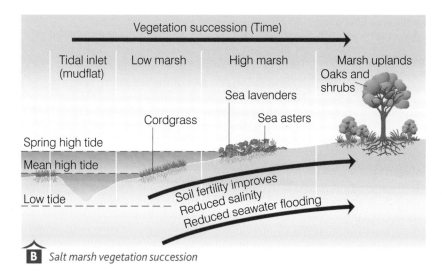

B *Salt marsh vegetation succession*

Keyhaven Marshes, Hampshire

Keyhaven Marshes is an area of salt marsh formed in the lee of Hurst Castle spit (map **A**, page 156). It supports a range of habitats including grassland, scrub, salt marsh and reed beds. This variety of habitats accounts for a rich diversity of wildlife in the area (table **C**).

C *Common wildlife species found at Keyhaven Marshes*

Species detail	Image	Species detail	Image
Plant: cordgrass – spiky, untidy-looking grass that grows fast on mudflats		Bird: ringed plover – feeds intertidally and nests on the salt marsh	
Plant: sea lavender – attractive, colourful flowers attract wildlife		Butterfly: common blue – resident butterfly commonly found on higher marshes	
Bird: oystercatcher – feeds and nests in salt marshes		Spider: wold spider – clings for hours to submerged stems of cordgrass waiting for low tide and food	

In common with many areas of salt marsh in the UK, Keyhaven Marshes is under threat:

- The salt marsh is retreating by up to 6 m a year. Although the causes of this are not yet fully understood, further sea-level rise threatens a 'squeeze' of the salt marsh as it lies between a low sea wall built in the early 1990s and the encroaching sea.

- The salt marsh has been under threat from the breaching of Hurst Castle spit during severe storms. In December 1989 storms pushed part of the shingle ridge over the top of the salt marsh, exposing 50 to 80 m to the full fury of the sea. It was eroded in less than three months.

- Increasing demands for leisure and tourism have meant that increasing numbers of people wish to visit the marshes. Careful management is required to prevent damage by trampling, parking and pollution. The area is popular with mariners who use the many creeks to moor their boats.

- In 1996 rock armour and beach nourishment were used to increase the height and width of the spit in an attempt to stop breaching. Since the completion of the £5m sea defences, the spit has not been breached and Keyhaven Marshes seems safe … at least for the time being.

- Keyhaven Marshes has been nationally recognised as an important site for wildfowl and wading birds. The area is officially a Site of Special Scientific Interest (SSSI) and part of the salt marsh is also a National Nature Reserve. This means that the area is carefully monitored and managed to maintain its rich biodiversity. Access is limited and development restricted.

- For the future, with sea levels expected to rise by 6 mm a year, the big issue concerns the 'squeeze' between the low sea wall and the rising sea.

D *The salt marsh at Keyhaven*

AQA *Examiner's tip*

Make sure you learn some specific details about the salt marsh at Keyhaven, as you are required to know a case study. Understand how it is being managed in a sustainable way.

Activities

1 Study photo **A**, diagram **B**, table **C** and photo **D**.

a What is a salt marsh and what makes it a special habitat?

b Why do salt marshes often develop in the shelter of spits?

c Why is cordgrass well suited to be a pioneer species?

d How does cordgrass improve the salt marsh environment to enable other plant species to grow?

e What is meant by a vegetation succession?

f How do the environmental conditions of a salt marsh change as the vegetation succession proceeds?

g Suggest ways that people's actions might affect a natural salt marsh vegetation succession.

2 Study the OS map extract of Hurst Castle spit and Keyhaven Marshes (map extract **A**, page 156).

a What symbol is used on a 1:25,000 map to depict an area of salt marsh?

b Locate the large area of salt marsh in grid square 3090. Would you expect this area to be flooded at high tide? Explain your answer.

c Locate Keyhaven Marshes in grid square 3191. What is the evidence from the map that there is a variety of habitats here?

d What is the evidence of human activity in Keyhaven Marshes (3191)?

e Locate the area of salt marsh just to the south-east of the village of Keyhaven (306912). What is the evidence of a vegetation succession here?

f Imagine that a major breach destroyed most of the spit in grid square 3090. With the aid of an annotated sketch map, suggest the possible impacts of such a breach on the natural salt marsh.

∞links

An excellent case study based on Chichester harbour can be accessed at www.conservancy.co.uk/learn/wildlife/saltmarsh.htm.

8.1 How does population grow?

■ Exponential growth

The world's population has grown exponentially. This means that the rate of growth has become increasingly rapid. Between AD 1 and AD 1000 growth was slow, but in the last thousand years it has been dramatic. By 2000, there were 10 times as many people living as there had been 300 years before in 1700. Not only is population increasing, but the rate of increase is becoming greater.

Population grew especially quickly during the late 20th and early 21st centuries. Between 2006 and 2007, 211,090 people were added to world population *every day*. Growth is predicted to continue, but now the rate is slowing down. Population is likely to rise to 9.2 billion by 2050 and finally peak a century later in 2150 at 10 billion. This should be followed by a more stable period of **zero growth** or even **natural decrease**.

Population growth is usually shown as a line graph. **Exponential growth** produces a line that becomes steeper over time, taking the shape of a letter J. Today, growth rates are slowing down (although the numbers being added daily are still high), so the shape of the graph is levelling off into an S curve (graph **A**).

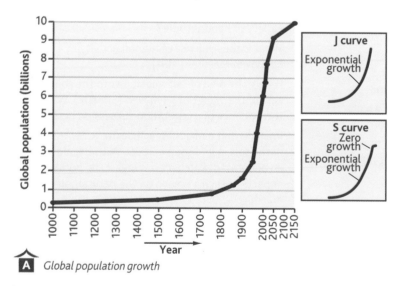

A *Global population growth*

■ Birth rate, death rate and natural change

Birth rate (BR) is the number of babies born alive per 1,000 people per year. Today, birth rates vary at between 5 per 1,000 per year and 40 per 1,000 per year, depending on the country concerned. In the past, figures have been as high as 50 per 1,000 per year, but availability of birth control means such high rates rarely occur now.

In this section you will learn

definitions of key terms involved in measuring population change

how to calculate population change by country

how to interpret population change calculations and compare countries.

Key terms

Zero growth: a population in balance. Birth rate is equal to death rate, so there is no growth or decrease.

Natural decrease: the death rate exceeds the birth rate.

Exponential growth: a pattern where the growth rate constantly increases – often shown as a J-curve graph.

Birth rate (BR): the number of babies born per 1,000 people per year.

Death rate (DR): the number of deaths per 1,000 people per year.

Natural change: the difference between birth rate and death rate, expressed as a percentage.

Natural increase (NI): the birth rate exceeds the death rate.

Life expectancy: the number of years a person is expected to live, usually taken from birth.

Death rate (DR) is the number of deaths per 1,000 people per year. Typically, death rates lie between 5 per 1,000 per year and 20 per 1,000 per year unless there is an epidemic, famine or war, which increase levels significantly.

Birth and death rates are expressed per 1,000 so that figures for countries of different sizes can be compared. Huge nations like China and India, with 1.3 and 1.1 billion people each, can be compared with smaller countries such as Singapore (4.6 million) and Luxembourg (480,222) (2007 figures).

Natural change is the difference between birth and death rates in a country. It is a useful measure of a population's growth or decline and, once we compare countries, we can to find reasons to explain their population statistics.

Until recently, all countries have been in a situation of **natural increase (NI)**, except in periods of epidemic, famine or war. In Europe between 1348 and 1353 the Black Death killed about a third of the population. Today, some countries have reduced their birth rates so much that they are now experiencing **natural decrease (ND)** and population is declining.

Data on birth rates, death rates and natural increase or decrease give us information on the level of development of a country. Whatever their stage of development, all countries now have low death rates. People often assume that DR in countries at lesser stages of development must be high because people are poor and **life expectancy** may not be long, but this is not true. Two factors affect the level of DR:

- health care has improved in poorer countries, lowering DR
- having so many people under the age of 15 reduces the chance of death.

Death rates in countries at further stages of development are often higher because their populations are older, causing DR to rise slightly. Birth rate is a better indicator of development. Although family size has reduced in most countries at lesser stages of development, parents in wealthier parts of the world still have fewer children.

Examples

Natural change calculations

UK

BR = 10.7 per 1,000 per year

DR = 10.1 per 1,000 per year

$NI = BR - DR$

$\quad = 10.7 - 10.1$

$\quad = 0.6$

Natural change is always expressed as a percentage, so the answer must be divided by 10.

Therefore:

NI = 0.06% per year

Czech Republic

BR = 9.0 per 1,000 per year

DR = 10.6 per 1,000 per year

$ND = DR - BR$

$\quad = 10.6 - 9.0$

$\quad = 1.6$

ND = 0.16% per year (sometimes written as –0.16%)

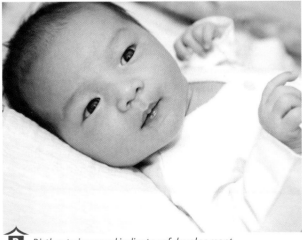

B   *Birth rate is a good indicator of development*

C   *All countries today have low death rates*

kerboodle!

D	European population, 1800–2005

Year	Population (millions)
1800	203
1900	408
1950	547
1975	676
1985	706
1995	727
2005	725

E	Birth and death rates for selected countries, 2007		

Country	Continent	Birth rate (per 1,000)	Death rate (per 1,000)
Afghanistan	Asia	46.2	20.0
Brazil	South America	16.3	6.2
China	Asia	13.4	7.0
Czech Rep.	Europe	9.0	10.6
Ethiopia	Africa	37.4	14.7
France	Europe	11.9	9.2
Germany	Europe	8.2	10.7
Hong Kong	Asia	7.3	6.5
India	Asia	22.7	6.6
Ireland	Europe	14.4	7.8
Italy	Europe	8.5	10.5
Nigeria	Africa	40.2	16.7
Pakistan	Asia	27.5	8.0
Romania	Europe	10.7	11.8
South Korea	Asia	9.9	6.0
UK	Europe	10.7	10.1
USA	North America	14.2	8.3

Key terms

Newly industrialising countries (NICs): these include the Asian 'tigers' as well as other emerging industrial nations such as Malaysia, the Philippines and China.

Asian 'tiger': one of the four east Asian countries of Hong Kong, South Korea, Singapore and Taiwan, where manufacturing industry grew rapidly from the 1960s to the 1990s.

Activities

1 Study table **D**.

a Use the data from the table to draw a growth curve for Europe. Plot the year along the *x*-axis and the population along the *y*-axis.

b Name the shape of the graph you have drawn. Why is this name used?

c Describe the population growth pattern that you have drawn. Give details of the graph's shape and quote figures from the graph to support your answer. Use key terms such as *exponential, J-shaped, S-shaped, levelling off, zero growth* and *population decline*.

2 Study table **E**.

a Choose five countries and calculate the natural change for each country. Set out your calculation clearly so that your teacher can follow each stage of your working.

b Classify all the countries in the table as one of the following:

■ country at a further stage of development

■ recent newly industrialising country (NIC)

■ Asian 'tiger' (or older NIC)

■ improving country at lesser stage of development

■ extremely poor country at lesser stage of development.

c Draw a summary table listing each country, its rate of natural change (in percent) and the type of country it is using your answers to questions 2a and b.

d Is there a relationship between population change (NI or ND) and level of economic development? Use your summary table to help you discuss this. Refer to figures from your calculations and your table to support your answer.

8.2 What is the demographic transition model?

Demography is the study of population. Transition simply means change. The **demographic transition model** (DTM) is shown in diagram **A**.

The model explains birth and death rate patterns across the world and through time. It includes the main period of a country's development and shows the links between demographic and economic changes. The diagram is divided into five stages, showing change from high birth and death rates in Stage 1 to much lower ones in Stages 4 and 5. Originally, the model was designed to explain population change in countries at further stages of development, but it has since been used to explain events in all countries. This allows us to compare different patterns of demographic and economic development.

In this section you will learn

the trends in the demographic transition model

how to interpret the demographic transition model and use it to compare the situations of different countries.

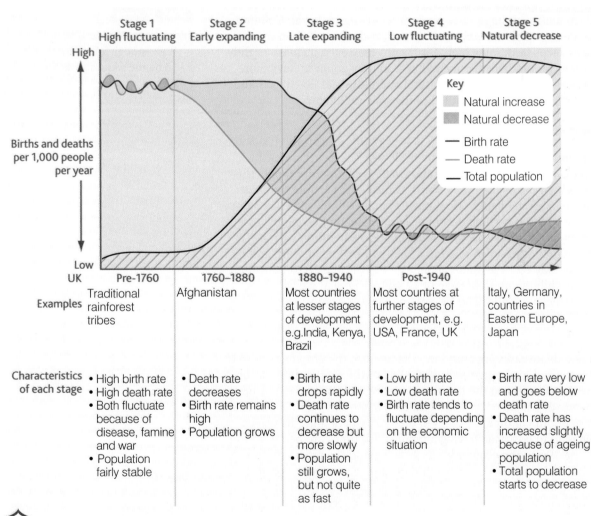

	Stage 1 High fluctuating	Stage 2 Early expanding	Stage 3 Late expanding	Stage 4 Low fluctuating	Stage 5 Natural decrease
UK	Pre-1760	1760–1880	1880–1940	Post-1940	
Examples	Traditional rainforest tribes	Afghanistan	Most countries at lesser stages of development e.g. India, Kenya, Brazil	Most countries at further stages of development, e.g. USA, France, UK	Italy, Germany, countries in Eastern Europe, Japan
Characteristics of each stage	• High birth rate • High death rate • Both fluctuate because of disease, famine and war • Population fairly stable	• Death rate decreases • Birth rate remains high • Population grows	• Birth rate drops rapidly • Death rate continues to decrease but more slowly • Population still grows, but not quite as fast	• Low birth rate • Low death rate • Birth rate tends to fluctuate depending on the economic situation	• Birth rate very low and goes below death rate • Death rate has increased slightly because of ageing population • Total population starts to decrease

Key:
Natural increase
Natural decrease
— Birth rate
— Death rate
— Total population

Births and deaths per 1,000 people per year

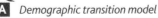 Demographic transition model

At first, only four stages were included in the model, but recent changes in parts of Europe have required the addition of a new Stage 5. Perhaps in the future we shall add yet another stage.

Characteristics of each stage

Stage 1

The high fluctuating stage occurs in societies where there is little medicine, low life expectancy and no means of birth control. Remote rainforest areas of Amazonia and Indonesia are the only locations where this stage might happen today. These are traditional societies, largely cut off from the rest of the world. The UK was at Stage 1 before around 1760.

Stage 2

The key factor that indicates the change from Stage 1 to Stage 2 is a decrease in death rate. Improvements in medicine and hygiene cure some diseases and prevent others. Life expectancy increases. The gap between birth rate and death rate results in population growth. The farther apart the lines are on the graph, the greater the rate of growth. Most economies in Stage 2 are predominantly agricultural. Children are needed to work the land, as they can produce more food than they eat. This keeps birth rates high. Birth rates may even increase slightly because fewer mothers die in childbirth so they go on to have more children. From 1760, new agricultural and industrial inventions and medical discoveries in the UK led to the start of Stage 2.

Stage 3

Death rate continues to fall, but more slowly. The key factor at the start of Stage 3 is a decrease in birth rate, which is often quite rapid. This is due to both the availability of birth control and economic changes, which mean people benefit from having smaller families. As a country develops, children become economic costs instead of economic assets – in other words, they cost the family money rather than earning it. When children have to go to school, they can no longer work and their education may cost the family money. In the UK, the 1870 Education Act made children up to the age of 12 go to school at a cost of one penny a week per child. The UK entered Stage 3 around 1880.

Stage 4

Birth and death rates are both low. The lines on the graph are close to each other and birth rate varies according to the economic situation. This is therefore called the low fluctuating stage. When the economy is growing and people have jobs and earn a good living, they are more likely to afford children. In times of unemployment and low wages, people tend to postpone having a family until times are better. In the UK in the 1960s, the economy was growing and the birth rate rose slightly, but in the 1970s, with world economic recession, the birth rate fell again. Overall, there is still population growth, but it is slow. It will be interesting to see whether the economic recession that began in 2008 will have an impact on birth rates.

Stage 5

Many Eastern and a few Western European countries are at Stage 5, but for different reasons. The UK remains at Stage 4. Death rate rises (as indeed it has in the UK too) because the population includes more elderly people. In Eastern Europe an uncertain economy discourages people from having babies, while Western European economies give young women so many career opportunities that they decide to be childless or to postpone motherhood.

Investigating countries

Case study

Stage 1: Traditional rainforest tribes

In parts of Indonesia, Brazil and Ecuador, small numbers of people live separately with little contact with the outside world (photo **B**). They retain high birth and death rates, making them the people closest to a true Stage 1 situation in the world today.

Stage 2: Afghanistan

Afghanistan is an extremely poor country, held back by political instability. In table **E** on page 172 it has one of the world's highest birth rates at 46.2 and a much lower death rate of 20.0. Natural increase is therefore 2.62 per cent per annum. Some 79 per cent of people are farmers, often nomadic (photo **C**), so need children to help with crops and livestock. Cities like Kabul, the capital, have even higher rates of natural increase because easier access to medical care reduces the death rate while high numbers of young adults (who have migrated to the cities) increase the birth rate.

Stage 3: Brazil

Brazil is a newly industrialising country (NIC). Although it is a country at a lesser stage of development, it is developing fast economically. Brazil's population will have almost doubled between 1975 and 2015, from 108 million to 210 million people. As a Roman Catholic country, it has a high birth rate, but rapidly improving standards of living mean people can see the benefits from having fewer children.

Stage 4: USA

The USA is the largest and most developed economy in the world. The world's third largest population (after China and India) with over 301 million people in 2007, its growth is quite high for a Stage 4 country – mostly due to immigration. Many immigrants come from Catholic Central America, but the USA is now encouraging a more highly trained Asian workforce who are likely to have lower birth rates.

Stage 5: Germany

Germany is almost as well developed as the USA, but it has moved a stage further in the demographic transition model. Women achieving high-powered positions at work, plus an ageing population, are factors that have brought Germany clearly into Stage 5. One of the first countries to enter this stage, Germany's birth rate is well below **replacement rate**. The government has to cope with the costs of a large elderly population and a declining workforce.

B   *Rainforest tribeswoman, Ecuador*

C   *Afghan nomads*

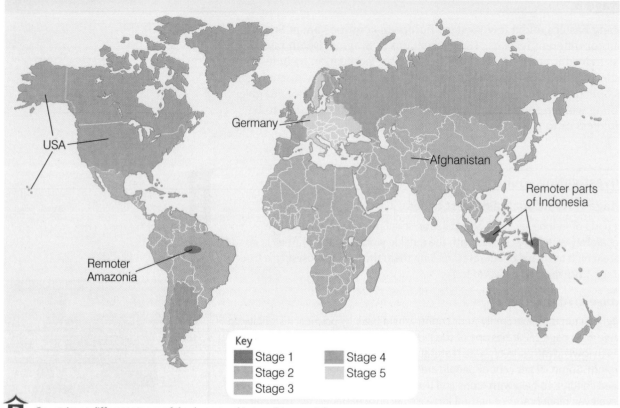

D *Countries at different stages of the demographic transition model*

Key
- ■ Stage 1
- ■ Stage 2
- ■ Stage 3
- ■ Stage 4
- ■ Stage 5

■ Major factors affecting population growth

Agricultural change

Changes in agriculture occur early in a country's development. Even at intermediate levels, technology improves yields and saves labour. This frees some workers for industry and more rapid economic growth. In the Industrial Revolution in the UK, factories needed a large workforce, so for a while larger families were a benefit. Soon, however, technological advantages reduced the need for labour, making smaller families more desirable.

Urbanisation

Rural-to-urban migration is common in poorer countries as cities are believed to have greater opportunities, and generally do. One major reason for such migration is to seek better educational opportunities for children. Children's labour is therefore of less value in cities than in rural areas.

Education

As levels of educational achievement increase, bringing improved standards of living, children become an economic disadvantage. Fewer children means parents have more money to be spent on each one, giving them better future chances. Many parents in poorer countries see education as their children's best chance in life.

Emancipation and status of women

As economies develop and education improves, opportunities for girls increase alongside those for boys. With development a larger workforce is required, so women must participate more in paid work outside the home. Reaching a good standard of living in a household requires two incomes.

Over time, prejudice against women holding more senior positions at work reduces. Equality increases and is perceived as not only acceptable but desirable. However, achieving highly in any career demands a large time commitment, leaving less opportunity for taking maternity leave or caring for children. Some women make deliberate choices regarding not having children or having them later, and these increase as an economy develops. Larger families, or even having a family at all, may be rejected. One in five women in the UK today is childless, compared with one in ten in their mothers' generation.

It must not be forgotten that most mothers in developed countries work. Most need childcare to enable this, which can be expensive. Many women deliberately do not achieve their maximum potential at work because they feel bringing up children is important. Countries like Sweden have changed the law to increase the proportions of women in management and government. At some point a balance should be reached.

E　*Women at work*

Activities

1　Study diagram **A**.

a　Describe how birth and death rates change through the five stages of the demographic transition model.

b　How do these changes affect population growth?

c　Why is Stage 1 really a thing of the past?

d　What is likely to happen to the number of countries at Stage 5 in the future?

2　Study map **D**.

a　Research the Internet for birth and death rates by country and suggest at least two more countries for each of Stages 3, 4 and 5.

b　Describe the distribution of countries at each stage of the demographic transition model.

3　Imagine you are a factory worker in an urban region of Brazil. Are you more likely to have a small or a large family? Explain your reasons, not forgetting that individuals may hold varying opinions.

4　Why might women in Germany and other Stage 5 countries have a very small family?

 links

You can find out more about birth and death rates at **http://encarta.msn.com**.

■ Population pyramids

A population pyramid is a type of bar graph used to show the **age** and **gender structure** of a country, city or other area. The horizontal axis is divided into either numbers or percentages of the population. The central vertical axis shows age categories: every 10 years, every 5 years or every single year. The lower part of the pyramid is known as the base and shows the younger section of the population. The upper part, or apex, shows the elderly.

Interpreting population pyramids tells us a great deal about a population, such as birth rates, to a lesser extent death rates, life expectancy and the level of economic development (or stage in the demographic transition model).

Population pyramids and the demographic transition model

In this section you will learn

- how to construct a population pyramid
- how to interpret population characteristics from a pyramid
- how to predict likely future changes in a population.

Stage 1

The Stage 1 pyramid has a very wide base due to its extremely high birth rate (up to 50 per 1,000 per year). However, **infant and child mortality rates** are high (only 50 per cent may reach their fifth birthday), so the sides of the pyramid curve in very quickly. Death rate is high in all age groups, so life expectancy is low. The result is a very narrow apex and the shortest of all the pyramids.

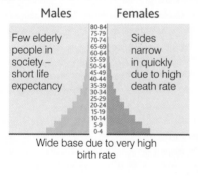

Males — Few elderly people in society – short life expectancy

Females — Sides narrow in quickly due to high death rate

Wide base due to very high birth rate

Key terms

Age structure: the proportions of each age group in a population. This links closely to the stage a country has reached in the demographic transition model.

Gender structure: the balance between males and females in a population. Small differences can tell us a great deal about a country or city.

Infant mortality: the number of babies that die under a year of age, per 1,000 live births.

Child mortality: the number of children that die under five years of age, per 1,000 live births.

Stage 2

Stage 1 and 2 pyramids are similar in shape. Death rate begins to fall, making the sides of the Stage 2 pyramid slightly less concave. The apex shows a few extra elderly people as life expectancy begins to rise.

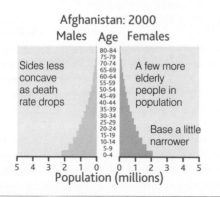

Afghanistan: 2000

Males　Age　Females

Sides less concave as death rate drops

A few more elderly people in population

Base a little narrower

Population (millions)

B　*A Stage 5 family*

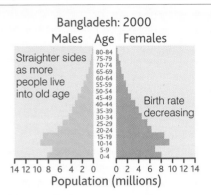

Bangladesh: 2000

Stage 3

The narrowing base shows the decrease in birth rate typical of Stage 3 countries. It also becomes straighter sided. Birth rate decreases quickly. Health improvements allow even more people to live into old age.

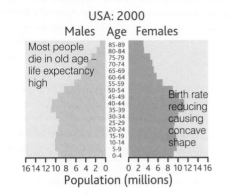

USA: 2000

Stage 4

This pyramid has become straight-sided, showing a steady low birth rate. High life expectancy allows most people to live into their 60s and 70s and a significant minority into their 80s.

Stage 5

Germany has been at Stage 5 since the 1970s – one of the first countries to reduce its birth rate to this extent (photo **B**). When today's middle-aged people become elderly, there will few adults of working age to support them. A Stage 5 population is not sustainable.

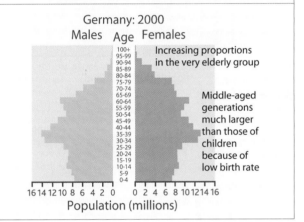

Germany: 2000

   Population pyramids and the demographic transition model

Future population change

Diagram **C** shows population pyramids for India in 2000, 2025 and 2050. Computer programs can predict future population patterns from today's statistics. Although India's birth rate has fallen, its base remains wide because there are so many young adults in their child-bearing years. By 2025 the number of babies born each year will have stabilised, reducing slightly by 2050. Increasing numbers live into old age, and the 90–94, 95–99 and 100+ age groups are included on future graphs. By 2050 most people will live into their 70s and India will have all the characteristics of a Stage 4 population.

Urban areas of many countries at lesser stages of development are predominantly male in all age groups up to 60–64. Usually more boys are born than girls, explaining the differences at the base (diagram **D**). Rural-to-urban migration in search of work remains common in countries at lesser stages of development. Men and older boys leave the women, younger children and elderly behind in rural areas. Cities offer greater opportunities to earn money, which can be sent back to improve the family's standard of living. Sometimes whole families make the migration.

AQA *Examiner's tip*

You are likely to be asked to complete a partially drawn population pyramid. Here are some tips:

- Read the horizontal scale carefully and mark the ends of the bars accurately.

- Use a sharp pencil and a ruler when drawing bars.

- Make your bars look as similar as possible to the ones already drawn, including the shading.

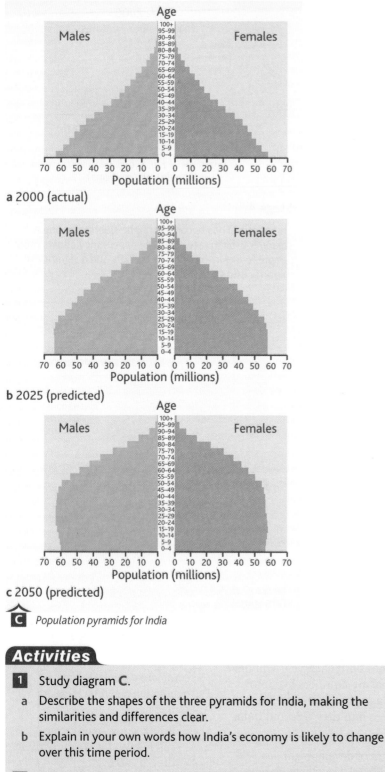

a 2000 (actual)

b 2025 (predicted)

c 2050 (predicted)

C *Population pyramids for India*

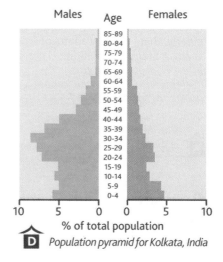

D *Population pyramid for Kolkata, India*

8.4 How can a population become sustainable?

A sustainable population is one whose growth and development is at a rate that does not threaten the success of future generations. Countries at Stage 4 of the demographic transition model, with low birth and death rates, are the most sustainable. The economy is stable or growing and the standard of living is maintained or improving.

Stage 5 populations are not sustainable because numbers are decreasing. In Japan, calculations were done to predict how long it would take the country to die out if current low birth rates continued well into the future. Although it is highly unlikely to happen, the fact that it was even thought about was worrying.

In this section you will learn

the reasons for the one-child policy in China from 1979

the severity of the rules imposed

how the one-child policy has changed over recent years and the outcomes of these changes.

Will China become a sustainable population?

The early days of the one-child policy

During the 1970s the Chinese government realised that the country was heading for famine unless severe changes were made quickly. Change to an industrial economy at the expense of farming had already caused a catastrophic famine from 1959 to 1961, with 35 million deaths. A 'baby boom' followed and population was growing too fast to be sustainable. The government stepped in to avoid another crisis. However, its methods have been considered too strict, even cruel.

Beginning in 1979, the one-child policy said that each couple:

- must not marry until their late 20s
- must have only one successful pregnancy
- must be sterilised after the first child or must abort any future pregnancies
- would receive a 5 to 10 per cent salary rise for limiting their family to one child
- would have priority housing, pension and family benefits, including free education for the single child.

Any couples disobeying the rules and having a second child were severely penalised:

- a 10 per cent salary cut was enforced
- the fine imposed was so large it would bankrupt many households
- the family would have to pay for the education of both children and for health care for all the family
- second children born abroad are not penalised, but they are not allowed to become Chinese citizens.

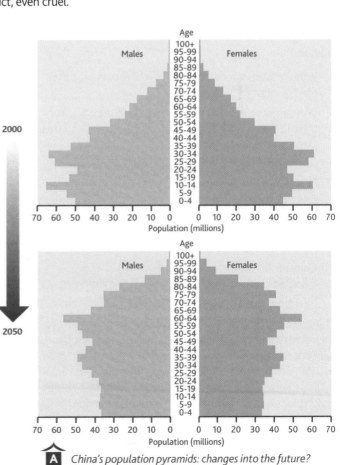

A China's population pyramids: changes into the future?

Case study

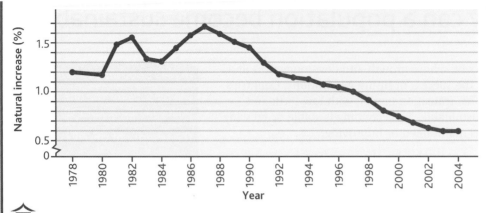

B *China's population growth through the one-child policy*

Pressure to abort a second pregnancy even included pay cuts for the couple's fellow-workers so they would make life unbearable. The 'Granny Police' – older women of the community entrusted with the task of keeping everyone in line – kept a regular check on couples of childbearing age, even accompanying women on contraception appointments to make sure they attended.

China is racially mixed but over 80 per cent of the population is of the Han race. Minority groups could become unsustainable under the one-child policy, so they were exempt. In rural areas, where sons are essential to work the family land, a second pregnancy was allowed if the first child was a girl, in the hope of getting a boy.

The problems and benefits of the policy

The one-child policy has been controversial for many reasons:

- Women were forced to have abortions as late as the ninth month of pregnancy.

- Women were placed under tremendous pressure from their families, workmates, the 'Granny Police' and their own consciences and feelings.

- Local officials and central government had power over people's private lives.

- Chinese society prefers sons over daughters. Some girls were placed in orphanages (photo **C**) or allowed to die (female infanticide) in the hope of having a son the second time round.

- Chinese children have a reputation for being over-indulged because they are only (single) children, hence the name 'Little Emperors' (photo **D**).

China's one-child policy has brought important benefits to the country, however. The famine previously forecast has not happened. Population growth has slowed down sufficiently for people to have enough food and jobs. It is estimated that 400 million fewer people have been born. Increased technology and exploitation of resources have increased standard of living for many. New industries have lifted millions out of poverty. This is partly the result of the one-child policy, but new technology from other countries has helped.

C *Children in a Chinese orphanage*

D *Only children are often spoiled*

Changes to the one-child policy in the 1990s and 2000s

Young couples who are both only children are allowed two children, but government workers must set an example and stick to one. Many agree with the government's policy during the 1980s but some prefer to choose for themselves. Young couples today face the problem of being responsible for four elderly parents. Having two children will share this burden in the future.

With less time needed for childcare, women have had the opportunity to concentrate on careers, so have achieved more. The attitude to having a daughter has improved. With increasing wealth, more people are able to break the rules, pay the fine and take the other consequences of having a second child. The policy is unlikely to relax any more because in 2008 China still had 1 million more births than deaths every five weeks and 600 million people – half the population – still live on less than $2 per day.

One major consequence of the one-child policy is gender imbalance. Some girls have been rejected, so there are now 60 million more young men than young women. Not all young men will be able to marry, which could present difficulties for Chinese society.

Did you know ??????

There are over 15 million orphans in China, most healthy girls, abandoned as a result of the one-child policy and Chinese society's economic and social preference for boys. Missionary-run orphanages are usually very good, but in state institutions the girls are neglected and sometimes treated cruelly.

Did you know ??????

Chinese couples had to apply for permission to try to become pregnant and take their turn. If they did not succeed in a six-month period, they had to wait for another opportunity.

We cannot just be content with the current success. We must make population control a permanent policy.

Adapted from the People's Daily *(China's Communist Party newspaper), 2000*

Beijing mother-of-one, Zhao Hui, who has a four-year-old daughter called Zhang Jin'ao, says she never wanted more than one child. 'One child is enough. I'm too busy at work to have any more,' says the 38-year-old.

*Adapted from BBC News website **news.bbc.co.uk**, 20 September 2007*

E *News items about the policy*

Activities

1 Study diagram **A**.
a Describe the shape of each pyramid.
b On copies of the pyramids, label the key features, paying particular attention to any similarities and differences.

2 Study graph **B**.
a Describe the shape of the graph.
b Do you think the one-child policy has been a success in controlling population growth? Explain your answer.

3 Study the extracts in **E**.
a Make a list of the controversial aspects of the one-child policy. Remember that the Chinese government and people may not see everything in the same way as outsiders.
b What were the successful outcomes of the policy during the 1980s?

4 Make a decision – was the Chinese government wise to operate the one-child policy? Were the benefits greater than the disadvantages? There is no right or wrong answer here – you simply need to justify your opinion clearly.

What alternative birth control programmes exist?

An alternative population policy in Kerala, India

India was the first developing country to launch a national family planning programme as early as 1952. This included not just contraception but many social changes. The south-western state of Kerala, with its socialist/communist government, has focused on social changes to create a society that encourages smaller families (photo **A**). Its decrease in birth rate has been the most dramatic in India.

A *The government of Kerala encourages smaller families*

B *Comparing Kerala*

Indicator	Kerala	India	Other low-income countries	UK
GDP/capita ($)	2,950	460	432	27,000
Adult **literacy rate** (%)	91	58	39	98
Life expectancy (years)	71	64	59	79
Infant mortality (per 1,000)	12	65	80	7
Fertility rate (per 1,000)	1.8	3.1	3.5	1.7

Kerala's 32 million people make up 3.4 per cent of India's population. At 819 people per km², population density is three times the Indian average. Fortunately, its growth rate is India's lowest. Kerala is one of the few regions in the developing world to have gone through the demographic transition as far as Stage 4, and it has social and demographic statistics close to those found in Europe and North America (table **B**). Its **gross domestic product (GDP) per capita** is low compared with richer countries (about a tenth of the UK's) but high against many other Indian states (table **B**).

Kerala's policy to reduce its previously high population growth rate has involved:

- improving education standards and treating girls as equal with boys
- providing adult literacy classes in towns and villages
- educating people to understand the benefits of smaller families

In this section you will learn

how all birth control programmes in the developing world do not take the same approach.

how to compare China's one-child policy and other birth control programmes.

Key terms

Gross domestic product (GDP) per capita: the total value of goods and services produced by a country in one year divided by its total population. Foreign income is not included.

Literacy rate: the percentage of adults in a country who can read and write sufficiently to function fully in work and society.

AQA Examiner's tip

To write an account contrasting two or more countries effectively, make sure you write a quick plan first, summarising the main points you are going to make. In an exam, this can be done in the spare space on the left-hand side of the lines in the answer booklet. Always support each idea with data from the diagrams given, as well as knowledge from your own notes.

- reducing infant mortality so people no longer need to have so many children
- improving child health through vaccination programmes
- providing free contraception and advice
- encouraging a higher age of marriage
- allowing maternity leave for the first two babies only
- providing extra retirement benefits for those with smaller families
- following a land reform programme.

Kerala's progressive education policy has been in action since the 1960s under the socialist/communist government. Today, more girls go to university than boys (photo **C**). The Right to Literacy programme organises reading and writing classes in villages, however remote, and they are always well attended (table **D**). Most villages have free libraries.

Land in Kerala was redistributed so that no one was landless. No family was allowed more than 8 ha and everyone could be self-sufficient. Obviously, larger families would be at a disadvantage.

C   *Keralan women enjoy higher status in society*

D   *Kerala: demographic and social facts*

Indicator	Value
Birth rate (per 1,000)	18
Death rate (per 1,000)	6
Natural increase (%)	1.2
Life expectancy (years)	73
Population older than 60 years (%)	11.2
Children born per woman (number)	1.66
Babies born in hospital (%)	95
Adult literacy rate (%)	89.9

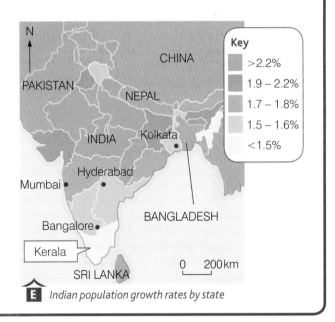

Key
- >2.2%
- 1.9 – 2.2%
- 1.7 – 1.8%
- 1.5 – 1.6%
- <1.5%

E   *Indian population growth rates by state*

Activities

1   Study table **B**.

a   What is Kerala's GDP/capita? How does it compare with those of the other regions in the table?

b   Compare Kerala's infant mortality and fertility rates with those of the other regions. Quote figures to make your answer clearer.

c   Write a paragraph to explain the advantages of living in Kerala compared with other regions of India.

2   Different countries tackle high population growth rates in different ways.

a   Read the case studies about China (page 181) and Kerala (above).

b   Which of the two policies has been dominated by birth control and which has used other methods?

c   Make two lists of the main similarities and differences in the policies of China and Kerala.

Richer world populations are ageing. Low birth rates and smaller families result in fewer children and adolescents. Better health care and more advanced medicines allow people to live longer, increasing the proportion of elderly people. Today's older people in countries at further stages of development tend to be wealthier, fitter and have wider interests, with the spare time to enjoy these. From the country's point of view, this is the age group with the most expensive needs, especially in terms of health care. The 85+ age group – the very elderly – is growing fastest, putting particular stress on health and social welfare systems (photo **A**).

A *The 85+ age group is growing fast. This couple is living in a care home*

In this section you will learn

In this section you will learn

- the ways in which an ageing population is different from younger populations
- the demands on a country that has an ageing population
- how, as the proportion of elderly people increases, the costs to the government increase dramatically
- different ways of solving the problem of an ageing population
- how to evaluate different approaches to coping with an ageing population.

■ The issues

Health care

The demand for health care increases because more illness occurs in old age. The elderly visit their GP (doctor) more often. They have more hospital appointments and more time in hospital than younger or middle-aged people. The government has to find more funding to support older people and this comes from taxation of present workers.

Social services

Elderly people need other services such as nursing homes, day-care centres and people to help them to care for themselves at home. These special needs put financial pressure on a country.

The pensions crisis

Life expectancy is higher in developed countries than in developing countries. In wealthier countries, people expect to be able to retire from work and have a pension (an income) for the rest of their lives. As there are more elderly people and the proportion of working people is decreasing, so the taxes must increase to pay the pensions bill.

The state pension began in 1908 when male life expectancy was 67 and retirement age was 65. The average person would therefore receive their pension for only two years. Today, the situation could not be more different. State pension still starts at the age of 65, but life expectancy is almost 80. Standards of living may go down.

Key terms

European Union (EU): a group of countries across Europe that work towards a single market, i.e. they trade as if they were one country, without any trade barriers.

Did you know ? ? ? ? ?

With the second lowest birth rate in Europe, Italy faces population decline in the 21st century. Projections suggest that 40 per cent of Italians will be over 60 and only 15 per cent under 20 by the end of this century.

Examples

An answer of 50 would mean that every 100 working people would be supporting themselves plus another 50 dependents. The lower the answer, the better things are economically for a country.

Activity

1 Study diagram **A** on pages 178–179.

a Describe how the numbers and proportions of elderly people change as a country passes through the demographic transition model.

b Look at the pyramids for Stages 4 and 5. What has happened to the gender ratio?

AQA *Examiner's tip*

Learn your EU case study (page 188) to give a detailed account of the problems one country has experienced in relation to an increasingly ageing population and the strategies used to tackle these problems.

■ The opportunities

The situation has a positive side too. Younger retired people (those in their later 60s and early 70s) contribute a great deal to the economy (photo **B**). They are relatively wealthy and have lots of leisure time. They spend money on travel and recreation, providing jobs in the service sector. Many do voluntary work and some still do paid work and therefore pay taxes.

B *Younger retired people contribute a lot to the UK's economy*

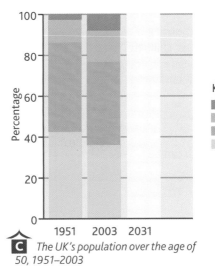

Key
■ 85 and over
■ 75 – 84
■ 60 – 74
■ 50 – 59

C *The UK's population over the age of 50, 1951–2003*

As soon as my hair turned white, people started to ignore me. It was as if I had become invisible.

Anna, 86, and still coping in her own home with no outside help

Britain's elderly are being neglected, poorly treated and marginalised by the country's health system. Inspectors found that many older patients were hungry because meals were taken away before they could eat them.

The Independent, 27 March 2006

It's horrid having to rely on other people for the most basic things. It's bad enough when it's your own family and friends, but when it's strangers from social services it's so much worse.

Barry, 95, who has four carers every day to get him up, washed, dressed and, later, back to bed

How bad is the UK's pension crisis? Every day seems to bring fresh warnings that Britons will not have enough money to live on when they retire.

Adapted from BBC News website news.bbc.co.uk, 23 September 2005

D *Quotes on being elderly, pensions and the health crisis*

Ageing in the EU

Table **E** shows **European Union (EU)** birth rates are very low – in fact, they have never been so low and they may decrease even more. Smaller families and later motherhood could soon result in a noticeable decline in population in some countries. In each generation there are fewer parents, so fewer children are born. In 2003, 1.5 babies were born in the EU for every woman, but 2.1 are needed for a population to be sustainable. Western European birth rates are higher than those in Eastern Europe, but Germany – a Western European country – has an even lower birth rate (table **E**).

E *Selected EU birth rates*

Country	Birth rate (per 1,000)
Ireland (W)	15.2
France (W)	12.7
UK (W)	12.0
Netherlands (W)	11.9
Poland (E)	9.3
Bulgaria (E)	9.0
Latvia (E)	8.8
Germany (W)	8.5

Note: (W) Western European country
(E) Eastern European country

Case study

France's solution

France is tackling its ageing population with a strong pro-natal policy – it encourages people to have children to produce a more favourable age structure and **dependency ratio**. This has had some effect, but has not been entirely successful as graph **F** shows.

Couples are given a range of incentives to have children:

* three years of paid parental leave, which can be used by mothers or fathers
* full-time schooling starts at the age of three, fully paid for by the government
* day care for children younger than three is subsidised by the government
* the more children a woman has, the earlier she will be allowed to retire on a full pension.

Nicole Falcou is 53. She lives in Muret, close to Toulouse in south-west France. She is married with three daughters aged between 14 and 22 and works locally with disabled children. She has three children of her own, so she is entitled to extra benefits from the government, including retiring in her early 50s on a full state pension if she chooses. Although a pension is always less than a salary, as her children grow up and become less financially dependent this is a tempting offer.

F *French birth and death rates, 1960–2007*

Activities

2   Study graph **C** on page 187.
  a   Copy and complete the graph, adding a similar bar for 2031 by using the data in table **G**.
  b   Describe the changes shown by the bars between 1951 and 2031.
  c   Explain why these changes are happening.
3   Study graph **F**.
  a   Describe the pattern of the birth rate line.
  b   List the problems affecting a country with a very low birth rate.

UK population aged over 50, 2031

G

Age group	%
50–59	28
60–74	44
75–84	20
85 and over	8

Key terms

Dependency ratio: the balance between people who are independent (work and pay tax) and those who depend on them. Ideally, the fewer dependents for each independent person, the better off economically a country is. Here is the formula (figures can be in numbers or percentages):

$$\frac{\text{number of dependent people}}{\text{number of independent people}} \times 100$$

8.7 What are the impacts of international migration?

■ Push–pull factors

People move home for many different reasons. Every individual's decision to move is the result of **push–pull factors**. Negative aspects of a person's home area push them away from it and make them look for somewhere better. Positive characteristics of new places, which attract people to move there, are called pull factors.

■ Impacts of international migration

The impacts of **migration** on both the countries of origin and the **destination** can be positive or negative.

Migrant workers often send money back to their **country of origin** to help their families. This means that money leaves the host economy – a disadvantage – but the country of origin can benefit enormously.

Finding accommodation can be difficult for migrants in the UK. Some have been helped by social services and this causes resentment from UK citizens who feel they are being treated as second-class citizens in their own country. Demand for housing has grown immensely in the UK during the early 21st century, fuelled by high levels of migration. The demand for housing is greater than supply, so property prices rose quickly in the early 2000s and immigration has contributed to this.

Migration brings labour and skills, and the economies of the UK and the EU have grown as a result. Most migrants are more successful than they would have been at home, although some are less fortunate. Exploitation does happen and not everyone earns as much as they had expected. Tragedies have occurred when gang masters, who often control large numbers of workers in agriculture and shellfish harvesting, have been negligent. The deaths of 23 Chinese cockle-pickers in Morecambe Bay in February 2004 was perhaps the worst example.

Too many migrants can be a burden. Schools taking many **immigrant** children may be under pressure. British parents sometimes feel this reduces opportunities for their own children because teachers are too busy with those whose first language is not English. On the other hand, cultural mixing is often seen as positive as long as racial prejudice does not become a problem.

```
◀ ▶  C  +  http://                                    Q Google
      News                        Home | Contact | Sitemap | News | Forum | Shop

  Home        The political controversy over immigration has intensified
              in Britain and other European countries. Many of those
  Contact     opposed to further immigration fear it may bring in its
              wake more crime. Opponents say that immigrants are
  Sitemap     taking the jobs of native workers, and lowering the wages
  News        of others. Supporters of immigration argue it has long-
              term benefits for the economy, providing needed skills
  Forum       and helping to boost growth.
  Shop
```

A *From the BBC News website*
Adapted from BBC News website **news.bbc.co.uk**, *17 June 2002*

In this section you will learn

- the concept of migration and people's reasons for moving home (push–pull factors)
- the positive and negative impacts of international migration
- who moves within the EU and why
- who comes to the EU and their reasons for wanting to live here
- the differences between voluntary economic migrants and refugees
- the benefits and difficulties of international migration for EU countries.

Key terms

Push–pull factors: push factors are the negative aspects of a place that encourage people to move away. Pull factors are the attractions and opportunities of a place that encourage people to move there.

Migration: the movement of people from one permanent home to another, with the intention of staying at least a year. This move may be within a country (national migration) or between countries (international migration).

Destination: the country where a migrant settles.

Country of origin: the country from which a migration starts.

Immigrant: someone entering a new country with the intention of living there.

Emigrant: someone leaving their country of residence to move to another country.

A Slovak girl working in Sussex

Jana Susinkova came to the UK in 2002 with her Czech boyfriend. She was only 18 and he a little older. She worked as a domestic cleaner, undercutting the level local women charged by at least £1 per hour. She had enough work to keep busy six days a week. Her boyfriend was a mechanic and odd-job man, and his job provided accommodation for them.

Late in 2007 Jana returned home to Slovakia. Her boyfriend had already left to find a job in the Czech Republic, where the growing economy offered increasing opportunities for skilled people. While in the UK they had saved enough money to buy the materials and labour to build a four-bedroom house in the Czech Republic, giving them an excellent start to their married life. Jana's English had become fluent, so she quickly found a well-paid job where she uses it every day.

B *Working as a cleaner*

C *Push–pull factors for a family in West Africa*

Family member	Situation
Father	Subsistence farmer – crops unpredictable due to climate.
	Part-time fisherman – catches are reducing because of overfishing.
Mother	Housewife with limited primary education.
Adult son	Secondary education completed.
	Would like the chance to go to university or obtain an interesting job.
Daughters	Part-way through school.
	Want to get as well qualified as possible.
	School resources are sometimes in short supply.

Key terms

Economic migrant: someone trying to improve their standard of living, who moves voluntarily.

Choropleth map: a map where areas are shaded to show a range of figures. The higher categories are shown in darker colours and the colours get lighter as the figures reduce.

Activities

1 Write the factors listed below into two groups of push and pull factors in terms of someone's possible migration:

job prospects   natural disasters

low income   housing shortages

health care   intolerance

high standard of living   high unemployment

improved housing   racial/religious tolerance

attractive environments   high wages

political or social unrest

educational opportunities   difficult climate

2 Study table **C**, which shows the factors affecting a family in West Africa who are considering moving to the EU. The family consists of two parents, an adult son and two younger daughters who are still at school.

a For each factor, say whether it is a push from the origin or a pull towards the destination. State clearly which members of the family are most affected.

b Taking all the factors and the situations of each family member into account, assess whether or not they should migrate to the EU. Give your reasons.

c Imagine similar families in the same situation. Would they necessarily make the same decision? Explain why.

Migration within the EU

There are two categories of migrants within the EU: those moving between countries and those coming in from beyond the borders. Wealthier countries usually receive immigrants searching for work and a better lifestyle. Poland and other Eastern European countries joined the EU in 2004. Since that date, many people have moved temporarily or permanently to the UK and other Western EU countries for work.

The UK received 600,000 Eastern European migrants between 2004 and 2006. The largest group are Polish (447,000 of the above figure). Most have found formal jobs with much better pay than they would receive at home. Poles in the UK earn on average five times as much as they would at home, while the UK cost of living is only twice that in Poland. Most migrants pay tax, which contributes to the UK's economy. However, some work in the informal economy – working for cash and not paying tax. Poles also use UK health and education services, which add to the government's costs.

Overall, the UK economy has benefited from the influx of migrants from Poland. More recently, as opportunities in Poland increase, many workers are returning home from the UK. Their international migration has been temporary.

E *Occupations of Polish workers in the UK, 2007*

Occupation	Total
Factory workers	95,865
Cleaners, maids and hotel room attendants	33,925
Farm workers, crop harvesters and fruit pickers	29,705
Kitchen and catering assistants, chefs	28,975
Warehouse operatives	25,215
Packers	24,130
Waiters/waitresses	15,840
Care assistants and home carers	12,610
Food processing operatives	11,325
Sales and retail assistants	10,535
Building labourers	10,525
Drivers (HGV and delivery vans)	6,315
Bar staff	6,030
Sub total	*310,995*
Others	77,270
Total	**388,265**

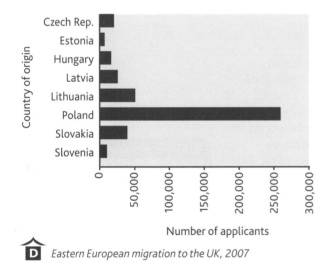

D *Eastern European migration to the UK, 2007*

Migration from outside the EU

Europe currently receives over 2 million immigrants from beyond its borders a year – more than any other world region. The ratio between current population and immigrants is higher for Europe than for the USA. European population is changing more in age and racial structure due to immigration than by changes in birth and death rates.

About 8.6 per cent of the EU's people are foreign-born, compared with 10.3 per cent in the USA and almost 25 per cent in Australia. The range of countries around the world from which migrants to Europe come has changed. Africa and Asia are the major sources of immigrants, but not the same regions as previously.

Activities

3 Study graph **D**.

a Using figures, describe the dominance of Poland in EU migration to the UK.

b Discuss the types of occupations Polish workers undertake. Are there any trends?

4 a List the benefits and difficulties of migration within the EU. Who benefits and who does not?

b If you know someone who has undertaken such a migration, interview them about their reasons for moving and their opinion of the success or failure of their move.

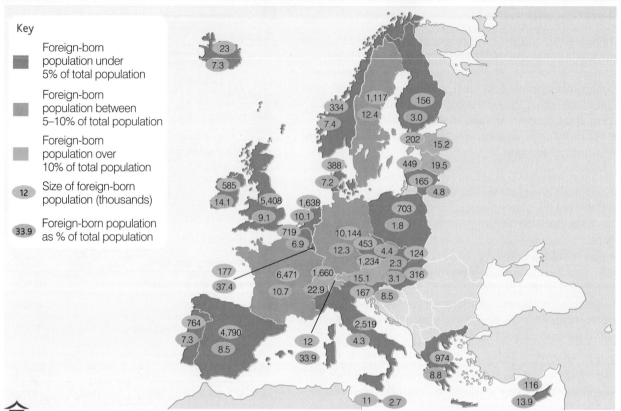

Key

Foreign-born population under 5% of total population

Foreign-born population between 5–10% of total population

Foreign-born population over 10% of total population

12 Size of foreign-born population (thousands)

33.9 Foreign-born population as % of total population

F *Migration into and within the EU, 2005*

Labour migration

Cheaper travel and more information attract skilled and unskilled labour to Europe. Many EU residents would like the flow of migration to reduce, but the United Nations predicts that immigration into the EU will rise by 40 per cent over 40 years. Immigration is a subject of political debate in all EU countries. Spain's immigrant population grew by 400 per cent in 10 years in the early 21st century. Italy expects 100,000 Romanians in the years following Romania's joining of the EU. Although Italy needs workers, not everyone is happy with such a large influx of new people.

Europe needs immigrants because of its falling birth rate and the resulting lack of workers. Highly skilled workers often come to the EU to take temporary jobs in areas of shortage such as teaching, nursing and high-tech computer jobs. About 20 per cent are graduates. Nevertheless, many people see immigrants as a problem rather than as an opportunity.

Activities

5 Study graph **D** on page 191.

a On a sheet of plain paper, draw proportional circles to scale to represent the values show in the graph. Cut them out carefully.

b On a blank outline map of Europe, stick each symbol on or beside the country it represents.

c Give your map a title and explain the scale of your circles in a key.

d Describe the pattern shown in your completed map.

Case study

International labour migration to the EU: Senegal to Italy

Senegalese children love football and support home teams, but many also support Lazio, AS Parma and other Italian teams. Many of these children's fathers and brothers already work in Italy and they aim to follow. Patterns of emigration from Senegal are well established.

G   *Many Senegalese fishing villages have been abandoned*

People aged from their teens to their 40s leave Senegal, mostly males. Those who return often bring enough money build a house. Money is sent home to support children, but those children suffer badly without their fathers at home. The dry climate has limited Senegal's subsistence farming. Funds sent home to the village of Beud Forage have helped to set up water and electricity supplies, but houses left by **emigrants** lie empty and unemployment is high. There is little to attract migrants to return.

Refugee movements to the EU

Asylum seekers are people who are at risk if they stay in their own country. They become refugees when they settle in another country. One-third of EU immigrants claim to be asylum seekers. Since EU countries reduced the number of EU migrants they would allow in, some **economic migrants** have claimed to be asylum seekers, believing this would give them a better chance of being allowed to stay in the EU. Unfortunately, this has sometimes caused strong feelings against genuine asylum seekers.

The EU has been criticised by the United Nations for not taking enough genuine refugees, but it has a good past record in taking those displaced by war. The 1990s Bosnian war produced hundreds of thousands of refugees to the EU. Germany alone took 400,000, many of whom returned home once the situation was peaceful.

Today, the wars in Iraq and Afghanistan – in which EU forces are involved – provide most asylum claims. Two million Iraqis have already left the country, some for neighbouring countries and some to the EU. Another 1.8 million refugees live away from their homes in Iraq and many feel sufficiently threatened to want to leave. Christians are particularly persecuted.

Sweden is particularly generous to asylum seekers (table **H**). By 2007, 70,000 Iraqis already lived there – half of those coming to the EU. The Netherlands, Germany, Greece, Belgium and the UK have given homes to most of the rest. Asylum requests to EU countries from Iraqis increased to 38,286 in 2007 from 19,376 the previous year. They are the largest group currently seeking refuge in the EU and this migration stream is likely to continue, although some decrease in the rate is likely towards the close of the war.

H   *Annual asylum applications per 1,000 existing inhabitants*

EU country	Number of applications
Sweden	2.57
Netherlands	2.27
Belgium	2.16
Germany	1.94
Denmark	1.84
Ireland	1.07
UK	0.97
Spain	0.21

Activities

6   Study map **F**.

a   Use the statistics to find out which countries have the highest proportions of their populations born abroad.

b   Which countries are the top three for numbers of foreign-born population?

c   Do you think it is more useful to study percentages or raw data when looking at migration statistics? Give reasons for your answer. (Note that there is no correct answer here; you will be marked on the quality of the reasons you give.)

d   Describe the pattern shown in the map. Is it what you expect to find having worked through this section?

∞ links

To investigate further why Eastern European migrants are returning to the UK, visit www.migrationwatchuk.com.

9.1 What are the characteristics and causes of urbanisation?

In 2008, for the first time in history, over half of the world's population lived in towns and cities. **Urbanisation** – where there is an increasing proportion of city dwellers in contrast to those in the countryside – is a worldwide process. It began at different times in different parts of the world and occurred at contrasting paces, as it continues to do so today (graph **A**). The contrasting proportions of people living in towns and cities is shown in map **B**.

> ### In this section you will learn
>
> the process of urbanisation and how it varies throughout the world and over time
>
> the causes of urbanisation.

> ### Key terms
>
> **Urbanisation:** a process where an increasing proportion of the population lives in towns and cities (and there is a reduction living in rural areas).
>
> **Rural–urban migration:** a process in which people move from the countryside to the towns.

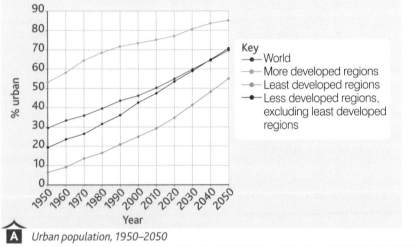

Key
- World
- More developed regions
- Least developed regions
- Less developed regions, excluding least developed regions

A *Urban population, 1950–2050*

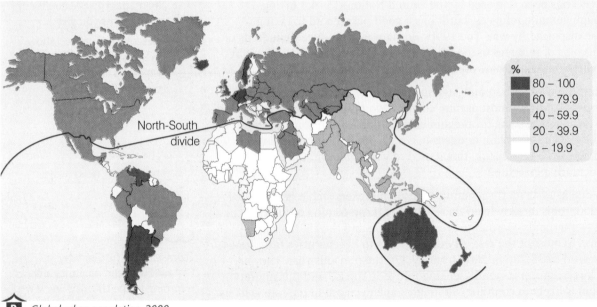

North-South divide

%
- 80 – 100
- 60 – 79.9
- 40 – 59.9
- 20 – 39.9
- 0 – 19.9

B *Global urban population, 2000*

Causes of urbanisation

There are two causes of urbanisation: **rural–urban migration** and natural increase. The initial cause of urbanisation is usually rural–urban migration. This is the result of push–pull factors, as shown in the photos in **C**. Reasons were similar in countries such as the UK in the 19th century. However, here mechanisation was the driving force in farming, which led to unemployment amongst farm workers, and the growth of large-scale production in factories in towns was the main attraction of the emerging industrial cities. The people that migrate into the towns are generally young and this results in relatively high levels of natural increase. The high proportion of young adults results in high levels of births. Falling death rates due to improved medical care mean more babies are born than people dying, which fuels the increase in the urban population.

> **Did you know** ??????
>
> The largest city in the world (based on the population of the metropolitan area) is Tokyo, with a population of 28,025,000, followed by Mexico City (18,131,000) and Mumbai (17,711,000). London is ranked 25th, with a population of 7,640,000.

> 66 Recent droughts have meant yields have been lower and lower, so I've struggled to provide enough food for my family to eat. I have only a small amount of land and can't afford to buy any fertilisers. 99

> 66 I came here because I thought we would be better off. I believed we would have a better house and I would have a good job and some money. My children would be able to go to school and we would get better medical help if we are sick. 99

C *Reasons why people leave the Ethiopian countryside*

Activities

1 Study graph **A**.

a Describe the trends shown by each of the lines. Support your answer with evidence from the graph.

b Give examples of how the process of urbanisation has been important at different times and has occurred at different speeds.

2 Study map **B** and a political map in an atlas. Describe the pattern shown by the map. Try to select areas where rates of urbanisation are very high, high, average, low and very low. Give a clear overview of the locations of places that generally fall into such categories. Exceptions may also be noted.

3 Study the photos in **C**. Imagine you are a resident of the rural area shown on the photo. Relatives have come back to your village and described the city shown in the second photo to you. Explain why you really want to move there.

links

You can find out more about urbanisation at www.esa.un.org/unup.

> **AQA** *Examiner's tip*
>
> Ensure that you can explain the process of rural–urban migration with specific reference to push and pull factors.

In all towns and cities the **land use** varies. In some areas, shops and services may be dominant; in others housing; elsewhere industry or recreation. Some areas have mixed land use where the **function** varies. In British cities, certain areas in similar locations tend to have similar characteristics. For example, the central area tends to be the shopping area or **central business district (CBD)**; the area around this is likely to have originally been the oldest part (although more recent building may have changed this) and this is referred to as the **inner city**. Newer areas are generally found on the outskirts, often known as the **outer city or suburbs**. The photos in **A** show the key areas and their features.

> **In this section you will learn**
>
> how urban areas have a variety of functions
>
> the characteristics and locations of some urban areas.

a CBD

b Redevelopment

c Suburbs

d Inner city

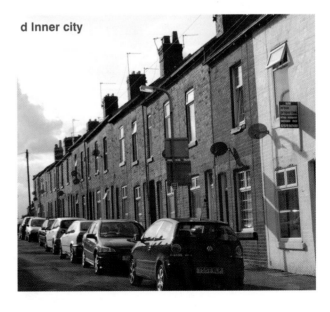

A   *Characteristics of different parts of Sheffield*

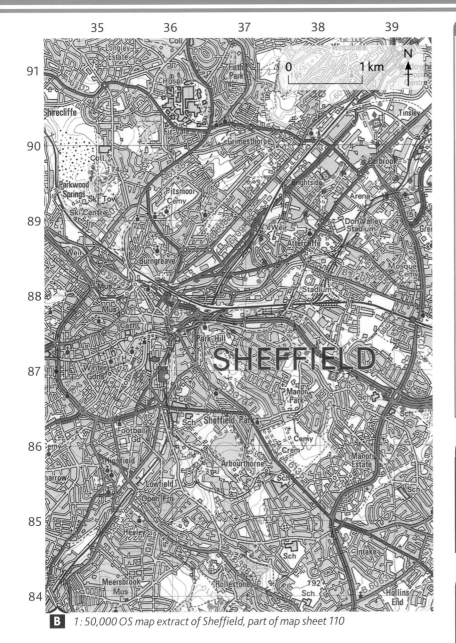

B | 1 : 50,000 OS map extract of Sheffield, part of map sheet 110

Land use: the type of buildings or other features that are found in the area, e.g. terraced housing, banks, industrial estates, roads, parks.

Function: the purpose of a particular area, e.g. for residential use, recreation or shopping.

Central business district (CBD): the main shopping and service area in a city. The CBD is usually found in the middle of the city so that it is easily accessible.

Inner city: the area around the CBD – usually built before 1918 in the UK.

Outer city or suburbs: the area on the edge of the city. Many suburbs were built after 1945 and get newer as they reach the edge of the city.

AQA Examiner's tip

Practise drawing a simple sketch map showing typical land uses in a British city. You may have to explain why each land use is found where it is.

Did you know ??????

The Burj Dubai is the world's tallest building, with 141 storeys and 693 m in height. The Freedom Tower at the World Trade Center will be 82 storeys and 417 m high when it is completed in 2011.

Activity

1 Study the photos in **A** and map **B**.

a What is the evidence that photo **a** shows a CBD?

b Describe the CBD (roughly in the centre of the area within the ring road) from the OS map. You should include reference to its size, shape and characteristics, including facilities present and how its extent can be recognised.

c Using the photo of the inner city, draw a labelled sketch to show the characteristics of the houses and the environment.

d Provide evidence to support the idea that land use in a city varies. You should try to give five different points.

⊙⊃links

Visit **www.multimap.co.uk** to see a complete range of maps at different scales. You can overlay them with satellite images to see how land use varies in Sheffield or a city near to where you live.

What are the issues for people living in urban areas in richer parts of the world?

There are many issues in towns and cities, including those relating to housing, traffic, services and provision for a mixed community. The photos in **A** highlight some of the issues.

 Issues in urban areas in richer parts of the world

In this section you will learn

the range and nature of issues that face people living in urban areas related to housing, inner city, traffic, CBD and multicultural societies

the success of strategies introduced to ease the problems.

Did you know ??????

Three blocks of high-rise flats in Everton, Liverpool were nicknamed 'The Piggeries' because of the living conditions there.

Activity

1   Study the photos in **A**. Work in pairs.

a   For one of them, write five questions to identify key features shown in the photo. Your partner should do the same for the other photo. Your questions should be designed to obtain clear, detailed and thoughtful answers.

b   Swap questions. Answer the questions your partner has written on the photo. You should give full and detailed answers to the questions.

c   Swap your answers and read and discuss them. Do you agree with the answers given? How good were the questions asked? How could they be improved?

d   Together, summarise the issues you think are shown.

Key terms

Household: a person living alone, or two or more people living at the same address, sharing a living room.

Brownfield sites: land that has been built on before and is to be cleared and reused. These sites are often in the inner city.

Greenfield sites: land that has not been built on before, usually in the countryside on the edge of the built-up area.

Urban Development Corporations (UDCs): set up in the 1980s and 1990s using public funding to buy land and improve inner areas of cities, partly by attracting private investment.

City Challenge: a strategy in which local authorities had to design a scheme and submit a bid for funding, competing against other councils. They also had to become part of a partnership involving the local community and private companies who would fund part of the development.

■ Issue 1: housing

Population in the UK has increased by 7 per cent since 1971 and this rate of growth is predicted to continue, giving a population of 52.5 million in England by 2021. The number of **households** has risen by 30 per cent since 1971. Most of this increase is because more people live alone – some 7 million of the UK's population. New single-person households account for 70 per cent of the increased demand for housing. This is due to people leaving home to rent or buy younger than previously, marrying later, getting divorced and living longer. A third of single-person households are aged over 65.

The government target is to build 240,000 new houses every year by 2016 so that house prices do not spiral out of control as a result of a shortage. Many of these new homes will be built throughout existing towns and cities, with a target of 60 per cent to be built on **brownfield sites** – areas that have been previously built on, usually in the inner city. However, some housing will inevitably be built on **greenfield** sites – areas that have not previously been built on, usually on the edge of the city. Table **B** summarises the points in favour of each of these two alternatives. The photos in **C** show the different types of housing being built to meet the growing demand and contrasting needs of the population in different parts of the city.

B   *Advantages of building on brownfield and greenfield sites*

Advantages of building on brownfield sites	Advantages of building on greenfield sites
Easier to get planning permission as councils want to see brownfield sites used	New sites do not need clearing so can be cheaper to prepare
Sites in cities are not left derelict and/or empty	No restrictions of existing road network
Utilities such as water and electricity are already provided	Pleasant countryside environment may appeal to potential home owners
Roads already exist	Some shops and business parks on outskirts provide local facilities
Near to facilities in town centres, e.g. shops, entertainment and places of work	Land cheaper on outskirts so plots can be larger
Cuts commuting	More space for gardens

a Riverside apartments, Bridgewater Place, Leeds

b Gentrified housing, Cambridge

c Retirement bungalows, Scartho, Grimsby

d Family homes, Scartho Top, Grimsby

C   *Types of housing*

Key terms

Regeneration: improving an area.

Sustainable community: community (offering housing, employment and recreation opportunities) that is broadly in balance with the environment and offers people a good quality of life.

Quality of life: how good a person's life is as measured by such things as quality of housing and environment, access to education, health care, how secure people feel and how contented and satisfied they are with their lifestyle.

Park-and-ride scheme: a bus service run to key places from car parks located on the edges of busy areas in order to reduce traffic flows and congestion in the city centre. Costs are low to encourage people to use the system – they are cheaper than fuel and car parking charges in the centre.

Segregation: occurs where people of a particular ethnic group choose to live with others from the same ethnic group, separate from other groups.

Activities

2 Read the text and study table **B** on page 199.

a Why is the demand for housing likely to increase significantly in the next few years?

b What is meant by the term 'brownfield site'?

c What are the advantages and disadvantages of building new houses on a brownfield site?

d Identify a brownfield site that is being used for housing in your local area.

e What is a 'greenfield site'?

f Why are many people concerned about the use of greenfield sites for building houses?

g Is there a greenfield site near you that is being developed for housing? If so, has it caused concern among local people? Explain your answer.

3 Study the photos in **C** on page 199.

a In groups, take on the following roles:

- professional accountant working in a city-centre office, aged 26 and single

- a 68-year-old pensioner, widowed and living alone

- a couple seeking an old, modernised house as a first home together

- a family with two children aged 10 and 12.

b For each role, identify which of the four houses shown in **C** would be most suitable and explain the advantages of living there.

Issue 2: the inner city

Part of the demand for housing will be satisfied in inner-city areas. Successive governments have had a variety of strategies to improve living in inner cities since 1945 – most infamously the building of cheap, high-rise blocks of flats in the 1960s and early 1970s as a 'quick fix'. Over the years, strategies have changed and there has been a greater involvement of private funding and the local community.

Urban Development Corporations (UDCs) were a major strategy introduced in the 1980s, with London Docklands Development Corporation (LDDC) and Merseyside Development Corporation (MDC) being established in 1981. Eleven more UDCs followed. These were large-scale projects where major changes occurred with the help of both public and private investment. Although the LDDC ceased to exist after 1998, the area has continued to develop and change. During its lifetime, £1.86 billion of public money was invested along with £7.7 billion from the private sector (diagram **D**). Changes are continuing today, with further improvements in the transport system (such as the eastward extension of the Jubilee Line), and the building of further skyscrapers at Canary Wharf (Heron Quay, North Quay) and other locations.

Activity

4 Study diagram **D** and photo **E**. You could work in groups of three for this task, with each member of the group doing one of the strategies.

a For each of the three strategies (UDCs, City Challenge and sustainable communities), complete a fact file including the following information:

- location

- dates

- what was done

- who was involved

- where funding came from

- what has been done to improve housing, the environment and the community.

b Create a table to show the advantages and the disadvantages of each of the three strategies.

c Which of the three strategies do you think is best? Give full justification for your answer. You should stress its advantages and the disadvantages of the other two.

Strategy 1: UDCs

The 1980 Act requires an urban development corporation 'to secure the regeneration of its area, by bringing land and buildings into effective use, encouraging the development of existing and new industry and commerce, creating an attractive environment and ensuring that housing and social facilities are available to encourage people to live and work in the area.'

The London Docklands Development Corporation

The LDDC was at work for 17 years. In its final annual report in 1998 it headlined its achievements as follows:

- £1.86 billion in public sector investment
- £7.7 billion in private sector investment
- 431 ha of land sold for development
- 144 km of new and improved roads
- the construction of the Docklands Light Railway
- 2.3 km² of commercial/industrial floorspace built
- 762 ha of derelict land reclaimed
- 24,046 new homes built

- 2,700 businesses trading
- contributions to 5 new health centres and the redevelopment of 6 more
- funding towards 11 new primary schools, 2 secondary schools, 3 post-16 colleges and 9 vocational training centres
- 94 awards for architecture, conservation and landscaping
- 85,000 people now at work in London Docklands.

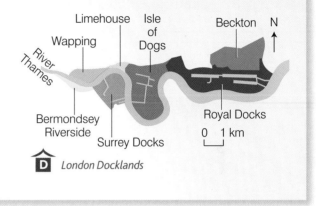

D *London Docklands*

City Challenge was a big initiative of the 1990s. It had a holistic approach to **regeneration**, where local authorities, private companies and the local community worked together from the start. The focus of projects varied. The Hulme (Manchester) City Challenge Partnership sought to improve the housing that had been built in the 1960s to replace the old terraces that had once stood there. Integral to this was an attempt to enhance the environment, community facilities and shopping provision (photo **E**).

Strategy 2: City challenge Hulme, Manchester

- Crescents were built in the 1960s and demolished in the 1990s
- Through City Challenge, Hulme received £37.5 million
- Some old buildings were retained
- Homes were designed to conserve water, and be energy efficient and pleasant
- There was a return to a traditional layout – Stretford Road (at the end of which is Hulme Arch) was rebuilt after demolition of crescents (original course was through the middle of these)
- Local schools and a new park have been built
- The views of local people were taken into account

E *Hulme, Manchester*

Sustainable communities allow people to live in an area where there is housing of an appropriate standard to offer reasonable **quality of life**, with access to a job, education and health care. This initiative began in 2003 and one area affected by it is an area of east Manchester, formerly known as Cardroom and now renamed New Islington Millennium Village. As diagram **F** shows, it seeks to provide for an appropriate quality of life in inner-city Manchester in the 21st century.

Strategy 3: Sustainable communities

What's coming to New Islington?

New homes
- 66 houses, 200 ground-floor apartments
- 500 two- and three-storey apartments
- 600 1- and 2-bed apartments
- 34 urban barns
- workshops
- refurbishment of Ancoats hospital and Stubbs Mill
- new office space

Waterways
- 3,000 metres of canalside
- 12 bridges
- 3 giant canopies
- 50 moorings for narrow boats and canalside facilities

Urban amenities
- 10 new shops
- 2 pubs, 2 restaurants, cafés and bars
- Metrolink stop in 10 minutes' walking distance
- New bus lines and bus stops
- 200 on-street and 1,200 underground car-parking spaces
- a safe Old Mill Street

Parks and gardens
- 300 new trees
- 2 garden islands, an orchard, a beach
- play areas and climbing rocks
- secured courtyard gardens
- private gardens and patios

Community facilities
- a primary school and play areas
- a health centre with 8 GPs
- 2 workshops
- a crèche
- an angling club and a village hall
- a football pitch

Sustainability Agenda
- boreholes will provide up to 25 litres per second of naturally filtered water
- central heat and power to generate 600 kW electrical energy and 1,000 kW thermal energy
- recycling collection points that allow occupants to recycle 50% of domestic waste

F *Living in New Islington, Manchester*

Issue 3: traffic

As we demand greater mobility and accessibility with flexibility, the number of cars has increased, as has the problem of traffic congestion. More people have more money and welcome the door-to-door service that comes with having a car. Many households (27 per cent in 2002) have more than one car, while 45 per cent have one car. Photo **G** shows environmental problems this creates, relating to standing traffic and congestion, air and noise pollution and an adverse impact on buildings and environmental quality generally. Diagram **H** shows strategies designed to reduce the use of cars by encouraging cycling, making public transport more attractive, introducing **park-and-ride schemes** and congestion charging.

- Air pollution from vehicles
- Noise from heavy vehicles
- Buildings discoloured
- Impact on health – respiratory conditions, asthma
- Unsightly

G *Environmental problems resulting from traffic congestion*

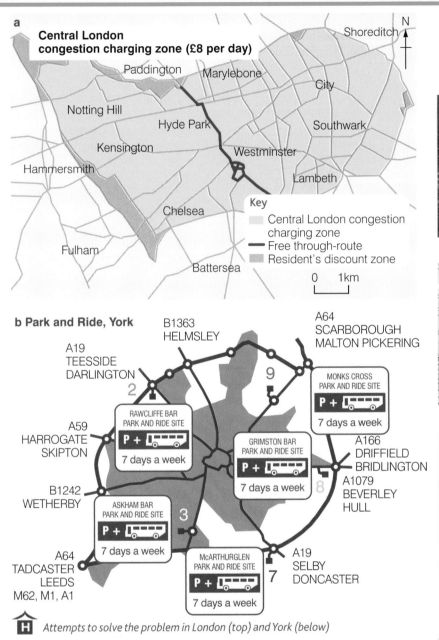

a

Central London congestion charging zone (£8 per day)

Shoreditch

N

Paddington Marylebone

City

Notting Hill

Hyde Park

Southwark

Kensington

Westminster

Hammersmith

Lambeth

Chelsea

Key

Central London congestion charging zone
Free through-route
Resident's discount zone

Fulham

Battersea

0   1km

b **Park and Ride, York**

B1363
HELMSLEY

A64
SCARBOROUGH
MALTON PICKERING

A19
TEESSIDE
DARLINGTON

2

9

MONKS CROSS
PARK AND RIDE SITE

P +

7 days a week

A59
HARROGATE
SKIPTON

RAWCLIFFE BAR
PARK AND RIDE SITE

P +

7 days a week

GRIMSTON BAR
PARK AND RIDE SITE

P +

7 days a week

A166
DRIFFIELD
BRIDLINGTON

B1242
WETHERBY

ASKHAM BAR
PARK AND RIDE SITE

P +

7 days a week

3

8

A1079
BEVERLEY
HULL

A64
TADCASTER
LEEDS
M62, M1, A1

McARTHURGLEN
PARK AND RIDE SITE

P +

7 days a week

7

A19
SELBY
DONCASTER

H *Attempts to solve the problem in London (top) and York (below)*

Issue 4: multicultural mix

Despite the apparent racial mix in many cities, a significant number of immigrants choose to live with people from similar areas and away from others with different ethnicity and culture. This represents **segregation**.

Here are some of the reasons why people choose to cluster in the same area as their fellow immigrants:

■ Support from others

People feel safe and secure when they can associate with other people from the same background. There is a sense of belonging and protection from racial abuse.

■ A familiar culture

In a strange country, there is comfort from being with people who have similar ideas and beliefs and speak the same language.

Activities

5 Study diagram **H**.

a Produce a poster to show the good and bad points of the options available to solve the environmental impact of traffic.

6 Study the photos in **I** on page 204.

a Select a strategy for revitalising the image of the CBD as shown here.

b For each strategy, explain how it works.

- Specialist facilities

In many areas these are provided so that, for example, Sikhs can worship in a *gurdwara* and Moslems in a mosque. Familiar foodstuffs will be available in shops.

- Safety in numbers

People have a stronger voice if they are heard as a group, rather than individually.

- Employment factors

Immigrant groups tend to do low-paid jobs or have a high rate of unemployment. They have limited money and so can only afford cheaper housing in certain parts of the city, usually inner city areas.

There is an attempt to integrate different ethnic groups and reduce segregation. In Leeds, this has involved the following measures:

- increasing children's achievement by improving educational provision and opportunities in deprived areas; and seeking to improve literacy in areas where English can be a second language
- increasing employment through initiatives to ensure basic skills and access to information and training
- increasing community involvement by ensuring that the needs of minority groups are understood and met
- providing facilities that encourage meetings of all sections of a community rather than separate ethnic groups.

I *Revitalising the image of Leeds CBD*

Issue 5: the central business district (CBD)

During the 1960s to the early 1980s, the CBD struggled to attract businesses. Out-of-town shopping areas and regional shopping centres became more favourable destinations as they offered pleasant shopping opportunities with ample parking. In contrast, city centres appeared busy and crowded. The air quality was poor, with the smell of diesel and lead concentrations in certain areas a cause for concern.

However, there have been significant changes in CBDs and their image is now, once more, a positive one. They have become vibrant and pleasant places as a result of a number of initiatives (photos in **I**).

⚭ links

Find out more about Leeds CBD and its plans for the city and Chapeltown at **www.leeds.gov.uk**.

9.4 What are the issues for people living in squatter settlements in poorer parts of the world?

The speed of the urbanisation process in many poorer areas of the world results in **squatter settlements** being built and the evolution of an **informal sector** of the economy. The pace of rural–urban migration is too fast to allow the time needed to build proper houses and for the economy to grow to provide jobs. People find unoccupied areas of land and materials and begin to build their own makeshift shelters. As there are few official jobs available, people create their own employment: selling items; making and repairing things on a small scale; becoming couriers, cleaners, gardeners; taking in laundry.

In this section you will learn

why squatter settlements have developed, their characteristics and effects on people's lives

different strategies to try to improve squatter settlements and evaluate them

how to apply general concepts relating to squatter settlements to a case study.

Key terms

Squatter settlements: areas of cities (usually on the outskirts) that are built by people of any materials they can find on land that does not belong to them. Such settlements have different names in different parts of the world (e.g. *favela* in Brazil) and are often known as shanty towns.

Informal sector: that part of the economy where jobs are created by people to try to get an income (e.g. taking in washing, mending bicycles) and which are not recognised in official figures.

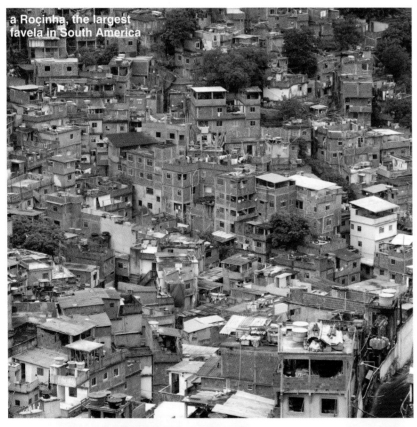

a Roçinha, the largest favela in South America

b Close-up of Roçinha

c Mumbai

A *Typical shanty towns*

Living in squatter settlements

No preparation for the building of these areas is done, so the houses are not provided with basic infrastructure such as sanitation, piped water, electricity and road access. The photos in **A** (page 205) include views of Roçinha, a *favela* in Rio de Janeiro populated by 100,000 – an image of squatter settlements that has become iconic. Close-up views give a clearer impression of conditions in the *favela*. The houses are made of any materials available nearby – corrugated iron, pieces of board – haphazardly assembled to provide a basic shelter. There is a simple layout that may have a living area separate from a sleeping area. Parents and large families inhabit a small shack in an area that is very overcrowded. There are no toilets, water must be collected from a nearby source – often at a cost – and carried back. Rubbish is not collected and the area quickly degenerates into a place of filth and disease. The inhabitants tend to create poorly paid jobs where the income is unreliable or they work in the less well-paid part of the formal sector. Quality of life is poor; the housing and environment are largely responsible for this. So too is the economic circumstances in which the residents find themselves, as a lack of money makes improvement difficult. Crime is a problem, children often do not go to school, the family lives on top of each other, there is no privacy, disease is rife and life is one of trying to survive from one day to the next.

Strategies to improve living conditions

Improvement by residents involves the residents seeking to 'do up' their original shelters (photo **B**). This means replacing flimsy, temporary materials with more permanent brick and concrete; catching rainwater in a tank on the roof; and obtaining an electricity supply (often illegally by tapping into a nearby source). Such improvements are slow and individual – not all the problems of poor living conditions can be solved.

Self-help occurs where local authorities support the residents of the squatter settlements in improving their homes. This involves the improvements outlined above, but it is more organised. There is cooperation between residents to work together and remove rubbish. There is also cooperation from the local authority, which offers grants, cheap loans and possibly materials to encourage improvements to take place. Standpipes are likely to be provided for access to water supply and sanitation. Collectively, the residents, with help from the local authority, may begin to build health centres and schools. Legal ownership of the land is granted to encourage improvements to take place, marking an acceptance of the housing.

B *DIY improvements in Roçinha*

Site and service schemes are a more formal way of helping squatter settlement residents. Land is identified for the scheme. The infrastructure is laid in advance of settlement, so that water, sanitation and electricity are properly supplied to individually marked plots. People then build their homes using whatever materials they can afford at the time. They can add to and improve the structure if finances allow later.

Local authority schemes can take a number of different forms. There may be large-scale improvements made to some squatter settlements or new towns may be constructed. In Cairo, new settlements such as 10th of Ramadan City were built to reduce pressure on the city (photo **C**). High-rise blocks of flats were built, together with shops, a primary school and a mosque. Industries were also planned to provide jobs for the new inhabitants.

C *10th of Ramadan City, Cairo*

Activities

1   Study the photos **A** and **B**. Imagine you are a 14-year-old resident of a squatter settlement such as Roçinha. Your name is Eduardo/Camila. You were born in Roçinha and have lived there all your life. You are being interviewed by a well-known journalist from a national paper. The reporter wants to know all about your house, your surroundings, your life, what you think the future holds for you and what you think are the good and bad points about living in Roçinha. Write down what you say. Include a labelled photo of a shanty town.

Kibera, a squatter settlement in Nairobi, Kenya

Map **D** shows the situation of Kibera in the capital city of Nairobi in Kenya. Some 60 per cent of Nairobi's inhabitants live in slums; over half of them in Kibera. Specific facts about Kibera are uncertain. It is believed that between 800,000 and 1 million people live in the shanty town, in an area of 255 ha. This gives extremely high densities, with people only having 1 m² of floor space each. Over 100,000 children are believed to be orphans as a result of the high incidence of HIV/Aids. Photos **E** and **F** show the squatter settlement. The homes are made of mud, plastered-over boards, wood or corrugated iron sheeting. The paths between the houses are irregular, narrow and often have a ditch running down the middle that has sewage in it. Rubbish litters the area as it is not collected. The area smells of the charcoal used to provide fuel and of human waste. A standpipe may supply water for up to 40 families; private operators run hosepipes into the area and charge double the going rate for water. Crime is rife and vigilante groups offer security – at a price. Police are reluctant to enter. However, there is a community spirit; homes are kept clean and the residents welcome visitors.

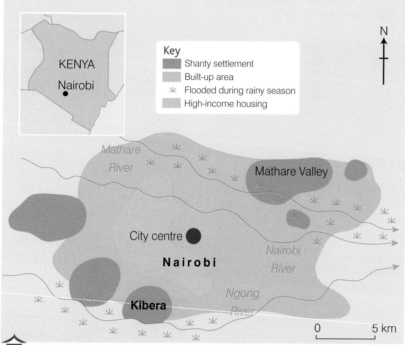

Key
- Shanty settlement
- Built-up area
- ⅄ Flooded during rainy season
- High-income housing

KENYA
Nairobi

Mathare River
Mathare Valley
City centre
Nairobi
Nairobi River
Ngong River
Kibera

0   5 km

N

D *Location of Kibera in Nairobi*

E *General view of Kibera*

Finding solutions

There are signs that things are improving. Practical Action, a British charity, has been responsible for developing low-cost roofing tiles made from sand and clay and adding lime and natural fibre to soil to create blocks used for building that are cheaper than concrete. These allow self-help schemes to progress. The United Nation's Human Settlement Programme (UN-Habitat) has provided affordable electricity to some parts at 300 Kenyan shillings per shack. There are two mains water pipes – one provided by the council and the other by the World Bank – at a cost of 3 Kenyan shillings for 20 litres. Improving sanitation is more difficult and progress is slow. Medical facilities are provided by charities. Many well-known charities support the improvements. Gap-year students are encouraged to go to Kibera to oversee the spending and to help coordinate efforts.

There are other schemes on a grander scale. In a 15-year project that began in 2003, plans are to re-house thousands of residents of Kibera. This is a joint venture between the Kenyan government and UN-Habitat. In its first year, 770 families were rehoused in new blocks of flats with running water, toilets, showers and electricity. Residents have been involved in plans and funding of 650 million Kenyan shillings had been set aside for the first year. It is hoped funding will be provided by charities and by setting up private loans.

F *Close-up view of Kibera*

AQA Examiner's tip

You may be asked to write a detailed answer covering many issues relating to squatter settlements, including their characteristics and improvement schemes. Practise this with reference to a specific squatter settlement.

Activities

2 Study map **D** and photos **E** and **F**. Produce a fact file on Kibera by completing the following tasks:

a Draw a labelled sketch map to show the location of Kibera in Nairobi.

b Write a summary of the key points about the Kibera squatter settlement.

c Draw a labelled sketch from one of the photos to show the characteristics of the squatter settlement and the living conditions there.

d Summarise the attempts that have been made to improve Kibera.

e In your opinion, is Kibera a slum of hope (where things are getting better) or a slum of despair (where there is no improvement and no hope for the future)? Justify your views.

Did you know ???????

A special report by the BBC in 2002 described Kibera as 'Six hundred acres of mud and filth, with a brown stream dribbling down the middle. You won't find it on a tourist or any other map. It's a squatter's camp – an illegal, forgotten city – and at least one third of Nairobi lives here.'

∞ **links**

Investigate Kibera at **www.bbc. co.uk** and **www.mojamoja.org**

9.5 What are the problems of rapid urbanisation in poorer parts of the world?

Examples of problems

During the early hours of 3 December 1984, the world's worst industrial accident unfolded in the Indian city of Bhopal. Poisonous gas escaped from a chemical plant and killed 3,000 people (photo **A**). Unofficial estimates put the number between 8,000 and 10,000 people. Around 50,000 suffered permanent disabilities and more died later. This is one example of how rapid urbanisation and **industrialisation** can lead to environmental problems in poorer parts of the world, which can often result in many deaths and have long-term effects on the health of populations. Expanding cities lead to problems of air and water pollution and **disposal of waste**, including toxic waste from plants like the one at Bhopal. Non-existent or poor regulations and a lack of planning for an environmental emergency make problems worse.

Creation of electronic waste is another major problem in a rapidly industrialising country like India. The country imports more than 4.5 million new computers a year, plus many second-hand ones with shorter lifespans. Computer waste is known as electronic waste or e-waste. In the cities, India's poor scrape a living by breaking down PCs and monitors (photo **B**). They boil, crush or burn parts in order to extract valuable materials like gold or platinum. But what they do not realise is that the toxic chemicals inside like cadmium and lead can pose serious health risks. India's hospitals are starting to see patients with ten times the expected level of lead in their blood. Dumping and unsupervised recycling of e-waste is literally leading to a brain drain.

The Ganges River contains untreated sewage, cremated remains, chemicals and disease-causing microbes. Cows wade in the river. People wash their laundry in it and drink from it (see Photo **E**).

In Shanghai, the construction boom is creating 30,000 tonnes of waste per day. Industry there is responsible for 70 per cent of the country's carbon dioxide emissions. Some 73 per cent of electricity is

A Deaths caused by the industrial accident in Bhopal

produced by coal-fired power stations. These factors are responsible for 400,000 deaths annually. Shanghai's Huangpu river is the main water supply for the city, and in the last 10 years water quality has fallen as 4 million m^3 of untreated human waste enter it daily.

B   *Breaking down PCs and monitors in India*

Reducing the problems

In order to seek to reduce the environmental problems resulting from rapid urbanisation and industrialisation, there need to be guidelines to indicate what is allowed and what is not. Limits must be monitored and enforced to ensure that industries, for example, do not exceed the stated limits.

Waste disposal

Waste provides a resource and a means of making a living for many shanty dwellers in poor countries. In São Paulo, Brazil, two huge incinerators burn 7,500 tonnes of waste a day, resulting in a problem caused by a management strategy. There were only two **landfill** sites in 1990. Children and adults alike scavenge and extract materials and then reuse or resell them. For example, car tyres may be made into sandals and food waste is fed to animals or used as a fertiliser on vegetable plots. In Shanghai, China, an effective solid waste disposal unit has been installed in most households and the waste is used as a fertiliser in surrounding rural areas. Toxic waste and its safe disposal is a key issue in areas where the manufacturing industry is increasing. In the aftermath of the Bhopal accident in 1984, the site was covered in toxic waste. This could not be disposed of safely in India. This meant that the waste was packed up and sent to the USA so that it could be disposed of safely. It is important that cities in poorer countries are not seen as areas where toxic waste can be disposed of more easily. Large companies need to take responsibility for safely disposing of electrical goods in areas such as Bangalore in India, where there are many call centres.

The large amount of e-waste in Bangalore is covered by one enforcement order, which is inadequate. There are not enough people employed to make sure the law is obeyed. Greenpeace believes that the high-tech companies that create the products should take responsibility for the waste created. This would involve extracting dangerous chemicals from the equipment at the end of its life. **Recycling** plants provide the way forward.

Air pollution

Air pollution is a real issue. Most industrial production is in the biggest cities and there is a need to encourage the use of new technologies that can reduce emissions of sulphur dioxide and nitrogen oxide. Switching to cleaner, alternative sources of energy is an option. However, given the plentiful supplies of coal in countries such as China (where 80 per cent of electricity is from this source) and India, this may need the introduction of a carbon tax to induce a change. In Shanghai, China, industries use low sulphur coal to try to reduce pollution. Greater monitoring and safety checks are essential if disasters such as Bhopal are to be avoided. Limits need to be set and enforced on emissions, and companies, including **transnational corporations (TNCs)**, must be monitored to ensure that emissions of carbon dioxide and sulphur dioxide are reduced. Transport also needs to be considered and strategies such as allowing odd-numbered cars into Mexico City on one day and even-numbered on another day can reduce traffic in towns. Other strategies include improving public transport, limiting the number of cars and introducing congestion charging to discourage car owners from entering city centres.

Water pollution

As with air pollution, limits relating to **water pollution** need to be identified and enforced if quality is to be improved. In 1986, the Ganga Action Plan sought to introduce water treatment works on the River Ganges in India, which it did successfully. However, the increasing population was not taken into account and water quality has since deteriorated. Such attempts have been replicated in other countries. In Shanghai, the Huangpu and Suzhou rivers have been the target for improving water quality. A World Bank loan of $200 million was granted to this cause in 2002.

⬭ links

For information on Union Carbide, the Bhopal disaster and toxic e-waste, go to **www.bbc.co.uk**.

For reducing air pollution in Asia and the Pacific, visit **www.unep.org**.

Look into the dumping of toxic waste in poor countries such as Ivory Coast at **www.npr.org** and research more about e-waste at **www.bbc.co.uk**.

C   *A recycling centre in Beijing, China*

D   *The River Ganges, India*

Activities

1

a   Describe the sources of pollution in the rivers Ganges and Huangpu.

b   Explain why they present a hazard to health.

c   Produce a fact file to summarise the events at Bhopal on 3 December 1984.

d   What is your view of the events of that night? Give reasons for your opinion.

e   Why do many poorer countries use coal as a source of energy?

f   Describe how some people living in cities in India make a living from discarded computers.

g   Explain the dangers of making a living in this way.

2   Study the text on the strategies to reduce the environmental effects of rapid urbanisation.

Produce a diagram to summarise the strategies adopted to manage the environmental problems in poorer parts of the world. Use case studies and illustrations as part of your work.

9.6   How can urban living be sustainable?

A **sustainable city** has certain characteristics that relate to its long-term future, which is ideally problem-free. The environment is not damaged; the economic base is sound with resources allocated fairly and jobs secure; there is a strong sense of community, with local people involved in decisions made. Some earlier parts of this chapter have included information that, you will see, could equally have been included here, such as the need to provide appropriate housing in rich and poor areas of the world, the need to deal with traffic problems and the need to address the issue of pollution in poorer areas. Photos **A** and **B** show two contrasting urban scenes in Los Angeles and Belfast.

In this section you will learn

attempts made to ensure that city life is environmentally and socially sustainable

the characteristics of a sustainable city.

A *Los Angeles: non-sustainable urban living?*

Key terms

Sustainable city: an urban area where residents have a way of life that will last a long time. The environment is not damaged and the economic and social fabric, due to local involvement, are able to stand the test of time.

B *Belfast: sustainable urban living?*

Activity

1   Study photos **A** and **B**.

a   Describe the characteristics of the urban scenes from the photos.

b   In which of the photos do you think life is sustainable? Explain reasons for your choice.

■ Seeking environmental sustainability

Conserving the historic and natural environment

The Liverpool Maritime Mercantile City provides an example of conserving an area of previous industrial use and historic commercial and cultural areas. The Liverpool waterfront and areas associated with its development were designated a World Heritage Site in 2004. The award recognised the importance of the area as a port and associated buildings of global significance during the heyday of the British Empire in the 18th and 19th centuries. Many of the buildings are architecturally as they were then, although their function has changed. The photos in **C** show some of the many faces of the sites that gained recognition, conserving an environment so rich in history and heritage.

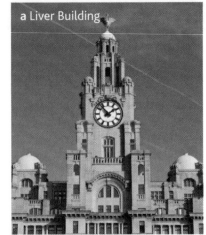

a Liver Building

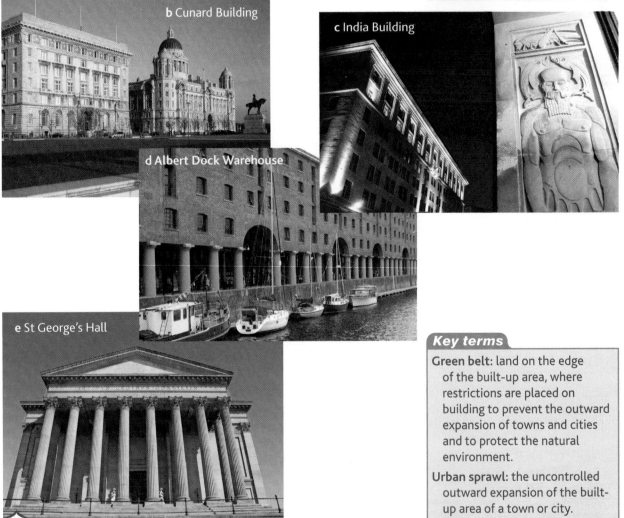

b Cunard Building

c India Building

d Albert Dock Warehouse

e St George's Hall

C *Liverpool maritime mercantile city*

Activity

2   Study the photos in **C**. Produce a leaflet for Liverpool City Council or another city near where you live. Produce an illustrated report to show how the city has used and conserved its industrial heritage in a sustainable way. Use the Internet to help you with your research.

Key terms

Green belt: land on the edge of the built-up area, where restrictions are placed on building to prevent the outward expansion of towns and cities and to protect the natural environment.

Urban sprawl: the uncontrolled outward expansion of the built-up area of a town or city.

Sustainability: development that preserves future resources, standards of living and the needs of future generations.

Incineration: getting rid of waste by burning it on a large scale at selected sites.

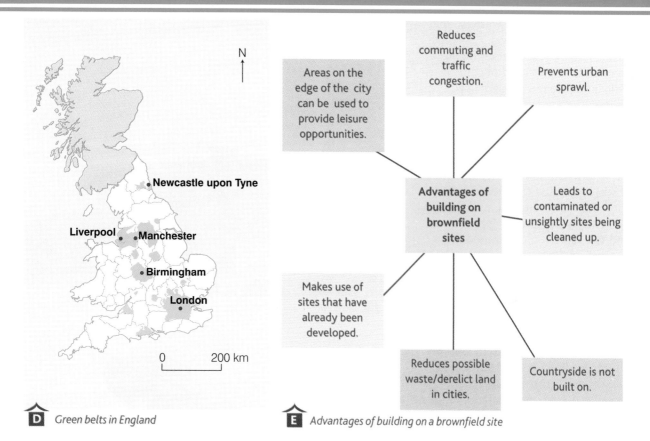

D *Green belts in England*

E *Advantages of building on a brownfield site*

The natural environment can be conserved by reducing, or even stopping, development on the edge of the existing built-up area and by encouraging development to take place on sites that have been previously used in the inner city or other areas. **Green belts** exist around many large towns in England or in towns where growth is occurring (map **D**). These were set up to prevent **urban sprawl** and to ensure that the surrounding countryside is protected from development. This often provides (and the policy intended this) recreational open space for urban residents. Limiting available sites on the edge of the city means that alternative locations for development must be offered if growth is to continue. This means that building on brownfield sites is simultaneously encouraged. In addition to limiting the growth beyond the city, as sites are available within the current built-up area, there are other advantages of building on brownfield sites that benefit the environment and encourage **sustainability** (diagram **E**).

Reducing and safely disposing of waste

By 2000, the UK was producing 330 million tonnes of waste each year – enough to fill the Royal Albert Hall in London *every hour*. Much of this was from mining and quarrying, but 30 million tonnes was from households, many of them in cities. There is a need to reduce the amount of waste produced. The government has a target of recycling 40 per cent of household waste by 2010. This is an ambitious target as only 18 per cent was recycled in 2004. However, 20 per cent of household waste is garden waste, a further 18 per cent is paper and cardboard and 17 per cent is kitchen waste. A note of caution is needed, however, as the cost incurred in transporting and reprocessing some of the products needs to be considered.

Did you know ??????

In the UK we produce 517 kg of waste per person every year.

Did you know ??????

Before many retailers introduced charges for plastic carrier bags and offered 'bags for life', UK shoppers used about 10 billion new carrier bags a year. Seven leading supermarkets have committed to reducing carrier bag use by 50 per cent by May 2009.

It is important to reduce waste so that fewer plastic bags are used. Consumer pressure could reduce packaging in general – do apples need to come in plastic bags? Do red peppers need individual packaging? Packaging can be made so that it can be returned and reused, such as milk bottles and 'bag for life' carrier bags.

Even with maximum effort, some waste will still be created that needs to be disposed of. There are two main options: **incineration** and landfill (photos **F** and **G**). The UK has favoured the latter option (73 per cent of household waste is disposed of in this way), but this is not without its problems. One significant issue is that we are running out of appropriate sites, with capacity available until 2015 before a shortage of sites begins to occur. Incineration only accounts for 9 per cent of household waste disposal. This has proved an unpopular option and created a range of issues.

Providing adequate open spaces

The presence of official green belts or areas where local authorities choose to restrict buildings around cities offers open space for recreation purposes. In addition, many areas within cities have designated areas of open space in the form of parks, playing fields and individual gardens. Map **H** shows the distribution of open space in Greater London and some of the types of open space available.

F   *Incineration: Sheffield energy recovery facility*

G   *A landfill site in Liverpool*

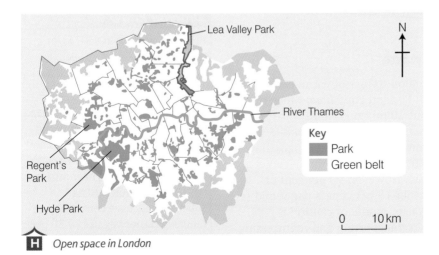

Lea Valley Park

River Thames

Key

Park

Green belt

Regent's Park

Hyde Park

N

0   10 km

H   *Open space in London*

Activities

3 Study map **D** and diagram **E**. Imagine you are a town planner on a visit to a local school where you have to explain how green belt policy and the use of brownfield sites can help in conserving the natural environment. Design a presentation using Powerpoint to outline the advantages of green belts and using brownfield sites for development.

4 Work in pairs. Produce a radio advert for your local council encouraging residents to 'reduce, reuse and recycle'. Make the advert clear and informative, and try to introduce a catchy slogan to promote a more responsible attitude to waste.

5 Study photos **F** and **G**. Summarise the advantages and disadvantages of:
   ■ incineration
   ■ landfill
   as a means of waste disposal.

6 a What materials are collected by the council from your home?
   b What other items do you recycle at home?
   c How do you reuse items at home?
   d Describe how you try to reduce waste at home.
   e What efforts are made at school to reduce, reuse and recycle?
   f How effective do you think efforts are at home and at school? In what ways could they be improved?

7 Study map **H**.
   a Describe the distribution of the different types of open space in Greater London.
   b London was designed to have many central parks. How do you think these parks are used by people?
   c Do you think cities should have open spaces such as parks? Justify your answer.

Involving local people

If people have ownership of ideas and feel involved and in control of their own destiny, they are much more likely to respond positively and care for the building and environment in which they live. Consulting people at planning stages – before decisions are made – is essential. Planners increasingly survey opinions before putting forward plans and consult after they have been produced. Residents form associations to give them a stronger collective voice.

Where improvements are planned, asking what residents want and providing it means that the people are happy in their homes and take better care of them. This can involve apparently minor things such as colour schemes for paint and new bathroom suites. Having meetings in local halls where people are invited to see what is planned gives people the opportunity to give their views so that they feel included, not excluded.

Providing an efficient public transport system

The volume of cars as a means of private transport is a problem and a barrier to a city being sustainable. London has sought to make parts of the city unattractive to drivers via congestion charging. However, an alternative needs to be offered. This means a public transport system that is efficient, reliable and comfortable. The mayor of London is keen to ensure the provision of a public transport system which the capital can be proud of and one that is sustainable (extract **I**). This inevitably means a focus on the Underground and improvements in buses and rail links. The 2012 Olympics has added impetus to improvements. The Tube is undergoing extensive upgrading, not just to the lines, but also to the trains and stations. London overground links will be extended to form a complete circuit around London – the railway equivalent to the M25. Buses are being improved – bendy buses will be abandoned

Did you know ??????

Between three and a half and four million journeys are made on the Underground (Tube) in London every weekday. The Tube is the oldest of its kind in the world – some parts dating back to the 1860s.

The mayor's transport strategy for London

The single biggest problem for London is the gridlock of our transport system. At the start of the 21st century, traffic speeds in central London have fallen to less than 10 miles an hour (16 kph) with knock-on effects on the speed and reliability of the bus system. Congestion is growing in outer London town centres. Rail services are in unprecedented crisis. The Underground is more and more overcrowded and unreliable, and stations are blighted by broken escalators and lifts.

The strategy sets out an integrated package of policies and proposals to take forward the entire transport system of this city and develop London as an exemplary sustainable world city.

To support the vision of London as an exemplary sustainable world city, the transport strategy will increase the capacity, reliability, efficiency, quality and integration of the transport system to provide the world-class transport system the capital needs. The top five key transport priorities that flow from this are:

- reducing traffic congestion

- overcoming the backlog of investment on the Underground so as to safely increase capacity, reduce overcrowding, and increase both reliability and frequency of services

- making radical improvements to bus services across London, including increasing the bus system's capacity, improving reliability and increasing the frequency of services

- better integration of the National Rail system with London's other transport systems to facilitate commuting, reduce overcrowding, increase safety and move towards a Londonwide, high frequency 'turn up and go' Metro service

- increasing the overall capacity of London's transport system by promoting: major new cross-London rail links including improving access to international transport facilities; improved orbital rail links in inner London; and new Thames river crossings in east London.

I *London's public transport: problems and priorities, July 2001*

in the interests of safety and buses are to be more frequent to reduce overcrowding and to make them more attractive to travellers. Faster journeys and greater frequency were key factors in encouraging bus use. By the end of 2008, all buses had CCTV to increase feelings of security and bus shelters were added at more bus stops. Buses have improved in quality – over 75 per cent have low-floor access. The extension of bus lanes has led to quicker journeys and cash fares have been frozen. Schemes such as the Oyster card, which allows for the advanced purchase of up to £90 worth of journeys on a swipe card, offer journeys at reduced rates.

Activity

8 Study extract **I** and read the mayor of London's transport policy at the link opposite. Imagine you work in the mayor of London's office. A key part of transport policy is to get people on reliable, affordable public transport. Design a poster encouraging the public to use the Tube and buses. It must be informative and eye-catching, have slogans and illustrations, and make people feel they really must 'give it a go'. The poster could show the negative impact of using cars on the environment. It should be produced to be displayed on the Tube and/or on buses.

AQA **Examiner's tip**

Know the characteristics of a sustainable city and the measures used to conserve the historic and natural environments of that city.

CO links

Investigate Liverpool Maritime Mercantile City at www.liverpoolworldheritage.com.

You can discover the mayor of London's transport policy at www.london.gov.uk/mayor, together with details of all London's transport systems, at www.tfl.gov.uk.

Lots of information on Curitiba can be found at www.pbs.org/frontlineworld. Type 'Curitiba's Urban Experiment' into the search box.

Sustainable urban living in Curitiba

Curitiba is the capital city of the Brazilian state of Parana. It is the seventh largest city with a population of 1.8million. The city is seen as a role model for planning and sustainability in cities worldwide. As long ago as the 1940s, a plan was commissioned to allow the controlled expansion. However, a later plan developed during the 1960s that sought to control urban sprawl, reduce traffic in the city centre, develop public transport and preserve the historic sector was adopted in 1968 – the Curitiba Master Plan. The emphasis has been on ensuring an appropriate quality of life for the residents of Curitiba with concern for the environment and the need to leave a suitable area for future generations living in the city.

The Bus Rapid Transport (BRT) System

Curitiba was the first Brazilian city to have dedicated bus lanes. The BRT system has four elements:

- direct line buses, which operate from key pick-up points
- speedy buses, which operate on the five main routes into the city and have linked stops
- inter-district buses, which join up districts without crossing the city centre
- feeder mini-buses which pick people up from residential areas.

Housing in Curitiba

In Curitiba, COHAB, the city's public housing programme, believes that residents should have 'homes – not just shelters'. They have introduced a housing policy that will provide 50,000 homes for the urban poor.

J *The innovative bus system in Curitiba*

Activities

9 Study diagram **J**, the information in this case study and research Curitiba at the weblink provided.

a Describe how Curitiba's bus system works and assess how successful it is.

b Explain why it was a good idea to preserve part of the old area of Curitiba.

c In Curitiba, 'planners come up with creative and inexpensive ways to go about solving universal problems for cities'. Provide evidence to support this statement.

10 Working in pairs, summarise the characteristics of a sustainable city.

10.1 What pressures are being placed on the rural–urban fringe?

The **rural–urban fringe**, sometimes abbreviated to the 'rurban fringe', is a transitional zone (area of change) or **rural–urban continuum** between countryside and city. The edge of a town or city is not bordered by a clear line. Density of land use changes gradually. The city can still have a huge impact on the surrounding countryside. Equally, people living in the countryside are prepared to travel to use urban services.

Demands for land in the rural–urban fringe

The rural–urban fringe is under intense pressure due to the growth of urban areas (**urban sprawl**) and the increasing mobility of the population because of car ownership. The main demands on the land are for retail developments, leisure (golf courses, horse-riding, etc.), residential (growth of villages) and transport (ring roads, airports, etc.).

Retail outlets

The areas of land on the outskirts of cities are prime land for the development of out-of-town **retail parks**. A retail park is a huge area of shops and leisure facilities such as cinemas and restaurants. The shops are often under cover and are served by large car parks. Most of the shops are well-known stores such as Next, Marks & Spencer, and WHSmith, but there may be parades of smaller shops and markets.

Commuting/commuter villages

Commuting is increasingly common. More people are prepared to spend time, money and energy travelling a greater distance to work to be able to live in the countryside. Rural areas not only depend on the urban zone for jobs, but also for services and administration. Rural-based businesses thrive because of the larger urban market close by. Garden centres and market gardens are examples.

Leisure

The rural–urban fringe serves the city as a leisure area. Box Hill, in the Surrey North Downs, lies in the commuter belt of London. Small towns like Leatherhead and Dorking have expanded and villages like Bookham house commuters. Local people use Box Hill for walks and cycling, and people come for days out from London, placing great pressure on a small green area (diagram **A**).

In this section you will learn

what is meant by the rural–urban fringe and know why people are attracted to live there

the impacts of out-of-town retail outlets and leisure provision

how transport networks have expanded in the rural–urban fringe

what is meant by commuting and its impacts on landscape and settlement

the characteristics of a suburbanised village, how they grow, their characteristics and the type of people who live there.

Key terms

Commuting: the daily movement of people travelling between home and work and back again.

Rural–urban fringe: an area around a town or city where urban and rural land uses mix and compete.

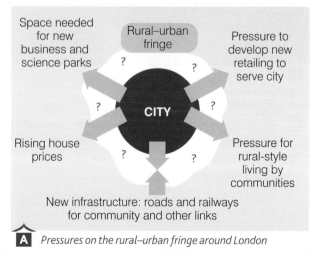

A *Pressures on the rural–urban fringe around London*

Bluewater Park, Kent

Bluewater Park, the largest and most recent of the five out-of-town **regional shopping centres** built in the UK, is located in Kent just beyond the M25 close to its major junction with the A2 (map **B**). It is a **high-access location** within London's rural–urban fringe. It opened in March 1999. Some 12 million people live within two hours' drive of Bluewater and 4 km from it is the recently opened Ebbsfleet station on the Channel Tunnel rail link, which may attract customers from France.

Bluewater was built in a disused chalk quarry. Many would consider this development an improvement on what was there before. Close by, just to the north of the River Thames, is Lakeside, a similar shopping development. Surrounding it is a huge area of retail parks, which include large stores like IKEA. The buildings themselves and the necessary infrastructure place immense pressure on the rural–urban fringe. This part of Kent is much more urbanised and congested than it used to be. Massive engineering works took place in 2008 to improve the link from the M25 to Bluewater.

Since 1999 three other large centres have been categorised as being on a regional scale: the Trafford Centre in Manchester, Braehead in Glasgow and the Westfield Centre in west London. However, these are within the existing urban area. Due to recent planning restrictions, it is unlikely that any more edge-of-town developments like Bluewater will be built in the rural–urban fringe.

Key terms

Urban sprawl: the spreading of urban areas into the surrounding rural/rural–urban fringe areas.

Retail parks: large warehouse-style shops often grouped together on the edge of a town or city, aiming to serve as many people as possible.

Regional shopping centre: a major indoor shopping centre with a large car parking area, located close to large urban area at a high access point, such as a motorway junction, so having millions of customers within two hours' driving time.

Rural–urban continuum: a graduation from rural to urban areas.

High-access location: an area with excellent transport infrastructure, making it easy to reach for people and goods.

C Bluewater Park regional shopping centre

AQA   Examiner's tip

Be clear about the pressure put on the rural–urban fringe. Ensure that you can explain some of the conflicts that arise as a result of these pressures.

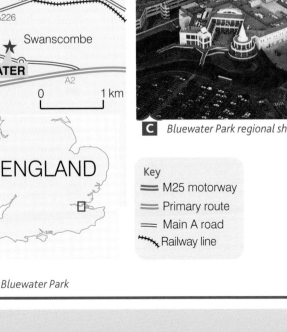

Key
≡ M25 motorway
≡ Primary route
≡ Main A road
⊬⊬ Railway line

B The location of Bluewater Park

Case study

kerboodle!

1   Use the Internet to choose one or more photos of the rural–urban fringe. Place a copy of the photo centrally on a sheet of plain A4 paper. Annotate the photo with detailed labels illustrating the rural–urban fringe characteristics of the picture. Draw clear arrows with a sharp pencil and a ruler to locate your points accurately.

2   Study map **B** and photo **C**.

a   Describe the location of Bluewater Park.

b   What are the advantages of its location?

c   Why do you think that it is unlikely that any more large centres like Bluewater will be built in the rural–urban fringe?

Suburbanised villages

Villages that have grown in recent decades and have housing estates with more recent residents than long-term ones are called **suburbanised villages**. They are most likely to be found in the rural–urban fringe as the reason they grow is because of access to both the city and the countryside. They are part of the process of **counter-urbanisation**, i.e. moving out of the city. These villages develop in stages (diagram **D**):

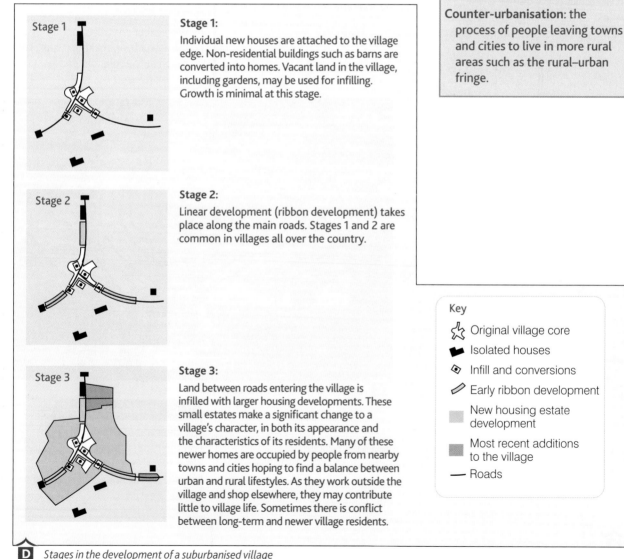

Stage 1:
Individual new houses are attached to the village edge. Non-residential buildings such as barns are converted into homes. Vacant land in the village, including gardens, may be used for infilling. Growth is minimal at this stage.

Stage 2:
Linear development (ribbon development) takes place along the main roads. Stages 1 and 2 are common in villages all over the country.

Stage 3:
Land between roads entering the village is infilled with larger housing developments. These small estates make a significant change to a village's character, in both its appearance and the characteristics of its residents. Many of these newer homes are occupied by people from nearby towns and cities hoping to find a balance between urban and rural lifestyles. As they work outside the village and shop elsewhere, they may contribute little to village life. Sometimes there is conflict between long-term and newer village residents.

Key

⭐ Original village core

◤ Isolated houses

◈ Infill and conversions

▱ Early ribbon development

▢ New housing estate development

▢ Most recent additions to the village

— Roads

D   Stages in the development of a suburbanised village

Chalgrove, Oxfordshire

Chalgrove lies on the edge of the rural–urban fringe between Oxford and London. Its residents can commute to work in either city. Its population has risen to 11,349, which means this village has grown into a small town. Although some Oxfordshire villages are growing, others are declining. Location and planning regulations are key factors. Reasons for the growth of Chalgrove include the following (map **E**):

- The village is close to the M40 motorway, allowing access to London.
- It has A-road access to Oxford (for much of the route).
- Mainline railway stations are accessible in nearby Thame and Wallingford.
- These towns provide a range of local services, including health and education.
- New job opportunities are growing in the Didcot–Thame development axis.
- Chalgrove is outside both the green belt and the Area of Outstanding Natural Beauty, so is not limited by their planning restrictions.
- Since 1993 the South Oxfordshire Development Plan has actively encouraged the further development of larger villages like Chalgrove.

Chalgrove's typical 'suburban' residents:

- are highly mobile (90 per cent of households have cars)
- work in the professions or in management (40 per cent)
- have lived in the village for less than 10 years.

Did you know ??????

The rural–urban fringe should serve the needs of both city and countryside. It should include parks and continuous green corridors with limited development. Public transport should be more efficient and attractive to allow access and daily travel. The rural–urban fringe should serve the city for recreation and sustainable waste management. More woodland could be planted.

∞ links

Find out more about Chalgrove at **www.chalgrove-village.org.uk**.

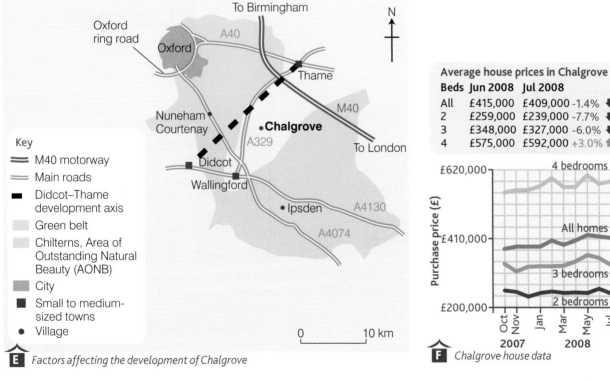

E *Factors affecting the development of Chalgrove*

Average house prices in Chalgrove			
Beds	Jun 2008	Jul 2008	
All	£415,000	£409,000	-1.4% ⬇
2	£259,000	£239,000	-7.7% ⬇
3	£348,000	£327,000	-6.0% ⬇
4	£575,000	£592,000	+3.0% ⬆

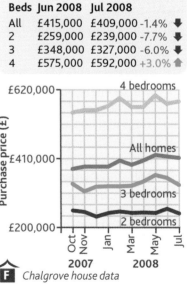

F *Chalgrove house data*

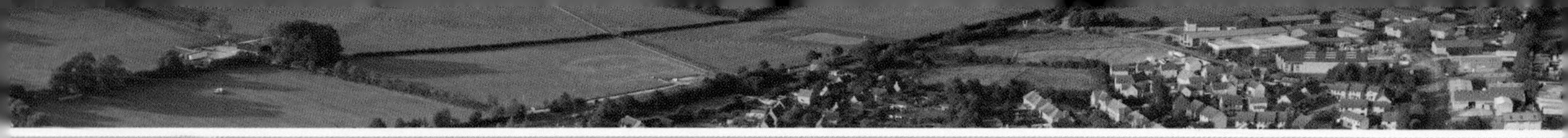

Transport developments

http://

Google

News

Home | Contact | Sitemap | News | Forum | Shop

- Home
- Contact
- Sitemap
- News
- Forum
- Shop

Gridlock fear for growing village

People living in a Bedfordshire village earmarked for more than 1,000 new homes are demanding a bypass. Residents in Shortstown are worried the development on the former RAF Cardington base will bring gridlock. Villagers running a 'No' to more cars campaign say the planned development could double the size of the community.

G *From the BBC News website*

Adapted from BBC News website ***news.bbc.co.uk****, 3 June 2004*

Rail and road networks are of good quality in the rural–urban fringe, so transport is relatively quick and efficient. As people move into the rural–urban fringe, infrastructure must increase to serve their needs. More people leads to more cars and therefore an expanded road network. The M25 motorway around London was primarily built to help movement around the city and within the rural–urban fringe. It is said, however, that traffic volume expands to fill any new roads developed.

Did you know ??????

Car pooling is becoming more common, sharing with someone going in the same direction. This cuts costs, is sociable and environmentally friendly. Councils and companies are increasingly running sharing schemes.

Activities

3 Write a short definition of the term 'suburbanised village'.

4 Study diagram **D** and map **E** on pages 222-223.

a With the aid of simple sketches, describe the characteristics of a village at each of the three stages of suburbanisation.

b What characteristics does Chalgrove have that make it a suburbanised village? Refer to the three stages of suburbanisation in your answer.

c Would you prefer to live in a suburbanised village like Chalgrove or a village that has not grown in this way? Give your reasons.

5 Study graph **F**.

a Using the data in the graph, describe the house price variations in Chalgrove between October 2007 and July 2008.

b During summer 2008 demand for houses fell across the UK due to an economic recession. Is there any evidence of this trend in Chalgrove? Use the table to help you.

6 Study extract **G**.

a Why are some villagers in Shortstown concerned about the new housing development? Make a list of three points.

b Do you agree with their concerns? Explain your answer.

10.2 What social and economic changes have happened in remote rural areas?

Locations of remote rural areas in the UK

There is a close correlation between uplands, hill farming zones and remote rural areas (map **A**). Hill sheep farming uses land that is too poor for most other agricultural activities. It is **marginal land** with severe physical limitations. Snowdonia's climate is harsh, with low temperatures, and high rain and snowfall. Glacial troughs have steep rock sides with bare rock or thin, stony soil. The landscape is both dramatic and physically attractive, but these remote areas can make people's lives extremely difficult.

Remote farming areas, usually used for hill farming

North-west Highlands
Grampian Mountains
Southern Uplands
Lake District
Pennines
SNOWDONIA
Cambrian Mountains
Brecon Beacons
Exmoor
Dartmoor

0   200km

N

A   Remote rural farming areas in the UK

> **In this section you will learn**
>
> the reasons for rural depopulation and the impacts it brings
>
> how village characteristics change during decline
>
> the growth of second homes and the impact this has on village decline.

> **Key terms**
>
> **Marginal land:** land that is only just good enough to be worth farming. It may be dry, wet, cool, stony or steep.
>
> **Rural depopulation:** people leaving a rural area to live elsewhere, usually in an urban district.

Rural depopulation in Snowdonia

Rural depopulation is the net emigration of people from an area, i.e. more people move out than move in. The 2001 Census showed a decline of 3 per cent in the number of residents in Snowdonia compared with 1991. During the 1980s there had been a 6.2 per cent growth, but the 1990s saw zero growth (static population since newcomers were moving in and locals were leaving).

As one of the highest and steepest regions in the UK, Snowdonia has never been densely populated, but today it continues to lose population. Those leaving are local people whose families have lived there for generations. People leave for many reasons.

Decline in hill farming

The number of hill sheep farmers is decreasing nationally, mainly due to economic pressures. In the 1990s lamb prices were considerably below the costs of production, so making a living became impossible.

Average farm income dropped as low as £8,000 per annum at that time, while fuel and other costs were rising. Farm vehicles need diesel and in such remote rural areas people often have no alternative but to travel by car. Many were simply driven out of business. It is impossible to bring up a family on this low amount. Some persevered, living on overdrafts, but that usually led to bankruptcy.

> **Did you know**
>
> The 1990s saw the worst agricultural recession for almost 100 years in the UK. The average family farm suffered a 90 per cent drop in income over five years.

Competition from abroad

Competition from abroad was also a problem. New Zealand can produce cheaper lamb and ship it round the world to undercut Welsh lamb in British supermarkets. To increase income, hill farmers have tried to diversify by working in other rural industries such as tourism and forestry. This has worked well in the Lake District because it is an important tourist and leisure destination. However, Snowdonia is an even more remote rural area and so has fewer tourists, providing less opportunity for **diversification**. Some farms run B&B businesses, but this is not a secure income. At the same time, other traditional rural industries such as mining and quarrying were suffering decline. Snowdonia's local unemployment rate is therefore high compared with Wales or the rest of the UK. Competition for jobs is intense as many people are chasing limited opportunities.

Snowdonia's farmers who have survived into the 21st century found that the new EU single farm payment system would be of less help than in the past. Area-based payments favour large farms. Hill farms benefited from the previous system where they were paid according to numbers of livestock. Wales and Scotland therefore opted to 'de-couple' (separate) from the new system and stick to the traditional payments.

Job losses

Agriculture is not the only sector experiencing job losses. Quarrying has reduced its labour force and the closure of Trawsfynydd Nuclear Power Station in 1993 caused a disastrous 500 job losses. For any chance of re-employment, people were forced to move to the towns.

Better job opportunities elsewhere

Emigration of young adults for higher education, training and employment has long been a trend in remote rural areas like Snowdonia. Once educated, the chances of a good job opportunity within the home region are small. The National Park employs some graduates of suitable subjects, as do health and education services, but most must look to urban areas for suitable careers.

Housing shortage

Snowdonia National Park Authority does not allow planning permission for building new homes within the park in order to preserve the traditional landscape. This limits housing availability and inflates prices. Local young people cannot afford to buy in their home village. Farmers leaving the land cannot find suitable family accommodation. They are forced into the towns to buy or rent. Developers are more prepared to build in towns, so places like Dolgellau and Bala can offer a wider choice of homes at better prices.

Loss of services

The trend towards loss of services in remote rural areas accelerates. With decreasing local population, services close. Those who remain find they cannot access what they need. Village shops, post offices, banks, primary schools and doctors' surgeries close. Communities are destroyed, so the people left can feel even more pressurised to move out. Unless new sources of employment and income can be put in place, as well as new housing being developed, many villages have little chance of being regenerated.

B *Snowdonia*

Growth in ownership of second homes

The average figure across Wales for houses that are **second homes** is 1.2 per cent. Within Snowdonia National Park it is a huge 13.7 per cent (photo **B**). A second home is a local property bought by someone whose main home is outside the region. Used for holidays and weekends, the houses' occupants are rarely part of the community, so they are not well accepted. They tend not to use local services, often bringing all they need with them.

News   Home | Contact | Sitemap | News | Forum | Shop

Home
Contact
Sitemap
News
Forum
Shop

Controls on second homes reviewed

Radical plans which would require second home buyers to apply for planning permission could be introduced in Wales to protect rural communities. If the proposals are approved the authority [Snowdonia] would become the first national park in the UK to stop outsiders from buying a holiday home in the region due to the sharp rise in property prices. 'It raises very interesting questions about civil rights and people's rights to live anywhere in the UK.'

C *From the BBC News website*

*Adapted from BBC News website **news.bbc.co.uk**, 5 September 2001*

Rhyd, Snowdonia

Rhyd is a small Snowdonian village linked to other settlements by a minor road. Local jobs used to be in farming, lead mining and clay works, but all have been in decline. In 2003 the village contained only 15 homes with 37 residents. Five were holiday homes. Rhyd has no services and no jobs within the village, whereas in the past it had a primary school, chapel, post office and shop. The only service now is two buses passing daily.

D *Distances to essential services for Rhyd's population*

Service	Distance (km)
Primary school	3
Pub	3
Doctor's surgery/ pharmacy	7
General store	11
Clothes shop	11
Bank	11

Case study

E *The village of Rhyd*

AQA **Examiner's tip**

Learn your UK case study of a remote rural area so that you can describe some of the social and economic changes it has undergone.

Activities

1. Define these terms:
   a. marginal land
   b. rural depopulation
   c. diversification
   d. second homes.

2. Write a list of the characteristics of declining villages in remote rural areas.

3. Study extract **C**. Does Snowdonia National Park Authority have the right to prevent outsiders buying cottages as second homes? Discuss your ideas in a group and either:

   a. present your argument to the class, or
   b. hold a class debate.

   Remember, your ideas and opinions must be justified.

4. a. Write a paragraph to describe the difficulties faced by the population of Rhyd in Snowdonia.
   b. Living In Rhyd has its advantages too. Consider whether you would like to live there or not. Explain the reasons for your answer.

How can rural living be made sustainable?

Conserving resources and protecting the environment

To live sustainably involves conserving resources. If we waste what we have, and do not protect our environment (an important resource in itself), not only will our economy and standard of living suffer, but so will the generations that follow us.

There are several government schemes and initiatives aimed at protecting the rural environment.

In this section you will learn

how rural environments can be protected and conserved

ways in which the needs of the rural population can be supported

government initiatives aimed at supporting the rural economy and environment.

1 Environmental Stewardship Scheme

This agri-environment scheme is open to all farmers to improve their sustainability. Run by the Department for Environment, Food and Rural Affairs (Defra), it aims to:

- conserve wildlife
- increase biodiversity (photo **A**)
- improve landscape quality
- provide flood management
- promote public access to the countryside.

Annual payments from UK government to farmers vary from around £30 to £60 per hectare or more, depending on the level of involvement of the farmer. There are 50 land management options.

2 Rural Development Programmes

The England Rural Development Programme (ERDP) was designed to support sustainable farming methods and greater environmental responsibility between 2000 and 2006. £1.6 billion of EU and UK government money was put into a wide variety of schemes.

A Meadow land developed under the Environmental Stewardship Scheme

The new Rural Development Programme for England (RDPE) replaced this in 2007 and will run until 2013 to distribute £3.9 billion to make rural life more sustainable. Most funding (£3.3 billion) will help farmers to manage their land more sustainably, and encourage biodiversity and good water quality while combating climate change. Money became available from January 2008.

The remaining £600 million will make agriculture and forestry more competitive and sustainable, as well as increasing rural job opportunities. Three types of project grants exist:

- the Forestry Commission can support sustainable woodland schemes
- Natural England operates the Energy Crops Scheme
- the Regional Development Agencies deliver new economic and social funding in rural areas.

Supporting the needs of the rural population

People need to be able to get to work and to access services as easily as possible. Some schemes have been put in place to help improve the quality of rural life.

1 Rural Transport Partnership

Rural transport is a difficult issue. Distances are longer and customers fewer. Making public transport networks pay can be almost impossible. The Devon Rural Transport Partnership began in 1999 to address these problems in the county. Financial grants support and advise local transport groups on the practicalities of setting up services.

In rural east Surrey there is support for students aged over 16 who need to travel to college courses. The moped loan scheme allows them travel where public transport routes do not serve. Safety helmets and clothing are provided. Student discounts on bus, coach and train fares are also supported by this funding.

Buses4U is a minibus scheme open to all members of rural communities in east Surrey. Journeys need to be booked at least an hour in advance and the minibus routes are then planned (photo **B**).

B *The Buses4U scheme*

2 Village Shop Development Scheme

This grant scheme aims to help village and farm shops and other local services to provide for their communities more efficiently and discourage people from getting in their cars and driving to the nearest town. This is important because:

- people save time and energy
- carbon emissions are lowered
- those without cars retain local accessible services
- a sense of community is enhanced.

Local shops, post offices and pubs provide local employment. More money is spent within the local community rather than going outside. Loyalty schemes encourage local people to support local services. Communities become more sustainable economically, environmentally and socially. Shops need funding to widen the goods and services they offer. Banking is an essential service usually found only in towns, but village post offices can increase the breadth of their services, which is more convenient for customers.

Some village businesses are taken over by community groups and may even use church and village halls for premises. The village of Redmire in Wensleydale, North Yorkshire, lost both its shop and post office.

Minimal support allowed both to re-open, based in a garage and run by community volunteers. High Etherley in the Wear Valley, County Durham has taken part in the Sainsbury's Assisting Village Enterprise (SAVE) scheme. The aim of the scheme is to increase the range of products available in the village shop and the number of customers.

Case study

Harting village store, West Sussex

With their shopkeepers anxious to retire and no new buyer in sight, the villagers of Harting set up an action committee. They formed a Village Shop Association, which gave them access to government funding and raised £60,000. A new shop owner was then able to pay the rest of the value of the premises, so Harting retained its services and social lifeline (photo **C**). The store is increasingly being used by people from the surrounding area. Environmentally it is also a success. People drive less often to towns, saving 'shopping miles'.

C Harting village store

Supporting the rural economy

There are two important programmes aimed at supporting the local economy in rural areas.

1 Rural Challenge

Rural Challenge involved 11 rural regeneration schemes in the 1990s to generate jobs, improve **living standards** and tackle social disadvantage, which can happen as easily in rural areas as in urban ones. £75 million created 3,000 jobs and 18,000 person weeks of training for young people to give them marketable skills. Around 200 new community facilities were set up. One example is in Bishop's Castle, Shropshire, which has 37 small business units with an IT resource centre and childcare facilities. Grants for arts, tourism, transport and social projects are available.

Key terms

Living standards: people's quality of life, mostly measured economically but also socially, culturally and environmentally.

Common Agricultural Policy (CAP): a policy to support and control farming in the EU.

2 Objective 1

EU funding is focused on the most deprived regions, where average incomes are 75 per cent or less of the EU average. This is the largest area of EU spending after the **Common Agricultural Policy (CAP)**. Money comes from four funds:

- the European Regional Development Fund
- the European Social Fund
- the Agricultural Guidance and Guarantee Fund
- the Financial Instrument for Fisheries Guidance.

AQA *Examiner's tip*

Be able to summarise at least two government initiatives aimed at supporting the rural economy and environment.

These funds must be matched by the national government concerned.

The aim is to revive failing traditional industries and community economic and social problems, which go hand in hand. Employment opportunities should increase and the workforce become more skilled and flexible. West Wales benefited in 2003. Some 6,000 jobs were created and another 7,000 protected. Cornwall and the Scilly Isles have benefited similarly.

D *Austwick, Yorkshire*

The Objective 1 programme in South Yorkshire from 2000 to 2006 was a £2.4 billion investment programme funded between the EU and UK public and private sector resources. South Yorkshire is classed as one of Europe's poorer areas. For its population size, it had fewer businesses, fewer people in work and lower wages compared with other regions. People could apply for support for projects to benefit their community, so increasing sustainability. For example, rural transport has been improved such as more school buses in rural communities. Plans have been made to restructure the regional economy, urban as well as rural.

Activities

1. a   Name two ways in which the UK government promotes protection of the environment.

    b   Choose **one** of the schemes named in a and explain briefly how it works.

2.   If you live in a rural area, this question applies directly to you. If you are an urban person, put yourself in the position of someone living in the countryside.

    a   What are the most difficult things about living in a rural area? Consider shopping, getting to school or college, transport in general and finding a job.

    b   Choose any of these problems. How might the projects discussed on pages 229–31 help you with this difficulty?

    c   Are there any ways in which you would improve the existing schemes?

Factors affecting farming

All agriculture is affected by both physical and human/economic circumstances. Farmers make decisions on land use based on quality of land, climate, supermarket pressure and people's choice of what to buy.

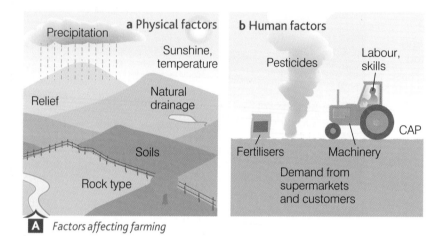

a Physical factors

Precipitation
Sunshine, temperature
Relief
Natural drainage
Soils
Rock type

b Human factors

Pesticides
Labour, skills
Fertilisers
Machinery
CAP
Demand from supermarkets and customers

A *Factors affecting farming*

Key terms

Intensive farming: high inputs of capital and/or labour to achieve maximum productivity.

Livestock farming: rearing animals.

Mixed farming: farming both crops and animals.

Arable farming: growing crops.

Inputs: anything entering the farm system, e.g. climate, soil, seed, labour.

Processes: jobs done on the farm to produce outputs.

Outputs: products leaving the farm system, usually for sale.

Activities

1 Define the following terms:
   a intensive farming
   b livestock farming
   c mixed farming
   d arable farming.

2 Draw a flow diagram to show the **inputs**, **processes** and **outputs** of a commercial mixed farm.

3 Describe the outputs typical of an East Anglian farm. Support your answer with facts and figures.

AQA Examiner's tip

You can achieve a much better answer to a question if you use facts and figures to support what you say. Using numbers like this gives a much clearer picture and allows you to compare more accurately.

links

Research in your school library and on the internet into different types of farming areas in the UK. Try the NFU's FACE website at **www.face-online.org.uk**. Look particularly at Lynford House Farm, Cambridgeshire.

Commercial farming in East Anglia

As a farming region, East Anglia covers Norfolk, Suffolk, Essex and Cambridgeshire. The landscape varies from undulating grazing land to fertile fenland, which is farmed **intensively**. In fact, East Anglia is one of the most productive agricultural landscapes in the world. Three-quarters of its land area is used for farming.

Crops

The region is well known for its cereal crops, growing more than a quarter of England's wheat and barley. Almost one-third of East Anglia's 1.4 million hectares is used for wheat alone. Well over half the UK's sugar beet crop and a third of all potatoes come from Norfolk, Suffolk and Cambridgeshire, much being processed locally. Market gardening and horticulture are also important: peas, beans, salad crops, strawberries and other fruits, as well as flowers, are cultivated.

Livestock

In terms of **livestock**, the UK's pig and poultry industry centres on East Anglia. Animals are fed directly on local grain. Some 2.2 million eggs and 25 per cent of chickens consumed weekly in the UK are produced in the region, mainly indoors under intensive systems. More than 1 million pigs live on 1,900 farms. Beef and dairy cattle herds and sheep flocks are small compared with other regions that specialise in these, such as the south-west peninsula. Nevertheless, grazing land is still an important part of the East Anglian landscape. Raising crops, feed and livestock is known as **mixed farming**.

Physical factors

East Anglia's climate is ideal for intensive **arable** production. Low rainfall (around 650 mm per year) comes mainly in spring and summer, when it proves most useful to growing crops. Warm summers (July = 18°C), with plenty of sunshine hours, ensure rapid ripening.

Human factors

Large farm machinery is essential to production on this scale. Most farms are hundreds of hectares in size – some of the largest in Europe. Farms often share the most expensive equipment such as combine harvesters, so during harvest time in August farmers work all day and at night using floodlights.

Many farms are owned by large companies that employ a manager to run one or more farms. Financial inputs are high. Much is spent on chemical fertilisers, pesticides and weed killers.

B   *The location of East Anglia*

C   *Sugar beet factory in East Anglia*

What are agribusinesses in the UK?

Agribusiness is large-scale farming of crops or livestock, aiming for high yields. Products usually supply major supermarkets. Farms are highly mechanised and **capital intensive**. Little labour is needed. Farms may be owned by companies and several farms can be operated by one company.

■ Impacts of agribusiness

Agribusiness is sometimes criticised for damaging and polluting the British landscape because of the following reasons.

1 Use of chemicals

Environmental groups oppose this system of farming. Chemical usage is high and much criticised.

- Pesticides are used to control pests, diseases and weeds. Without them, yields would be 45 per cent lower. This is why yields on **organic farms** are always lower. However, chemicals kill important wildlife such as bees, which are essential for crop pollination.

- Artificial fertilisers are mineral compounds that increase plant growth. Most soils do not contain enough nutrients for the high yields we want. Problems include cost (most are oil-based) and leaching of chemicals into rivers and the water table, leading to **eutrophication**.

- Phosphates released from farm slurry (animal manure) pollute water supplies and can be up to 100 times as damaging as household effluent.

2 Damage to ecosystems

Intensification of agriculture has caused the loss of important wildlife **habitats**, such as:

- wetlands
- hedgerows.

Wetlands act as sponges, soaking up excess rainwater and preventing floods. When drained, they make high-quality farmland, but flooding has increased as a result.

Although hedgerows are manmade (they were mostly planted in the 18th century), they are an important wildlife habitat (photo **A**). Large machinery is more efficient in big fields, so less turning is necessary. Hedgerows also take up valuable cropland. Between 1945 and 1990, over 25 per cent of UK hedges were destroyed, but in the most commercial arable areas like East Anglia, the figure was over 60 per cent. However, attitudes have now changed and hedges are being replaced, even on very large farms. They act as windbreaks and their roots bind soil together, which reduces soil erosion. Trees within hedges act as a windbreak and provide added wildlife habitat.

In this section you will learn

what is meant by the term 'agribusiness' and what its impacts are

the demands of the supermarket chains and food processing firms on farmers.

Key terms

Agribusiness: running an agricultural operation like an industry. Inputs and outputs are both high.

Capital intensive: farming to achieve maximum production via inputs of money to allow purchase of fuel, fertilisers and buildings that will allow maximum output.

Organic farm: a farm that does not use chemicals in the production of crops or livestock.

Eutrophication: pollution of fresh water from agricultural waste or excess fertiliser run-off.

Habitat: the home to a community of plants and animals.

Genetically modified (GM) crops: involves putting genes from other species (sometimes animals) into a crop to give it certain characteristics that increase yield.

A *A hedgerow in winter*

3 Use of genetically modified (GM) crops

Agribusiness farms are more likely to use **genetically modified (GM) crops**, which aim to prevent food shortages and reduce costs for the customer. Their use in the UK is still experimental, but is much opposed by environmental groups. If the added genes transfer into the natural environment, ecosystems could be irreversibly altered. GM development is controlled by multinational companies with a great deal of power.

> **AQA** *Examiner's tip*
>
> Be able to demonstrate a thorough understanding of agribusinesses. Ensure that you can define the term and use specific terminology to describe the impacts of agribusinesses on the environment.

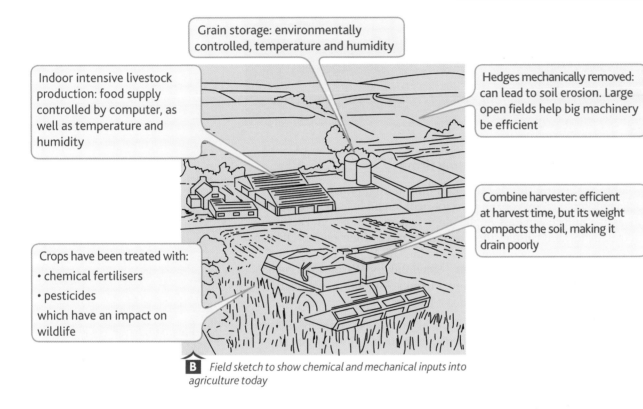

Grain storage: environmentally controlled, temperature and humidity

Indoor intensive livestock production: food supply controlled by computer, as well as temperature and humidity

Hedges mechanically removed: can lead to soil erosion. Large open fields help big machinery be efficient

Combine harvester: efficient at harvest time, but its weight compacts the soil, making it drain poorly

Crops have been treated with:
• chemical fertilisers
• pesticides
which have an impact on wildlife

B *Field sketch to show chemical and mechanical inputs into agriculture today*

■ The influence of supermarkets on UK farming

The major supermarkets have considerable power and control over the whole food supply system. Almost all our food is purchased in supermarkets. One-tenth of all spending (and not just on food) is in Tesco alone, now one of the world's largest retail outlets. If farmers want to sell their produce, they often have to take what the supermarkets offer them.

UK dairy farmers receive the lowest price for their milk in the EU. In 1996 they received 25p per litre; by 2006 it was only 18p. Milk is a daily basic food and people do not want to pay more than they have to, giving supermarkets an incentive to reduce the price. Supermarkets put pressure on dairy food processing companies for lower prices, so they in turn pay farmers less for milk too.

People who feel strongly about these issues have demonstrated outside and inside supermarkets (photo **C**). Bad publicity can have a great influence. Some companies now offer farmers a better deal and they use this in their advertising.

> **AQA** *Examiner's tip*
>
> Labelling a field sketch is similar to labelling a photo. Labels should be placed around the sketch rather than on it, for clarity. Link the label to the relevant part of the sketch with a clear, ruler-drawn arrow. Make labels detailed.

C What can the customer do?

The influence of food processing firms

Food processing transforms raw foods from farm production into packaged, often ready-cooked food for people and livestock. The food processing industry in East Anglia alone is worth £3 billion. Factories are usually located close to their raw materials or their market. British Sugar processes all sugar beet grown in the UK. Its five factories are all within the growing region (map **D**). Shaw Bakery, Warburton's bread factory in Oldham, Lancashire, manufactures 750,000 loaves every week. All are sold within 48 km of the factory.

Food processing companies are put under pressure by supermarkets to produce food at a certain price. To maintain their profit margin, they try to buy their raw materials as cheaply as possible from farmers. Everyone wants to make a profit. Farmers are at the end of the chain and so find themselves having to accept low prices rather than not sell their produce at all.

Growing region
● Factories
● Head office

0 ———— 200 km

Newark
Wissington
N
Peterborough
Cantley
Bury St
Edmunds

D British Sugar factory locations

Activities

1 Draw a mind map (concept map) to show the main features of agribusiness.

a Write 'agribusiness' in the centre of your page and define it. Use a photo from the Internet too.

b Around the edges, add information using text and photos to describe the main features and impacts of agribusiness.

2 Write at least three paragraphs to discuss the impacts of supermarkets and food processing companies on farmers and customers.

3 a Write lists of advantages and disadvantages of using chemical fertilisers and pesticides on farms.

b Consider the environmental issues involved.

c Make a decision: When is it right to use these chemicals, and when not so?

10.6   How can we reduce the environmental effects of high impact farming?

The importance of organic farming

Organic farming is a form of agriculture relying on **crop rotation**, green manure, compost and biological pest control. Animal welfare issues are a priority. Organic farming rules out manmade chemical fertilisers and pesticides, livestock feed additives and genetically modified (GM) crops. It is generally considered to be more sustainable because it looks after the soil and natural habitats, as well as producing healthier food for people. However, yields per hectare are significantly lower than in non-organic farming, making produce more expensive to buy.

Since 1990 the market for organic produce has been growing by 20 to 25 per cent each year, despite prices being higher. In 2005 organic produce was worth £33 billion in the UK. The amount of organic land is also growing: 2 per cent globally and 3.6 per cent in the UK.

A certain level of pest damage is accepted as organic methods are less effective. Beneficial organisms are introduced, such as ladybirds to eat aphids. Bees and other insects thrive without the chemicals that kill them, in turn improving pollination. Weeds are controlled mechanically, which is more time-consuming, and by mulching (adding a layer of organic material on top of the soil to prevent weed germination).

Some people argue that organic foods are not what they seem – they are an indulgence the world cannot afford. Organic pesticides such as copper stay in the soil for ever and are toxic. This may be worse for the environment than some modern, biodegradable pesticides.

Dairy production is a major source of greenhouse gas emissions, mainly methane, which is 20 times as powerful as carbon dioxide. Organically reared cattle produce twice the amount of methane of conventional cattle because of their diet. Organic crops produce less per hectare but cost the same amount of energy to harvest.

The Department for Environment, Food and Rural Affairs (Defra) estimates that organic tomato production in the UK takes 25 per cent more water than non-organic production. On the other hand, organic wheat uses less water and less energy per kilogram harvested than non-organic wheat.

In this section you will learn

government policies to reduce environmental effects of high impact farming

how to compare and contrast intensive and organic farming systems.

Key terms

Crop rotation: changing the use of a field regularly to help maintain soil fertility.

A *Organic produce in a supermarket*

How do government policies affect farming today?

UK farmers have to compete in global markets. Local products like lamb and wheat can be imported cheaply from abroad. The UK exports some of its agricultural produce, much to other EU countries. However, this is not always a reliable market. Foot-and-mouth disease in the UK in 2001 meant that other countries stopped buying our produce for some time.

Farmers are the main people working to conserve the countryside. The UK government aims to support them in several ways, including paying grants under the Environmental Stewardship Scheme. By December 2006, 3.7 million hectares of British farmland were included in this scheme. Farmers restore wildlife habitats by:

- replacing hedgerows and keeping them well trimmed
- leaving a strip of land along both sides of a hedge to increase wildlife habitat
- planting trees and managing woodland
- looking after ditches and ponds (photo **B**)
- maintaining hay meadows and grassland.

The National Farmers' Union (NFU) has calculated that farmers carry out £400 million of unpaid work in maintaining the British landscape.

The change from the subsidy and quota system to the single farm payment affects farmers differently. The end of quotas means farmers are no longer limited in the amount of food they produce. In a world of rising demand and rising prices, this is a positive change. However, because single farm payments are based on area under cultivation, larger farms inevitably benefit most.

B Ponds are an essential wildlife habitat on farmland

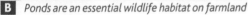
Activities

1 Study photo **A** on page 237.

a Can you identify some organic produce being sold?

b What is meant by 'organic farming'?

c How do organic farmers cope with pests?

d Why is organic food often more expensive than intensively grown food?

e Would you buy organic food? Explain your answer.

2 a How does the Environmental Stewardship Scheme help farms to restore wildlife habitats?

b Suggest reasons why wildlife habitats should be preserved.

3 Research some of the problems in the UK farming industry such as BSE and foot-and-mouth disease. These are often associated with farming too intensively. Since demand for food is increasing, farming is likely to become even more intensive. Discuss some of the problems this may bring and consider some possible solutions.

10.7   How do countries change from subsistence to cash crop production?

Most poor countries are in debt. One solution is to turn land currently used for subsistence crops to cash crop production. Cash crops can be exported for foreign currency to help repay debt and some of the profit might pay for development benefits for the country.

In this section you will learn

how cash crop cultivation impacts on subsistence food production.

Cash crop production in Kenya

Poorer countries like Kenya are committing more land to cash crop production for export to generate hard currency income (map **A**). The land involved was previously used for growing subsistence crops for local people or belonged to small-scale producers of tea and coffee. Many benefits and costs result for the communities involved and the country as a whole.

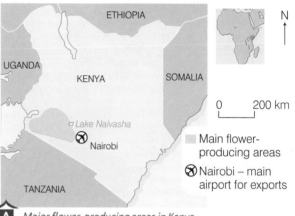

A   *Major flower-producing areas in Kenya*

ETHIOPIA
UGANDA
KENYA
SOMALIA
Lake Naivasha
Nairobi
TANZANIA

N

0   200 km

Main flower-producing areas

Nairobi – main airport for exports

B   *Kenyan roses on sale in a UK supermarket*

BLOSSOM & BLOOM
FAIRTRADE LONG
STEM KENYAN
ROSES (P)   163/L
Price
£4.99
Guaranteed for 7 Days
FAIRTRADE

Kenya's flower industry is the oldest and largest in Africa. It became important as Israel reduced its production. Kenya's main flower crops are roses and carnations; no other country exports as many roses as Kenya. Just look at your local supermarket – they will almost certainly have Kenyan roses (photo **B**). By 2003 their export value was £77 million, 8 per cent of export value and third after tea and tourism. Business has grown rapidly. The busiest time of year is the first half of February because of Valentine's Day in Europe. Some 20 per cent of annual business takes place then and hours worked are at a maximum. Millions of roses are flown from Nairobi to Europe every year. The USA is also an increasing market.

C   *Worker in Kenya's flower cash crop industry*

Kenya has some of the largest flower farms in the world, even compared with the Netherlands – a country famous for its flower production. Even though Kenya has a good climate for flower growing, most roses are grown in greenhouses to protect them from rain, which rots them, and wind and hail, which damage them. Flowers are grown close to the Rift Valley lakes because they are a reliable source of water and cultivated flowers need vast quantities of water.

The business employs tens of thousands of workers (photo **C**). Two-thirds of them are casual labourers with no job security or benefits – pregnancy usually means loss of employment. Wages of £1 per day – Kenya's minimum wage – are just above the globally accepted poverty level of $1 per day.

AQA   *Examiner's tip*

There are always at least two points of view to any issue. You need to put your own opinion to one side initially and write about all sides of the situation. In your conclusion, state your own opinion but it must be backed up with reasons.

∞ links

See p.292 for more on flower production in Kenya. This will help inform your activity answers.

Activities

1   Study map **A** on page 239.

  a   Describe the location of the flower-growing area in Kenya.

  b   Explain the advantages of this location. Think about workforce, market and transport.

2   **a**   List the advantages and disadvantages of flower farming in Kenya for local labourers.

  b   How might their lives have changed working on flower farms rather than in subsistence production?

  c   What other types of job are created by new flower businesses? Who profits?

3   **a**   How is the land affected by flower production? Be careful to include both positive and negative impacts in your answer.

  b   Write a conclusion in favour of or against flower farm expansion in Kenya. Use your answers to earlier questions to justify your point of view.

10.8   How do forestry and mining impact on the traditional farming economy?

Traditional farming in the Amazon Basin

Shifting cultivation (also called 'slash and burn' agriculture) is the traditional form of rainforest subsistence farming. Native tribes have practised the system for thousands of years. Rainforest soils are poor quality; most nutrients in this rich ecosystem are stored in the vegetation. Soil takes at least 20 years to regenerate after use for only 2 or 3 years. This is extensive farming because so much land is needed per person. One family has three or four plots under cultivation each year and ten times as many **fallow** ones.

Tribal people often have no legal right to the land they cultivate; governments can sell it off to transnational corporations (TNCs). Once the forest is opened up, whether for logging or mining, the ecosystem breaks down. Vegetation is cleared and the soil quickly loses its structure. Heavy rain then washes it away. Once clearance has happened, traditional farming cannot restart: the most basic resource – the soil – has been ruined. Regeneration is almost impossible, so shifting cultivators cannot return to this land.

Rainforest destruction

There are four main causes of rainforest destruction.

Key terms

Fallow: land that has been left unseeded to recover its fertility.

Deforestation: the removal of trees and undergrowth.

A   *Satellite images of deforestation in the Brazilian Amazon. The red shows vegetation*

1 The impact of cash crops and ranching

Economic pressures force the governments of poorer countries to grow and export cash crops. Land is needed and rainforest is cleared. This is called **deforestation**. In the Amazon, huge areas have been cleared for soya production, mainly to feed beef cattle in the USA or Brazil itself to produce beef for export. In Indonesia, cassava – another fodder crop – is grown for export to the EU. Large areas of forest have been cleared for cattle ranching, especially in the Amazon Basin. On average one animal requires two hectares of land.

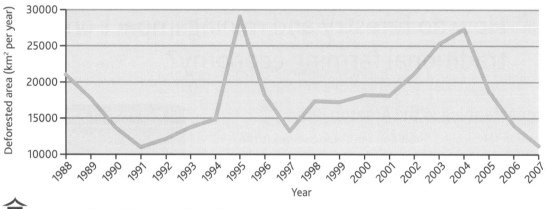

B *Brazilian rainforest deforestation, 1988–2007*

2 The impact of forestry

Logging has devastated local tribes in Indonesia such as the Moi people. Their way of life threatened, they tried to stand up to the major Intimpura Timber Company. The government sent in the army to support the company over its own people. Land, air and water pollution were the results of this over-exploitation. Rivers were silted up from soil erosion. Water and soil were polluted by machine oil.

3 The impact of mining

Carajas Iron Ore Mine is one of the largest operating open-pit mines in the world. Rainforest has had to be cleared on a large scale to make way for the mining operators and for transport. Iron-bearing rock has had to be dug out of the ground (photo **D**). Pollution is high for both water and land. Most companies are foreign-owned, although Brazilian investment is increasing.

4 The impact of growing infrastructure

Road construction in the Amazon has led to great losses too. Roads provide access to logging and mining sites, as well as allowing poor farmers to penetrate deeper into the forest. The Trans-Amazonian Highway was a huge project 3,200 km long bisecting the rainforest region. Some 100 tonnes of soil per hectare were eroded due to clearance for this highway. (Remember, a hectare is a square with sides of 100 m.)

Malaysia's controls on deforestation have been relatively strict. Nevertheless close to Kuala Lumpur, the capital city, rapid industrialisation is still causing increased deforestation.

C *A farmer burns the rainforest*

D *Large-scale mining in the Amazon Basin*

Rainforest clearance in Indonesia, 1997–1998

From September 1997 to June 1998, much of South-east Asia suffered major pollution caused by thousands of rainforest fires. An area as large as Western Europe had visibility reduced to as little as 50 m. Some fires were started by new farmers clearing land and they were largely blamed. However, 80 per cent were due to large forestry companies taking advantage of the dry El Niño conditions to clear land for large-scale mining and agriculture projects.

Indonesia has lost 20,000 plant species and 17 per cent of the world's birds. Thousands more species face extinction due to deforestation. From 1997 to 1998, rainforest fires polluted the whole region.

Case study

Did you know ??????

Between 1970 and 2006, over 600,000 km² of Amazon rainforest were destroyed, mostly by logging and some by mining concerns. Speed of clearance has increased. From May 2000 to August 2006, nearly 150,000 km² were lost, a quarter of the total losses in one-sixth of the time period. A peak clearance year was 2004. Since then there has been some evidence of a decline in annual deforestation.

links

Research deforestation in the Amazon Basin at www.mongabay.com/brazil.html.

E *Advantages and disadvantages of large-scale exploitation*

Advantages	Disadvantages
Timber and minerals produce foreign exchange for the country, which can pay for development projects	Local people are thrown off land they use for traditional farming activities and have nowhere else to go
TNCs bring expertise to the country – new skills are learned by local people	Businesses are often owned by TNCs and the profits go abroad
Jobs are provided and income for local people	Senior workers are brought in from outside and local workers are often low paid
Infrastructure is improved to provide access to exploited sites	New roads and railways are cut through the rainforest, leading to soil erosion

Activities

1 a Use the data below to draw a pie chart to show the causes of deforestation in the Amazon Basin. The diameter of your pie chart should be at least 6 cm for clarity.

Causes of deforestation in the Amazon Basin, 2000–2005	
Cattle ranching	60 per cent
Small-scale subsistence agriculture	30 per cent
Fires, mining, urbanisation, road construction and dams	4 per cent
Logging, legal and illegal practices	4 per cent
Large-scale commercial arable agriculture	2 per cent

b Why do you think cattle ranching is by far the main cause of deforestation?

c Why has rainforest been cleared for roads?

d Do you think the government should spend money stopping illegal practices? Explain your answer.

2 Study graph **B**. Do you agree that rainforest clearance has increased over the years? Use data from the graph to support your answer.

3 Study photo **C**.
a Why is the farmer burning the rainforest?
b Suggest some environmental consequences of what he is doing.

4 Study photo **D**.
a What is happening in the photo?
b What might be the environmental consequences of this mining operation?

10.9 What are soil erosion issues?

What is soil erosion?

Soil erosion occurs when the soil is either blown away by wind or washed away by rain (photo **A**). It is common in areas with steep slopes. Deforestation removes the roots that bind the soil together and the shelter that limits the impact of the wind. Poor farming techniques make the problems worse because they weaken soil structure and make it more likely to erode. Population increase puts pressure on farmers to grow more.

In this section you will learn

why soil erosion happens

methods for solving soil erosion problems.

What are the impacts of soil erosion?

Traditional agriculture has continued for centuries, so it ought to be in balance with its environment. This is not always true. Iraq was part of the Fertile Crescent, where agriculture began 10,000 years ago. It was the wealthiest and most technologically advanced region in the world. Today, it has the worst soil problems in the world: erosion, salinisation and deforestation. The USA, Australia and China also have serious widespread soil erosion problems. In the world as a whole, an area 10 times the size of the UK has become so severely degraded that it can no longer produce food – a serious problem as world population continues to grow.

A Severe gullying in East Africa

Where does soil erosion occur?

In Madagascar, the best farmland is used for export coffee production, so people have had to clear more and more forest for their subsistence crop production. Half the island's forest vanished between 1950 and 1985. Losses of topsoil are huge. Air and satellite photos show red staining in the Indian Ocean due to topsoil being washed out to sea (photo **B**). In Democratic Republic of Congo and Nigeria, increased population and food demand meant shifting cultivators could not leave soil plots long enough to regenerate between crops. They cleared too many areas too close together, which allowed heavy rain to wash the soil away. This whole farming system has broken down as a result.

B Satellite photo showing eroded soil washed into the Indian Ocean

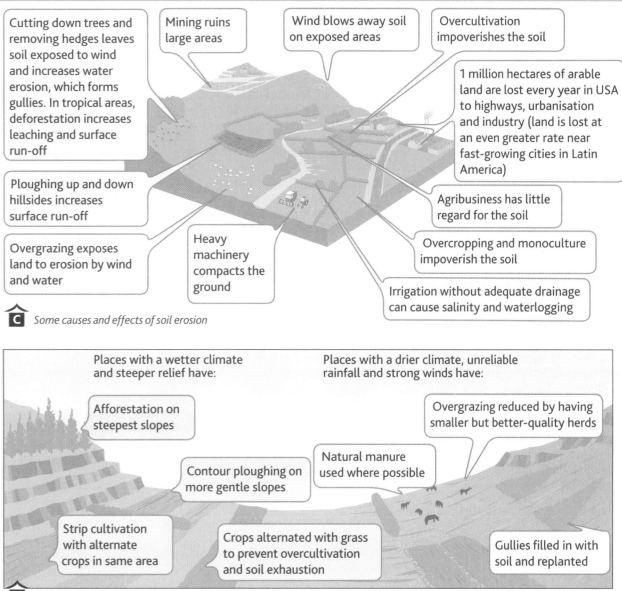

Cutting down trees and removing hedges leaves soil exposed to wind and increases water erosion, which forms gullies. In tropical areas, deforestation increases leaching and surface run-off

Mining ruins large areas

Wind blows away soil on exposed areas

Overcultivation impoverishes the soil

1 million hectares of arable land are lost every year in USA to highways, urbanisation and industry (land is lost at an even greater rate near fast-growing cities in Latin America)

Ploughing up and down hillsides increases surface run-off

Overgrazing exposes land to erosion by wind and water

Heavy machinery compacts the ground

Agribusiness has little regard for the soil

Overcropping and monoculture impoverish the soil

Irrigation without adequate drainage can cause salinity and waterlogging

C   *Some causes and effects of soil erosion*

Places with a wetter climate and steeper relief have:

Places with a drier climate, unreliable rainfall and strong winds have:

Afforestation on steepest slopes

Overgrazing reduced by having smaller but better-quality herds

Contour ploughing on more gentle slopes

Natural manure used where possible

Strip cultivation with alternate crops in same area

Crops alternated with grass to prevent overcultivation and soil exhaustion

Gullies filled in with soil and replanted

D   *Some attempts to reduce and prevent soil erosion*

Activities

1   Write a definition of the term 'soil erosion'.

2   Study photo **A**.

a   Make a list of annotations you could write for the photo.

b   Describe the photo, making clear what has happened to the landscape and its usefulness to people.

3   Study diagram **C**.

a   What are the causes of soil erosion?

b   Which causes do you think are more responsible for soil erosion, physical or human factors? Justify your answer.

c   Draw your own version of diagram **D**. Choose items from the list below to add extra labels:

■   terracing on steep slopes

■   stone lines laid to trap surface water run-off and soil

■   hedgerows replanted to hold the soil together and act as windbreaks.

How do irrigation schemes increase food production?

What is irrigation?

Irrigation is the artificial watering of the land. It has played an enormous role in increasing agricultural output worldwide. As global population is predicted to rise to 10 billion by 2150, food production must rise dramatically. More than half of rice and wheat are already grown using irrigation. One-third of the world's food is grown on 17 per cent of its land – but all that land is irrigated. This shows just how important irrigation is today and in the future.

What problems does irrigation bring?

California and Australia have employed irrigation schemes on a vast scale. Yields were dramatically increased, but after several years problems such as salinisation emerged. Up to half the irrigated land in these developed world regions suffers from this, putting it out of production until the problem can be solved – an expensive process. Human activities make the problem worse. Irrigation increases the likelihood of salinisation. This is a difficult problem to put right. Some 40 per cent of Australian soils are affected, especially in South Australia and Victoria where large areas are cultivated using irrigation. California has similar problems while Syria, a neighbour of Jordan, has 50 per cent of land affected. In Uzbekistan, a staggering 80 per cent of soils are damaged by salt.

Two solutions have been tried in Egypt's Nile Valley:

- Underground drainage systems take the saline water away, but they are expensive to build.
- Changing land use is a cheaper alternative. Some grasses are salt-tolerant, so turning arable land to pasture keeps it productive while the salt washes away slowly with the rain.

Irrigation in sub-Saharan Africa

Sub-Saharan Africa suffers from erratic, unreliable rainfall patterns and includes many of the world's poorest countries. Only 3.5 per cent of land is currently irrigated (photo **A**). Africa lags behind Asia and the rest of the world in all aspects of irrigation. In West Africa, 1 in 20 of Senegal's farmers has irrigation and this is one of the better-served countries. In Asia, one in three farmers has irrigation. The ratio of irrigated land to total cultivated land in Africa is one-fifth that of the global average. This means that opportunities for improving nutrition and standards of living are being lost. Calorie intake in the 24 sub-Saharan countries is only 78 per cent of the global average.

Due to lack of technology, most African irrigation systems utilise the water only once. It is not recycled within the system. Most technology improvements are funded by aid.

In this section you will learn

how irrigation schemes can change agriculture.

Key terms

Irrigation: watering of farm land by artificial means, mainly in dry areas or during dry periods.

Did you know ? ? ? ? ?

By 2025, 80 per cent of food is expected to come from irrigated land.

A *An irrigation scheme in sub-Saharan Africa*

Types of irrigation

For a drip irrigation scheme, pipes are laid across the fields. Water flows through them slowly and drips out through holes in the pipe. It soaks gradually into the soil over a long period of time (diagram **B**). With sprinkler irrigation, water is piped to one or more central locations within the field. Pressure sprinklers then distribute water to the crops (photo **D**). Countries at further stages of development, including the UK, favour this method but it is expensive to set up and run and is therefore less valuable in poorer countries.

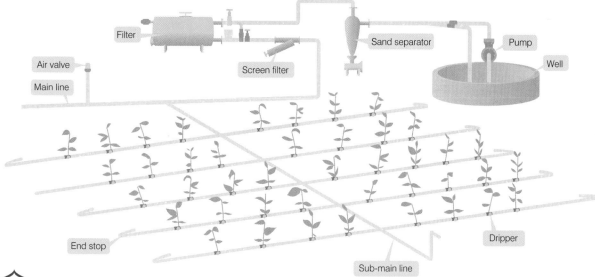

B *A drip irrigation scheme*

C *Advantages and disadvantages of drip and sprinkler irrigation*

	Drip	Sprinkler
Advantages	• Uses a small amount of water • Little chance for water to evaporate • Can be laid in the ground, so even less water is wasted • Plants have regular supply of water • Easy to control • Inexpensive to set up • Can use recycled water • Farmer can control soil moisture • Minimises soil erosion • Leaves stay dry; no rot	• Flexible – can be fed by hose from water source • Can be moved easily around a field • Can use recycled water
Disadvantages	• Only works with crops grown in rows • Needs gently sloping land to work efficiently • Pipes can become clogged with silt • Needs a pump to flow well on flat land	• Expensive to start up • May need a pump to run • Leaves get wet and might rot • Uses more water • Higher rate of evaporation

D *Sprinkler irrigation*

Water use in Jordan

Jordan is a particularly dry country in the Middle East. With very low rainfall and a rapidly growing population, Jordan has concentrated on using as much water for irrigation as possible. Only 6 per cent of the country is farmed. Irrigation is essential and 70 per cent of all water resources are already used for food production (photo **E**). Treated waste water is now used to provide extra irrigation water. Sixteen treatment plants put 60 million m³ of clean water directly into irrigation schemes or into the river systems of the River Jordan basin to be extracted for use further downstream. The World Bank financially supports the companies involved. Other countries might be able to learn from Jordan's experience.

E *Intensive production of tomatoes using pumped water in the Badia desert, Jordan*

Appropriate technology

Poorer countries need affordable technology at a level suitable for the purpose, which can be realistically maintained by the people using it in the environment in which they live. Sometimes in the past aid projects have introduced technology at too high a level to be practical in the circumstances. To repair machinery needs tools that must be available and those using them must have sufficient training. To run machinery may require fuel, which could be inaccessible and expensive. Just as small, handheld ploughs may be more appropriate in a remote part of rural Africa than tractors, so simple irrigation systems, like the drip method, may turn out to be more effective than more expensive and complex systems.

Activities

1.
   a. Write a short definition of 'irrigation'.
   b. Explain briefly why irrigation is important in world farming today.

2.
   a. Explain what is meant by the term 'salinisation'.
   b. Describe the appearance of a landscape suffering from salinisation.
   c. What problems does salinisation cause for the farmer?

3.
   a. Write a brief definition of 'appropriate technology'.
   b. Explain why any new irrigation scheme in sub-Saharan Africa would need to be based on appropriate technology.
   c. Write at least two paragraphs to explore ways in which irrigation in Africa could be much more effective.

4. Read the case study of water use in Jordan and look at the photo **E**.
   a. How is the expansion of irrigation in Jordan likely to help the country's people and economy?
   b. Does Jordan's new irrigation scheme involve appropriate technology? Give one piece of evidence to support your answer.

5. Imagine you are a consultant on a new irrigation scheme in a less well-off country. What advice would you give on the following points?
   - the best type of scheme to be used
   - the construction of the scheme
   - avoiding future problems of salinisation.

   Present your advice in the form of a short report using photos and diagrams.

10.11 How does rural-to-urban migration affect rural areas?

Reasons for rural-to-urban migration

Rural-to-urban migration is caused both by rural push factors and by urban pull factors. The rural push begins the process.

Poor rural standards of living make people want to do better for themselves and their children. Rural areas often lack opportunities for improvement. Agricultural systems have evolved over centuries and they work well as long as the population does not grow too large and people are prepared to remain at a subsistence level. As a country develops, more opportunities become available, but most of these are in the cities. People begin to look elsewhere for improvement. The pull factors, or attractions of the cities, become important at this stage in a person's decision-making process.

Rural people often believe there are more opportunities in the city than there really are because of the sources of their information, which include:

- what they hear on the radio and from newspapers
- what they learn from friends or relatives who have already moved. These people often exaggerate as they want to appear successful to others. This encourages others to follow them in the rural-to-urban migration.

Impacts of rural–urban migration on the rural area

Rural areas benefit from this migration flow. The main aim is to improve the economic situation of individuals and their extended family. Migrants send money home to those remaining in the villages. This can be used to improve farming – increasing livestock, buying new tools or better seed – or to help the family by paying school fees.

It also brings some problems. Sometimes whole families move, but often it is the main breadwinner – the husband and father. Wives, children and the elderly are less able to carry out the hard physical labour involved in farming. Crop yields may reduce and rural development may become limited. Children are brought up with less influence from their fathers. They may have to help more on the farm, which can affect their education. Although the money sent home helps, the family and farming system are put under pressure.

> **In this section you will learn**
>
> the reasons for rural-to-urban migration
>
> the impact rural-to-urban migration may have in the rural areas from which people move
>
> why some agricultural systems are failing.

> AQA **Examiner's tip**
>
> With reference to specific places, you need to be able to explain the main causes of soil degradation and possible solutions.

A *Poor pasture in Lesotho, Africa*

Crisis in Lesotho

Farming employs 57 per cent of Lesotho's labour force, mostly on small subsistence farms. This figure is low compared with other poor countries. The reasons vary:

- Lesotho's mountainous environment limits the amount of agricultural land available. Much of the land suffers from serious soil erosion, partly the result of drought.

- Some 35 per cent of adult males leave home to work in the cities and mines of neighbouring South Africa (map **B**). If they were all at home on the land, farm labour would be a much higher 86 per cent of the national workforce.

- It is difficult to manage efficient and productive farming in such circumstances. In addition, since the area of productive land has decreased (graph **C**), 40 per cent of families are landless.

Table **D** shows the main crops produced in Lesotho. Maize is the staple carbohydrate. Notice how along with other crops its production has declined since 2000/01. Crop production in Lesotho is high risk and low yield because of poor soil quality and the harsh, unpredictable climate (photo **A**). About 10 per cent of the country is arable land, but only 1 per cent is high quality. Lesotho has to import much of its grain as well as some other food. Livestock production is important to rural income because animals cope better with poor land than crops, although drought in the 1980s and 1990s limited the amount of pasture.

With so many men and older boys leaving to find paid work elsewhere, much of the farm labour is left to women, children and the elderly.

On a more positive note, Lesotho has one of the most advanced soil conservation programmes in Africa. It is beginning to solve some of the problems by terracing the land and introducing irrigation systems.

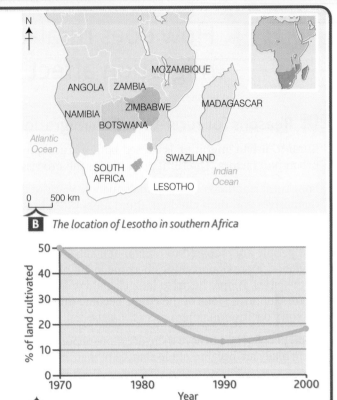

B The location of Lesotho in southern Africa

C Decline in agricultural land in Lesotho, 1970–2000

D Agricultural production, 2000–2005 (thousands of tonnes)

Crop	2000/01	2001/02	2002/03	2003/04	2004/05
Maize	158.19	111.21	82.08	81.00	74.98
Sorghum	45.35	11.92	11.95	10.30	10.32
Wheat	37.31	18.96	13.11	11.65	11.13
Beans	7.86	4.36	3.70	4.83	5.23
Peas	3.67	3.04	3.04	1.50	1.36

Activities

1 Study graph **C**. Briefly describe what has happened to agricultural land in Lesotho.

2 Study table **D**.

a Draw the data in the table as a line graph, with one line for each crop. Plot the years on the *x*-axis and agricultural production on the *y*-axis.

b What is the general trend shown by your graph?

c Suggest reasons for the trend you have described.

11 The development gap

11.1 What are the traditional ways of dividing up the world?

First, second, third and fourth worlds

This early method of dividing up the world was from a Western European perspective. Europeans saw themselves as the **first world**. The wealthier regions they colonised, such as North America and Australia, were referred to as the **second world**. Poorer countries were then grouped together as the **third world**. It then became clear that this was an insufficient number of divisions, as the variety between poorer countries was increasing. The poorest countries – those that were standing still or even declining in economic growth – were labelled the **fourth world**. This system had no clear place for Communist countries such as the former Soviet Union, which were quite well developed, and had never been colonies or part of Western Europe. They were added into the second world group, but this was always unsatisfactory.

North/South

The Brandt Report on the state of world development was published in 1971. It made a simpler division, contrasting economically wealthier and industrialised countries with poorer, less mature and largely agricultural ones. A line called the North-South divide was drawn on a world map to make this difference clear. The development indicator used was GNP (Gross National Product) per capita (map **A**) which is a measure of a country's wealth. This system has decreased in popularity because, as economies have become more varied, it is just too simple.

> **In this section you will learn**
>
> the traditional ways of dividing up the world in terms of development
>
> a modern division based on wealth.

> **Did you know** ??????
>
> Map **A** shows GNP per capita (per head) across the world. The type of map used to show this is called a choropleth map. It is a shaded map where the depth of colour increases as the value gets higher. Also, the shades are close to each other in the colour wheel. Choropleths give a clear visual impression of the pattern of values. We naturally associate darker colours with higher values and paler ones with lower values. The appearance of the map helps us to understand it easily and quickly.

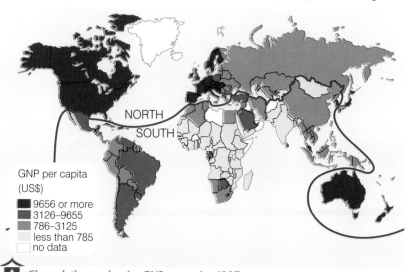

NORTH
SOUTH

GNP per capita (US$)
- 9656 or more
- 3126–9655
- 786–3125
- less than 785
- no data

A *Choropleth map showing GNP per capita, 1997*

Countries at different stages of development

Another simple twofold division involved the use of 'less developed country' (LDC) and 'more developed country' (MDC). This was not widely accepted because development is not only economic but also social and cultural. Many poor countries have a rich culture and society. It was negative to suggest that culture was at a low level just because the country was economically poor. LDC therefore became less economically developed country (LEDC) and MDC became more economically developed country (MEDC).

Since the late 20th century the world order has changed tremendously. Globalisation means there are more contacts and trade between countries than ever before. Some LEDCs are growing more rapidly than most developed economies. A new category had to be introduced to cover those developing fastest. They became known as newly industrialising countries (NICs).

Within any country there are increasing differences between people and regions. The middle class is growing in many poor countries. Their capital cities, along with their surrounding regions, are often well developed and almost indistinguishable from any MEDC city. The LEDC/MEDC division is therefore becoming much less useful.

The five-fold division based on wealth

The five categories recently suggested are based on wealth (map **B**):

1 rich industrialising countries
2 oil-exporting countries
3 newly industrialising countries
4 former centrally planned economies (those previously having a Communist political system)
5 heavily indebted poor countries.

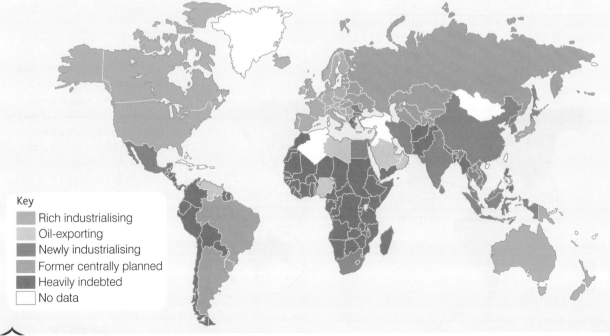

Key
- Rich industrialising
- Oil-exporting
- Newly industrialising
- Former centrally planned
- Heavily indebted
- No data

B *Newly suggested five-fold division based on wealth*

There is a clear and useful division between those poorer countries that are doing well and those that are not (numbers 3 and 5). One reason some countries find it difficult to move their economy forward is the burden of debt. They have borrowed large sums of money from other countries or from world organisations such as the World Bank, but earning enough to repay the debt with interest is taking funds that should be used for development projects. This division is similar to NICs and LEDCs, or third and fourth world countries.

The category of former centrally planned economies (number 4) covers a wide variety of countries today. Russia still has considerable Communist influence. Its old allies, such as Poland and the Czech Republic, are becoming more like Western European countries and are members of the EU. China still has its Communist government, but it is one of the world's fastest growing economies and its speed of industrialisation has been phenomenal. There is still a huge variety within this category.

Oil-exporting countries (number 2) have a great spread of wealth in their society. Huge amounts of money are made from exporting oil. Rich people and companies are investing their money in businesses abroad. Some UK football clubs have been bought by Middle Eastern oil billionaires (photo **C**); in 2008 some invested in struggling British high-street banks. Nevertheless, the majority of people remain poor in oil-exporting countries such as the United Arab Emirates (UAE) and Venezuela. Some profit is used for development projects, so eventually everyone may benefit.

The UK would be classed as a rich industrialising country (number 1). We are still wealthy, but our manufacturing industry has declined over recent decades.

The classification does not always fit well – no division will fit all countries neatly. We need to continue to improve the method of division to make it as useful as possible.

C   *Manchester City football club was bought by a Middle Eastern company*

Activities

1   Study map **A**.

  a   Use an atlas to help you name two countries with less than $785 per person per year and two with $9,656 or more.

  b   Describe the pattern of GNP per capita across the world in 1997. Identify clusters of similar countries.

  c   Describe the pattern made by the North/South divide line. Can you see anything odd about the shape of the line?

  d   Does the line separate countries of similar choropleth shading? Name some countries that are exceptions to the rule.

2   Study maps **A** and **B**. Write at least two paragraphs comparing and contrasting the patterns on the two maps.

3   Consider, using Internet research, the cultural richness (music, art, etc.) that exists in many of the poorer countries of the world.

How can we measure development?

■ The correlation between different measures of development

There are many indicators of development that tell us a great deal about a country and allow us to compare them. Any **development measure** that tells us about wealth, poverty or economic development should correlate with similar indicators. For example, a country with low GNP per capita is likely to have most of its people surviving on subsistence agriculture with low levels of income. Money is limited to pay for education and health, so living standards for most people are low.

Birth rate is closely correlated with level of development. The more developed a country, the lower its birth rate. The demographic transition model gives us evidence that birth rate decreases as countries become more developed (map **A**). A country at a further stage of development is likely to have a high **human development index (HDI)**, low **infant mortality** rate and widespread access to clean water. There are likely to be many doctors for the number of people and literacy rates will be high because the government has sufficient money to spend on health and education. Poorer governments do not have enough funds to provide high-level services. Often even the basics like clean water and a living wage are not possible (table **B**).

In this section you will learn

a range of measures of development and interpret what they tell us about a country

how these measures correlate.

Key terms

Infant mortality: the number of babies (0–12 months) dying per 1000 live births. There is an inverse relationship between the wealth of a country and its infant mortality.

GNI: Gross National Income – is a measure of a country's wealth.

GNP: Gross National Product – is also a measure of wealth but does not take account of some business taxes.

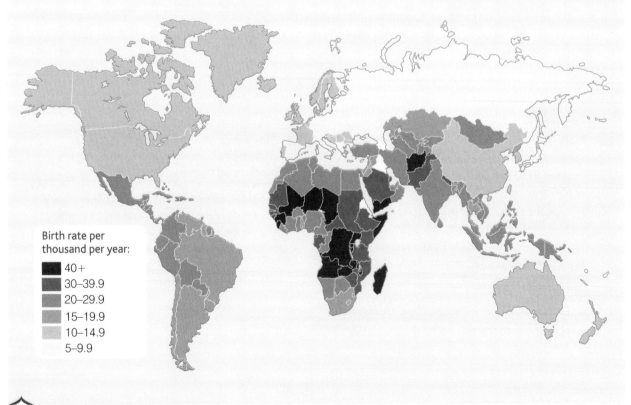

Birth rate per thousand per year:
- 40+
- 30–39.9
- 20–29.9
- 15–19.9
- 10–14.9
- 5–9.9

A Birth rate as a development indicator

B *A range of development indicators for selected countries*

Country	GNP per capita (US $)	GNI per capita (US $)	HDI	Birth rate per 1,000 per year	Death rate per 1,000 per year	Infant mortality per 1,000 live births per year	Number of doctors per 1,000 people	Literacy rate (%)	Percentage of population with access to clean water
USA	43,743	44,710	0.948	14.0	8.2	6	2.6	100.0	100
Japan	38,984	38,410	0.949	8.3	9.0	3	2.0	100.0	100
UK	37,632	40,180	0.940	12.0	9.9	5	2.3	100.0	100
Turkey	4,704	5,400	0.757	18.4	5.9	26	1.4	87.4	96
Romania	3,834	4,850	0.805	9.8	12.4	16	1.9	97.3	57
Brazil	3,455	4,730	0.792	19.2	6.3	31	1.2	88.6	90
China	1,736	2,010	0.768	13.1	7.1	23	1.1	90.9	77
Ivory Coast	843	870	0.421	35.3	15.4	118	0.1	48.7	84
Bangladesh	467	480	0.530	24.8	7.5	54	0.3	47.5	74

Graph **C** shows a clear correlation between two development indicators, although there are a few anomalies. An anomaly is a figure that does not fit in with the pattern. For example, Romania has a lower birth rate than expected, given its low GNI.

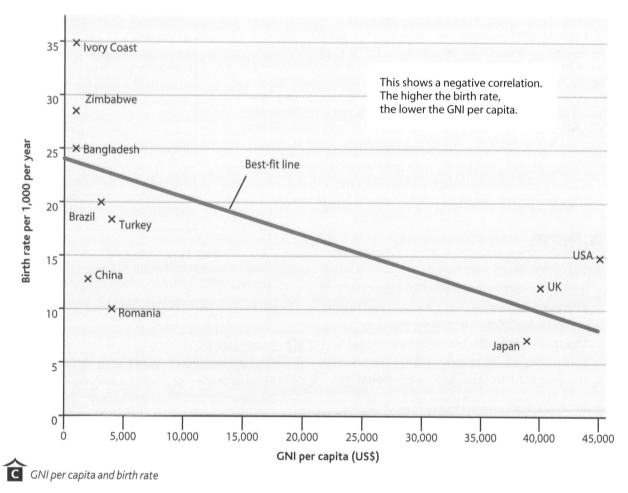

This shows a negative correlation. The higher the birth rate, the lower the GNI per capita.

C *GNI per capita and birth rate*

Limitations of using a single development measure

Birth rate is an excellent measure of development. In rich industrialising countries, women achieve high levels of education and career prospects. As it is difficult to pursue a high-flying career and bring up several children, having children is often sacrificed for success in the working world. Some countries such as France have tried to encourage career women to have children, using tax advantages and benefits.

Death rate is a poor indicator of development. Almost all countries have low death rates today. The more economically developed the country, the higher is the death rate. When birth rates fall, there are fewer young people. Improved health care allows most people to live longer. Death rate increases because there are so many elderly people.

Some measures of development can give a narrow or confusing picture. GNP or GNI per capita are only economic measures so give no clear indication of people's personal living standards. They do not tell us what people earn or how much that buys. Nor do they touch directly on how educated people are and the cultural quality of their lives.

All indicators of development are averages across a country. Within any society there are extremes of wealth and opportunity, which a single figure hides. Relying on one indicator increases this problem. The solution is to use a number of indicators together, giving a much broader view of economic and social development as well as living standards.

Quality of life or standard of living?

The terms 'standard of living' and 'quality of life' are often used interchangeably but they are not the same thing. Standard of living refers to the economic level of a person's daily life. Are they comfortably off or not? Do they fall below an income of $1 per day – the global measure of absolute poverty? During the 1980s a new measure was devised to describe people's quality of life. **Physical quality of life index (PQLI)** uses only social measures of wellbeing.

Ideally, the two measures should be used together to give a more complete picture. People may be quite poor yet they are educated, live to a good age and their children are healthy. In this situation, standard of living might be low, but PQLI would reflect a higher level.

Key terms

Development measure: statistics used to show the level of development, which allows countries to be compared.

Human development index (HDI): an index based on three variables: life expectancy at birth; level of education, including both literacy rate and years spent in school; income adjusted for purchasing power (how much it will buy). Maximum HDI = 1. Wealthy countries like Japan have an HDI of over 0.9, whereas poor countries are around half that figure or less.

Physical quality of life index (PQLI): the average of three social indicators: literacy rate, life expectancy and infant mortality.

AQA Examiner's tip

The relationship between development indicators can be shown clearly on a scattergraph. Ideally, the relationship is close enough for you to be able to draw a best fit line. A best fit line is a straight line that goes through or close to all, or almost all, the points on a scattergraph. Each axis shows a different development indicator. There are several combinations of data sets. It is interesting to test out pairs of data sets and see the results achieved.

Activities

1. Study table **B**.
   a. Which country has the highest GNI per capita?
   b. Find evidence to show how the country named in a uses its wealth to benefit its people.
   c. Which country has the lowest GNI per capita?
   d. In which ways is the poverty of the country named in c obvious from other development indicators?

2. Study graph **C**.
   a. Describe the relationship between GNI and birth rate.
   b. What is meant by the term 'anomaly'?
   c. Name two countries that are anomalous in graph **C**. Suggest possible reasons for these anomalies.
   d. Do you think the variables in graph **C** show a clear correlation? Explain your answer.

3. Study table **B**.
   a. Draw a scattergraph to see if there is a relationship between GNI (on the horizontal axis) and another development indicator of your choice.
   b. Try to draw a best-fit line for your graph.
   c. Discuss how closely your data sets correlate. Remember to point out any anomalies. Give reasons for the patterns you have identified.

11.3   What factors make global development inequalities worse?

Physical factors

The physical geography of some countries does not favour development. Africa has more landlocked countries than any other continent and this makes trade less easy. Tropical Africa, South America and Asia tend to be the poorest world regions and these areas suffer from more climate-related diseases than cooler parts of the world. Many people in the Tropics are debilitated by malaria and other diseases. HIV/Aids is a recent health problem from which sub-Saharan Africa has suffered particularly badly. People cannot improve their standard of living when they are so ill, and many children are orphaned.

Climatic hazards such as drought regularly strike Africa, especially Ethiopia, Eritrea and Somalia in the east. These limit future development and destroy what may already have been achieved.

Economic factors

Poverty causes poverty. Low life expectancy and low standard of living make an almost impossible base from which to develop and expand. The poorest countries are those in, or just emerging from, civil wars: Democratic Republic of the Congo, Burundi, Somalia and Sierra Leone are examples. Ethiopia had a successful history yet today it is one of the world's poorest nations due to famine and war.

∞ links

Find out more about the importance of clean water at www.wateraid.org.uk

---

In this section you will learn

different factors that contribute to development inequalities.

Did you know ???????

As a rule, Africa exports raw materials (mainly minerals and petroleum) and imports manufactured goods. The production of export goods employs around 2 million people over the whole continent, a fraction of the total population of over 934 million. Companies involved are transnational corporations (TNCs), so profits go to their bases abroad. African governments have sometimes squandered their profits, so trade has been of little help towards true development.

Key terms

Tariffs: government taxes on imported or exported goods.

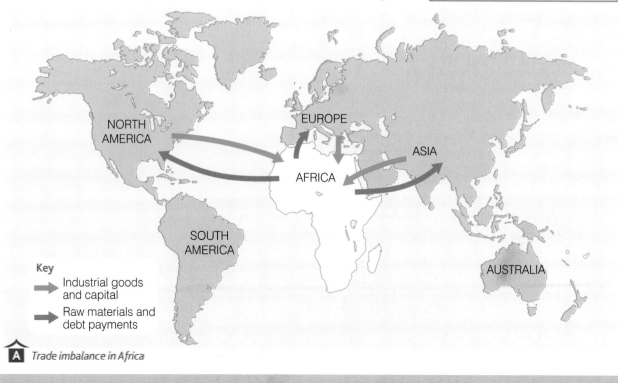

Key

→ Industrial goods and capital

→ Raw materials and debt payments

A *Trade imbalance in Africa*

Global trade policies have not favoured the poorest countries. Sometimes **tariffs** are placed on goods by the purchasing country. Africa is the least industrialised continent. Even though it has really cheap labour, processing is usually done in the purchasing country. Most African exports are primary goods (raw materials). There are several reasons why companies do not invest there:

- banks financing industry want political stability
- good infrastructure to move raw materials, finished goods and labour needs to be in place
- a reliable electricity supply must be available
- an educated workforce is necessary.

These factors are rare in Africa, whereas Asia has supplied these and industry has been attracted, pushing forward its development.

A country's income is primarily measured by its gross domestic product (GDP) per capita and the welfare of its people by the human development index (HDI). In general, when the income is low, so is the level of welfare (table **B**).

Environmental factors

People abuse the land, but there are sometimes understandable reasons for doing so. Farmers on the verge of starvation are unlikely to be concerned about the rainforest they are cutting down to try to feed their families. Throughout the Sahel region just south of the Sahara, deforestation and overgrazing have increased desertification. Wildlife have been sacrificed to poachers when they could have been the basis of a successful tourist industry. Poor governments have little spare to protect and develop environmental resources.

Social factors

Water quality

Poor water quality causes disease, which debilitates people and prevents economic development. Many tropical countries suffer from endemic malaria, yellow fever, bilharzia and river blindness. Diseases are carried in water and water quality is unreliable.

Reliability of water supply

Inadequate water supplies limit crop yields and therefore food supply. There is not enough water for irrigation to allow for yields to be increased.

Education

A poor country finds it difficult to fund education for all children to a good level. Investors are put off by a lack of an educated workforce.

Health

When Sierra Leone got its independence in 1963 it had a health system in place, but as a poor country it has not been able to maintain this. It is difficult for sick people to work hard, so Sierra Leone's economy has spiralled downwards. It is one of the most distressed countries in the world, with an HDI of only 0.048.

B GDP per capita and HDI compared, 2005

Continent/ country	GDP per capita ($)	HDI (out of 1)
Africa:		
Chad	1,749	0.388
Comoros	1,063	0.561
DR of Congo	264	0.411
Ethiopia	591	0.406
Ghana	1,225	0.553
Somalia	199	0.284
Sudan	2,249	0.526[1]
Zimbabwe	538	0.513[2]
Asia:		
India	490	0.590
Indonesia	790	0.682
Iran	1,500	0.719
Malaysia	4,130	0.790
Pakistan	400	0.499
Philippines	990	0.751
Latin America:		
Mexico	6,260	0.800
Peru	2,130	0.752

[1] Confusing, as some regions are now war zones.
[2] Much reduced due to major political disruption.

Activities

1 List two physical and two human factors that make global inequalities worse.

2 Study table **B**.

a Draw a scattergraph to show HDI and GDP per capita for the selected countries shown in the table. Put GDP on the horizontal axis and HDI on the vertical axis.

b Draw a best fit line and describe the relationship.

c To what extent do the countries cluster closely together? Does one continent stand out as being especially different from the others?

11.4   How do physical and human factors increase global inequalities?

Hurricane Ivan

Hurricane Ivan was one of the Caribbean's most powerful and destructive hurricanes, hitting several Caribbean countries as well as the USA. Ivan hit the island of Grenada on 7 September 2004 (map **B**). Winds of 200 km per hour from this Category 4 hurricane on the Saffir-Simpson scale caused major damage, though rainfall proved not as heavy as expected.

Hurricane Ivan affected the entire island of Grenada. The southern part was the worst affected. Trees were broken and uprooted, services and buildings were destroyed (photo **A**). Roads were seriously blocked by fallen trees, although the coastal defences did well against the storm surge. Thirty-seven people died from the effects of Ivan on Grenada, before it continued across Jamaica to the USA. Around 90 per cent of houses were damaged or destroyed. Most people were affected in some way, half being made homeless – almost all schools were damaged. Water, power and telecommunications systems were also disrupted – potentially a major health risk for the island. Water was prioritised as the most basic health issue and was efficient again within three weeks of the storm.

In terms of development, long-term damage is more important. In the short term, people lacked food, clean water and medical care, but these shortages were overcome. In the long term, agriculture, tourism and infrastructure were badly hit and took much longer to repair and replace, and at much greater cost. Even 10 years is not enough to catch up with what was destroyed by a major hurricane like Ivan, and poorer countries have a low tax base and individuals are usually without insurance to help them recover.

In this section you will learn

the impacts of a natural hazard on development (a physical factor)

the unfairness of world trade (an economic factor)

the effects of water quantity and quality on standards of living (a social factor)

how unstable and corrupt governments limit economic development (a political factor).

Case study

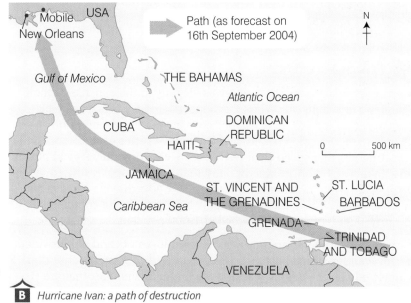

A *Hurricane Ivan's damaged landscape*

B *Hurricane Ivan: a path of destruction*

News

Home	Grenada's Prime Minister, Keith Mitchell, whose official residence was destroyed, told the BBC the island was '90 per cent devastated' and that he had declared a national disaster.
Contact	
Sitemap	The capital, St George's, was hit by 200 km per hour winds, flattening homes and disrupting power. The storm destroyed the city's operations centre, the main prison, many schools, and damaged the main hospital. Mr Mitchell said the hurricane had caused 'hundreds of millions of dollars of damage'. He said the country's key export crop, nutmeg, was likely to have 'taken a tremendous hit'.
News	
Forum	
Shop	

Adapted from BBC News website **news.bbc.co.uk**, *9 September 2004*

C *From the BBC News website*

Economic factors

World trade is often unfair. Richer countries want to pay as little as possible for their raw materials, many of which come from developing countries. There is often more supply than demand, which keeps prices low. Poorer countries find themselves competing against each other for business and have lowered their prices to attract buyers. The result is that poor farmers do not receive enough to support their families.

Exports by countries at lesser stages of development are mainly exports of primary goods (raw materials). Processing is done in richer countries, which get the benefits of the jobs provided.

Social factors

Clean water is essential for health and a good standard of living. It also means good crop production. If water is in short supply, people have to spend time searching for it and wasting valuable energy carrying it. If water is dirty and contaminated, people become ill and less able to support their families. Great improvements in standard of living have been experienced by villages whose water supply has been made secure and clean.

Political factors

Corrupt politicians enrich themselves illegally at the expense of their country's development. When this happens, money is not available for education, health services, roads, clean water and sanitation. People in the Solomon Islands and Zimbabwe suffer from corrupt governments that spend their resources by ensuring they stay in power. In Zimbabwe, highly productive land that was previously owned by large-scale white farmers was taken over by government officials and 'war veterans'. The country's economy has been almost completely destroyed and inflation has exceeded 1,000 per cent.

Corrupt governments are unstable. Foreign investors and aid providers are discouraged from putting funds into such a country because they cannot rely on the money reaching the target. Economies that are already weak cannot afford to miss out on this income.

AQA **Examiner's tip**

Be clear as to the differences between 'physical', 'economic', 'social' and 'political' factors in relation to global inequalities and be able to give at least one example of each.

Activities

1. Study map **B** on page 259.

   a. Make a list of the countries affected by Hurricane Ivan in September 2004.

   b. Give three reasons why these countries were so badly damaged.

   c. Explain why it was difficult for individuals, families and the Caribbean governments to recover from the physical impacts of Hurricane Ivan.

2. Undertake some Internet research into the economic and developmental impacts of another physical hazard on poorer parts of the world. Try earthquakes, volcanoes and river floods. Consider the impact of your choice on the future development of the country(ies) concerned.

11.5 How can international efforts reduce global inequalities?

Loans and aid

Loans are sums of money that at some time in the future have to be paid back with interest. Brazil borrowed huge sums in the 1970s. The aim was to build factories, manufacture goods for export and make a profit. Loans were to be repaid from that profit. This policy was successful to a great extent, but world recession caused some problems. Not all countries have been so lucky.

Aid is gifts of money, goods, food, machinery, technology and trained workers. The aim is to raise standards of living. True aid is not a loan that needs to be repaid. However, in the real world some 'aid' really is a loan because some form of payback is required. Many people in the richer world want to relieve the poorer world of its debt burden. There is popular support for the G7 proposals, shown by the marches that took place in London, Edinburgh and elsewhere (photo **A**).

> ### In this section you will learn
> the key differences between loans and aid
>
> how debt results from loans and how individuals and countries are affected
>
> the concept of fair trade, and its advantages and disadvantages for producers and consumers
>
> the different types of aid.

Loans and their impacts

A country in need of funds for development projects can borrow from other countries, world financial organisations (e.g. IMF, World Bank) or international banks. If the project is a success, the debt is repaid. However, often things do not go according to plan and the country defaults on the debt, which then has to be paid back over a longer period. People in the debtor country have to work hard to produce goods for export to fund the interest on the loan. Standard of living and level of development simply cannot improve.

> ### Did you know ??????
> Each year Africa sends more money to Western bankers in interest on its debts than it receives in foreign aid from these countries. Relieving at least some of the debt would help these economies grow, as long as they are well managed.

Debt relief and abolition

Poorer countries can be helped by **debt relief**, which means reducing the interest rate or the amount of the loan. Sometimes debts are abolished or written off. Debtor nations (those who have borrowed) benefit hugely as they can begin to improve life for their citizens. In the UK, the Midland Bank wrote off some debts owed by countries at lesser stages of development in the 1980s, but found its own finances suffered so much it was taken over by HSBC. In the 21st century there is a move from wealthier governments, like those in the G7 economic group, to write off all the debts of the poorest countries.

A Make Poverty History march, 2005

Loan solutions

Micro-enterprises may be the way forward. Non-profit groups in the USA have been lending money to individuals in poor countries. Sara Garcia in Lima, Peru borrowed $1,845 in 1984. She invested in equipment to make patterned handkerchiefs. She hired extra workers and family members to help. Output has risen from 20 to 500 items per day. She met her repayments on time. The technology used was appropriate, allowing the business to become sustainable and successful.

> ### Did you know ??????
> Loans from governments, world organisations and banks from countries at lesser stages of development total $1.2 trillion.

This is typical of what can be done on a small scale. Today, this system has expanded greatly. Small business is the basis of any economy and it employs a surprising number of people and supports their families. It is an important way forward and everyone involved gains.

Conservation swaps

In their attempts to develop, poorer countries are tempted to use every natural resource available to them, even if it is something valuable whose loss will result in future difficulties for the country itself and sometimes for the rest of the world. After some debate, some richer countries have realised that, although conservation of valued resources and landscapes are important, the poorer country should not lose out on development opportunities. New 'conservation swaps' are being developed to try to solve this problem.

Swaps are agreements whereby a proportion of a country's debts are written off in exchange for a promise by the debtor country to undertake environmental conservation projects. These were first set up by environment groups in the 1980s to reduce the debt problem of poor countries and to promote the conservation of important environments.

Between 1987 and 2001, 50 countries took part. Usually, areas of valuable land are set apart for protection, especially tropical rainforest. The first swap took place in Bolivia, South America. A North American conservation group took over $650,000 of Bolivia's national debt in return for the Bolivian government setting aside a large area of rainforest as a nature reserve. Other countries that have taken part in such schemes include Guatemala, Peru, Ecuador, Costa Rica and Poland.

Fair trade

Fair trade is an international movement ensuring producers in poor countries get a fair deal. They receive a minimum guaranteed price for their crop, which provides them with a living wage, long-term contracts for security and skills training to develop their businesses.

Table **B** shows the rapid growth in fair trade goods consumed in the UK between 1999 and 2007. The volume of goods produced increased by 150 per cent in 2006/07 alone. Global sales of fair trade goods in 2007 were £1.6 billion. Around 7 million farmers, farm workers and their families in 58 poorer countries benefit from the improved trading conditions brought about by fair trade. Recognition of the Fairtrade Mark has grown rapidly in the 21st century. Almost 60 per cent of adults in the UK know what it means.

Aid: positives and negatives

Short-term aid in the form of disaster relief can help to save lives and with the country's recovery. Long-term aid can build schools and hospitals, and invest in industry and agriculture. There are many successes, both large and small scale.

During the 1990s it was realised that many poorer countries were not improving with aid and many richer countries cut their contributions. Some aid is lost to corrupt governments and individuals. Lack of infrastructure can prevent development happening as planned, for example schools and hospitals cannot operate without roads and power. Trained staff may not be available. Too much aid has been tied, which means the donor country gains more than the receiving country.

B *Fair trade goods consumed in the UK, 1999–2007 (£ millions)*

Product	Year				
	1999	2001	2003	2005	2007
Coffee	15.0	18.6	34.3	65.8	117.0
Tea	4.5	5.9	9.5	16.6	30.0
Cocoa/chocolate	2.3	6.0	10.9	21.9	34.0
Honey	–	3.2	6.1	3.5	5.0
Bananas	–	14.6	24.3	47.7	150.0
Flowers	–	–	–	5.7	24.0
Wine	–	–	–	3.3	8.2
Cotton	–	–	–	0.2	34.8
Other	–	2.2	7.2	30.3	90.0
Total	21.8	50.5	92.3	195.0	493.0

The right kind of aid is essential, i.e. at a suitable scale, appropriate to the level of technology and local culture, as well as being sustainable.

Possible effects of the recent global recession

The global economic recession of 2009 has affected most countries in the world, though at differing rates. The problems of the wealthier countries will spread to all those with whom they trade or exchange loans or debts. Countries that normally donate aid will have less to give away in aid or to lend. As world prices decrease, poorer countries will receive less for their exports. These countries will no longer be able to repay the interest on their debts to richer countries and to global financial institutions. They may therefore go even further into debt.

Did you know ??????

The US government gave the Guatemalan government $15 million towards the cancellation of its debt. Conservation groups added more. The total contributed from all sources was $24 million. Threatened forests, including high-altitude 'cloud forests', were protected in this trade. Hundreds of rare and endangered species were assisted.

Did you know ??????

Fair trade chocolate in Ghana
The Kuapa Kokoo cooperative in the Ashanti region at the heart of Ghana's cocoa belt is working with fair trade organisations. It is helping 35,000 members to get their fair share of the profits generated by cocoa.

⊙links

More on Guatemala's debt for nature swap can be found at www. wildlifeextra.com/go/news/ guatemala-debt.html.

You can find out more about Kuapa Kokoo by entering the name of the organisation into the search box on www.fairtrade.org.uk to find out more.

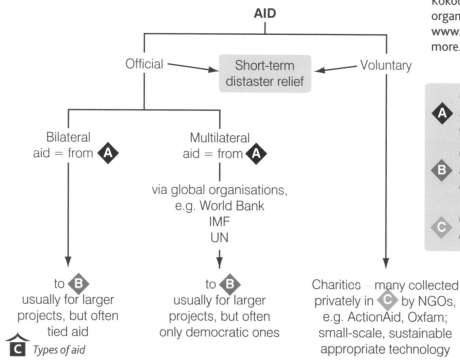

A Governments of countries at further stages of development

B Governments of countries at lesser stages of development

C Countries at further stages of development

C *Types of aid*

D *Advantages and disadvantages of aid for donor and recipient countries*

Type of aid	Donor countries		Recipient countries	
	Advantages	Disadvantages	Advantages	Disadvantages
Short-term aid	People give willingly in a disaster – 'feel-good factor'	None	Immediate help – lives saved Flow of aid may continue following publicity of the disaster	Occasionally, well-meaning governments and organisations fail to provide exactly what is needed
Long-term aid	Companies and individuals find satisfying and well-paid work in projects overseas Trade may continue into the future	None	New industries improve skills and employment Agricultural improvements – new and better crops New infrastructure such as schools and hospitals built Trade may continue into the future	Tied aid – makes recipient country reliant on donor country Local people may not have sufficient skills and training to reach senior posts Agricultural change may not be sustainable – level of technology too high Lack of money for fuel, spare parts, etc. May not be sufficient funding to maintain schools and hospitals to an adequate level Local people may lose their land to large-scale projects
Top-down aid	Coordinated by government or international organisations – makes donor feel in control	Projects swallow large amounts of money – donors may feel it is wasted	Capital-intensive – aims to improve country as a whole Large projects, e.g. dams, improve national infrastructure	Most ordinary people do not benefit directly
Bottom-up aid	NGO aid, so individuals give to charity – 'feel-good factor' Feeling of direct link between donor and recipient – e.g. sponsorship schemes like ActionAid	None	NGOs work with recipient communities, who have input into the project Money not lost to corruption Appropriate technology used, so projects are sustainable	Charity funds may reduce in economic recession

Activities

1 Study photo **A**.

a Make a list of possible reasons for going on this march.

b Several moves have been made to improve both debt and aid problems. Suggest why these are likely to be successful or unsuccessful.

2 Study table **B**.

a Draw a series of line graphs to show the consumption of Fairtrade goods in the UK. Notice that some of the lines will be longer than others! Use different colours and a key. Plot the years on the x-axis and consumption on the y-axis.

b Describe the overall trend shown by the lines and comment on differences between the lines.

c Do you and your family choose to buy Fairtrade products? Why?

11.6 How successful are development projects?

A large-scale aid project: the Cahora Bassa dam, Mozambique – an example of bilateral aid

The Cahora Bassa dam was begun by the Portuguese government of Mozambique in the 1960s, although it was only completed three decades later. Civil war (1977–1992) prevented development and use of the scheme, as well as damaging its infrastructure. Renewal work did not begin until 1995, ending in 1997.

It is the largest HEP scheme in southern Africa, with five huge turbines and a surface area second only to Nasser behind Egypt's huge Aswan dam. Three major dams have been constructed in the River Zambezi Basin: the Kariba on the Zambia–Zimbabwe border and the Itehzi–Tehzi on the tributary River Kafue, as well as Cahora Bassa dam. Cahora Bassa is the most recent and potentially the most important (fact file **A**).

Despite this huge resource, only 1 per cent of Mozambique's rural homes have a direct electricity supply and this level has hardly changed during the life of the dam. Most of the power is sold to South Africa, which makes money for the Mozambican economy but does little for its citizens. The Cahora Bassa dam has much greater potential than produces today. It could provide the whole of Mozambique with all the power it needs for the foreseeable future as long as some natural gas, solar and wind projects are also developed to serve the most rural areas.

Having three dams in one basin has caused environmental damage. River flow is low because so much water is held in reservoirs. The shrimp fishing industry in the lower valley has been almost destroyed. A new dam, the Mepuna Uncua, was planned in 2005 downstream from Cahora Bassa, even though the existing dam is not operating at anywhere near its full capacity. The environmental consequences of this are uncertain, but risky.

Has Cahora Bassa been a success as a development project? Certainly it has much greater potential than has ever been realised. Perhaps if it had concentrated on serving Mozambique, its success would have been much greater. Also, if it was to work in harmony with other dams, its total potential might serve the whole region of Mozambique plus its neighbouring countries. It was hoped that the Cahora Bassa dam would bring major river flooding – typical of the region – under control. Some success has been experienced, but with more careful control much more could be achieved.

In this section you will learn

detailed case studies of aid at different scales and from different donors

the advantages and disadvantages of aid for both recipient and donor countries.

Fact file

- Catchment area: 56,927 km²
- Length of lake: 292 km
- Maximum width of lake: 38 km
- Surface area of lake: 2,739 m
- Average depth of lake: 20.9 m
- Maximum depth of lake: 157 m

A *Cahora Bassa dam fact file*

B *The Cahora Bassa dam*

Case study

A medium-scale aid project: ActionAid, Kolkata – an example of voluntary aid

ActionAid is a UK charity working in local communities in the developing world to alleviate poverty. Its six target areas of relief are:

- HIV/Aids
- hunger and food
- women's rights
- the right to education for all
- the right to security
- the right to good government.

ActionAid works in the poorest districts of Kolkata such as Dharavi, known as the world's worst slum. This area has unimaginably high population densities and few services.

Some of ActionAid's work is done through sponsorship schemes. Donors are encouraged either to give monthly to general projects or to sponsor a child and his or her family or community. This is a successful approach. Donors like the idea of improving the life chances of an individual, especially that of a child.

ActionAid's donors feel they are doing something worthwhile for real people. Low technology is used. Local people benefit, both individuals and communities, but costs remain low. Environmental quality is improved as sustainability is an important aspect of ActionAid's work.

A small-scale aid project: Community Youth Empowerment Programme, Uganda

Student Partnership Worldwide (SPW) is a UK-based organisation placing gap-year students in development projects in countries like Uganda. They work with school pupils and farmers. Volunteer UK and Ugandan students work together to:

- raise awareness of the Aids risk in order to prevent as many infections as possible, especially among young people, through non-formal education techniques in school such as role-play and drama
- improve knowledge of environmental health concerns such as nutrition, sanitation and waste management
- teach energy conservation methods such as how to construct a fuel-efficient stove or how to start a tree nursery to provide precious fuelwood
- promote sustainable organic farming ideas.

Projects include the following examples:

1. In Kebager village, the three natural springs were polluted. SPW students constructed a covered water tank to keep pollution out. People took their supplies from this by tap, so water never lay open

C SPW works with school pupils in countries like Uganda

to risk. Appropriate technology has improved living standards as well as enhancing the environment.

2. In Bwanyanga village, both the primary and secondary schools received SPW students as volunteer teachers, who provided a number of lessons and workshops over the course of six months. These focused on sexual health awareness and improving life skills. These projects were well received but suffered setbacks due to issues with school fees affecting the attendance of pupils.

Activities

1. Study the case study on the Cahora Bassa dam.
   a. Who are the winners and who are the losers in this project? Make two lists to show who has benefited and who has suffered.
   b. Decide whether the dam should have been built, giving reasons for your answer based on your answer to a.

2. a. Why do you think medium- and small-scale projects are considered to be more appropriate than large-scale projects?
   b. Consider the short-term and long-term benefits of improving water quality and education in Uganda.

11.7   How do levels of development vary within the EU?

A *The 27 EU member countries*

In this section you will learn

about using indicators to contrast levels of development within the EU

the reasons for these development levels

how to contrast levels of development in two EU member countries

the policies the EU uses to try to even out these differences.

Key terms

Economic periphery: the edge of a country or region in terms of economics. It may not physically be the edge, but is a more remote, difficult area where people tend to be poorer and have fewer opportunities. A less well-developed area.

Economic core: the centre of a country or region economically, where businesses thrive, people have opportunities and are relatively wealthy. A highly developed area.

Sustainable development: this allows economic growth to occur, which can continue over a long period of time and will not harm the environment. It benefits people alive today but does not compromise future generations.

There are marked variations in levels of development in the EU (map **A**). Table **B** (page 268) shows three measures of development for all 27 EU countries. They have been selected to show you that there are many similarities and some differences. HDI is high for every member country. All are in the top 60 in the world. Life expectancy is always over 70 years, but it is significant that some are over 80, indicating an extremely high standard of living. GDP per capita is the indicator that really shows up the clearest differences. Luxembourg's GDP per capita is almost seven times higher than that of Bulgaria. The older members have higher GDP per capita than the newer ones.

Contrasting two EU countries: Bulgaria and Ireland

Bulgaria's population in 2008 of 7,640,238 was lower by almost 300,000 than in 1998. People are leaving to find better opportunities in the rest of the EU, which Bulgaria joined in 2007. Today it is a democracy, but previously it was politically and economically dominated by the USSR and was a Communist country. With the break-up of the Communist block in the 1990s, its standard of living fell by 40 per cent. Although Bulgaria is the second poorest country in the EU and on its **economic periphery**, new funding and projects are improving quality of life.

Ireland's 2008 population of 4,422,100 is growing slowly because of natural increase and migrant labour. Until it joined the EU in 1973, it was a poor country within Europe. Membership has benefited Ireland enormously, changing its focus from agriculture to a high-tech service economy. Its 10 per cent per annum economic growth rate from 1995 to 2000 earned it the title 'Celtic tiger'. Ireland is no longer on the economic periphery but very much part of the **economic core**.

B *EU comparative development indicators*

Country	HDI rank in the world	Human Development Index (HDI)	Life expectancy at birth (years)	GDP per capita ($)
Ireland	5	0.959	78.4	38,505
Sweden	6	0.956	80.5	32,525
Netherlands	9	0.953	79.2	32,684
France	10	0.952	80.2	30,386
Finland	11	0.952	78.9	32,153
Spain	13	0.949	80.5	27,169
Denmark	14	0.949	77.9	33,973
Austria	15	0.948	79.4	33,700
UK	16	0.946	79.0	33,238
Belgium	17	0.946	78.8	32,119
Luxembourg	18	0.944	78.4	60,228
Italy	20	0.941	80.3	28,529
Germany	22	0.935	79.1	29,461
Greece	24	0.926	78.9	23,381
Slovenia	27	0.917	77.4	22,273
Cyprus	28	0.903	79.0	22,699
Portugal	29	0.897	77.7	20,410
Czech Rep	32	0.891	75.9	20,538
Malta	34	0.991	79.1	19,189
Hungary	36	0.874	72.9	17,887
Poland	37	0.870	75.2	13,847
Slovakia	42	0.863	74.2	15,871
Lithuania	43	0.862	72.5	14,494
Estonia	44	0.860	71.2	15,478
Latvia	45	0.855	72.0	13,646
Bulgaria	53	0.824	72.7	9,032
Romania	60	0.813	71.9	9,060

■ EU policies to reduce different levels of development

The Common Agricultural Policy (CAP)

The CAP includes a system of subsidies paid to EU farmers. Its main purposes are to:

- guarantee minimum levels of production so that there is enough food for Europe's population
- ensure a fair standard of living for farmers
- ensure reasonable prices to customers.

Supporters of the CAP say it guarantees the survival of rural communities, where more than half of EU citizens live, and preserves the appearance of the countryside. Critics say that as only 5 per cent of EU citizens work in agriculture, which generates only 1.6 per cent of GDP, the CAP costs too much (extract **C**).

News

Home | Contact | Sitemap | News | Forum | Shop

Home
Contact
Sitemap
News
Forum
Shop

The Common Agricultural Policy is regarded by some as one of the EU's most successful policies, and by others as a scandalous waste of money.

A series of reforms has been carried out in recent years, the most significant being the Single Payment Scheme introduced in 2003, intended to break the link between farm aid and production.

In the latest reforms, announced in November 2008, farmers are to have more of their subsidies for food production diverted towards rural development. Milk quotas will be boosted before being scrapped in 2015.

Instead of paying farmers to produce more, the EU now makes payments conditional on farmers meeting environmental and animal welfare standards and keeping their land in good condition.

Adapted from BBC News website **news.bbc.co.uk**, *20 November 2008*

C *From the BBC News website*

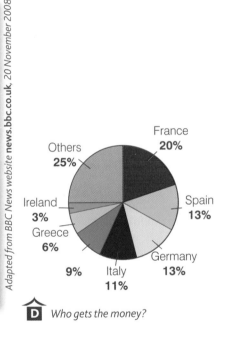

D *Who gets the money?*

The addition of 10 new member countries in 2004, together with Bulgaria and Romania three years later, brought another 7 million farmers into the EU, in addition to the 6 million already there. CAP cannot possibly maintain them at the previous Western European level.

Urban II fund

Most Europeans live in urban areas because they are the centres of economic activity and hold greater opportunities. However, all cities have concentrations of social, environmental and economic problems. Urban II fund money comes from the European Regional Development Fund and it is for **sustainable development** in troubled districts of European cities. It aims to provide economic and social regeneration. Any successful idea in one city is shared with others to try to improve living conditions as widely as possible.

Social and economic urban regeneration includes:

- improving living conditions (e.g. renovating older buildings)
- creating new jobs in services that benefit the whole population
- integrating less-favoured groups of people into education and training so that they can find satisfactory employment
- developing environmentally friendly transport systems
- making greater use of renewable energy
- using the most up-to-date ICT systems to make work more efficient and to improve people's skills and therefore their job prospects.

Urban II was preceded by Urban I (1994–99), which benefited 118 urban areas. Some €900 million was spent and 3.2 million people were helped. Urban II has 70 different programmes that affect 2.2 million people. Its budget was €728.3 million between 2000 and 2006. Money is divided according to need, which is measured by population numbers and unemployment rates. See table **E**.

The people of the town of Teruel in northern Spain have a new ring road, paid for by Urban II funds. It will reduce traffic flows through the town by at least 20 per cent, cutting congestion and improving travel times and air quality in the town. The new road also links previously isolated neighbourhoods. There are paths for cyclists and joggers. The project cost €16.6 million.

European Investment Bank (EIB)

The EIB's money comes from the member countries who own it. They contribute according to their size and wealth. In 2004 they contributed €163.6 billion. The bank borrows on the world financial markets.

Its main purpose is to invest in regional development. Some regions are suffering difficulties because of the decline of local industry or reduced farm incomes. Projects are usually locally based and funds are used to train people with new skills and to help set up new businesses.

Structural funds

Structural funds support poorer regions of Europe and improve infrastructure, particularly transport because that enables the economy in an area with difficulties to work more efficiently. Together with CAP, it makes up most of EU spending.

Regions whose GDP per capita is less than 75 per cent of the EU average are targeted – the aim is to accelerate economic development so they catch up with other regions. The budget for 2007–13 is €347.41 billion. In addition, the most deprived regions will receive extra money from other funds.

E *Urban II fund across Europe, 2000–06*

Country	Urban II fund (Euros)
Austria	8,400,000
Belgium	21,200,000
Denmark	5,300,000
Finland	5,300,000
France	102,000,000
Germany	148,700,000
Greece	25,500,000
Ireland	5,300,000
Italy	114,800,000
Netherlands	29,800,000
Portugal	19,200,000
Spain	112,600,000
Sweden	5,300,000
UK	124,300,000

Activities

1. a Which countries have the highest life expectancy in Europe?
   b Suggest possible reasons for this.

2. Use the Internet and the information in this section to carry out a comparative study of Bulgaria and Ireland. Try to identify how and why the two countries differ in terms of their development.

3. Describe the effects of the Urban II fund on the town of Teruel. Try to find some information and photographs from the Internet to support your study.

12 Globalisation

12.1 What is globalisation?

Globalisation has existed since the 1960s. Its meaning has evolved from one relating to a way of doing business in an international arena to the wide-ranging process that it has become today. Globalisation relates to the greater connectivity between different areas of our seemingly shrinking world and an increase in **interdependence** as a result. This is due to the fact that the manufacturing of goods and the provision of services takes place internationally and can include all continents on a worldwide basis. This has been made possible by the relaxation of laws allowing foreign investment in countries (which encouraged the rise of transnational corporations, or TNCs), the increased provision and speed of international transport and developments in communication with the use of fax, telephone and e-mail that negates the impact of long distances.

Nike is a TNC that manufactures footwear and clothing. It has 124 plants in China, 73 in Thailand, 35 in South Korea, 34 in Vietnam and others elsewhere in Asia. In addition, it has factories in South America, Australia, Canada, Italy, Turkey and the USA. Africa is the only inhabited continent not represented.

A *Nike shoes on sale in China*

In this section you will learn

what is meant by the term 'globalisation'

the increased interdependence and interrelationships that result from greater connectivity between the different countries.

Key terms

Globalisation: the increasing links between different countries throughout the world and the greater interdependence that results from this.

Interdependence: the relationship between two or more countries, usually in terms of trade.

AQA Examiner's tip

Ensure that you explain the meanings of the terms 'globalisation', 'interdependence' and 'transnational corporation' (defined on page 276).

The production of a Wimbledon tennis ball

Slazenger has supplied the All England Club since 1902. Dunlop Slazenger is responsible for making the 48,000 balls that supply the showpiece tournament in June each year. From the 1940s to 2002, the tennis balls were made at Barnsley, south Yorkshire. However, in an attempt to boost profits by cutting labour costs, production shifted to Bataan in the Philippines in 2002.

Table **B** shows the ingredients for making the tennis balls and their origin. Many of the components are used to vulcanise the rubber so that the balls have the correct amount of bounce. They are all transported to Bataan for manufacture, before being despatched to London.

B *The manufacture of a Wimbledon tennis ball*

Ingredient	Origin	Destination
Wool	New Zealand	UK
Cloth (made from wool)	UK	Philippines
Dyes	UK	Philippines
Silica	Greece	Philippines
Zinc oxide	Thailand	Philippines
Rubber	Malaysia	Philippines
Tins	Indonesia	Philippines
Clay	USA	Philippines
Magnesium carbonate	Japan	Philippines
Sulphur	South Korea	Philippines
Tennis balls (product)	Philippines	UK

C *A Wimbledon tennis match*

Activities

1 a Write down a definition of 'globalisation' and give one example to illustrate it.

   b Compare your answer with a partner's and agree on a definition.

2 Study table **B**.

   a On a world political map outline, draw flow lines to show the movement of materials to the Philippines and then the completed tennis balls to the UK.

   b Describe how your map illustrates the concept of 'globalisation'.

   c Why might it be said that the production of the cloth to cover the tennis balls is the best example of the globalisation process?

∞links

Find out more about globalisation by typing the word into the search box at **www.bbc.co.uk**.

12.2 / How has globalisation changed manufacturing and services worldwide?

■ The influence of developments in ICT

Advances in transport, communication and increased links underpin changes in manufacturing and services, enabling them to become a worldwide affair. Graph **A** shows the rapid decline of air and sea transport as well as the cost of a transatlantic phone call and satellite charges that enable media, mobile and satellite phone connections. Much of the advances made in communication are the result of developments in satellites. A satellite is an object that revolves around the earth following a particular path or orbit. It is usually built for a specific purpose. For example, weather satellites include TIROS and COSMOS satellites, as well as cameras for photos of the weather; communication satellites allow telephone and data conversations to be relayed via transponders within the satellite – this is like a radio that receives the conversation on one frequency and then transmits it back to earth on another frequency. Even cable television depends on satellites initially to allow transmissions.

In this section you will learn

how developments in ICT have encouraged globalisation

how specialist, localised industrial areas have developed with global connections, e.g. Motorsport Valley

why call centres have developed abroad.

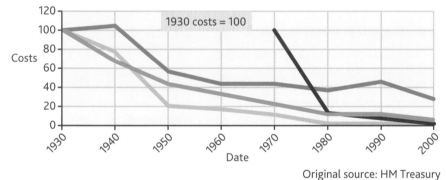

1930 costs = 100

- Transatlantic phone call [1]
- Sea freight [2]
- Air transport [3]
- Satellite charges

Original source: HM Treasury

[1] Cost of three-minute telephone call from New York to London
[2] Average ocean freight and port charges per short ton of import and export cargo
[3] Average air transport revenue per passenger mile

A *Falling transport and communication costs*

The development of submarine cables has been important in allowing global operations for both manufacturing and service industry. The extent of this network is shown in map **B** on page 274. One particularly important submarine cable system is the SEA-ME-WE cable linking South-east Asia, the Middle East and Western Europe, which was developed during the 1980s. Further developments have followed with SEA-ME-WE3 extending the original area and linking Western Europe to the Far East and Australia. This is 39,000 km in length and improvements in cable quality ensure increased capacity and quality of reception, especially over long distances. SEA-ME-WE4 was developed by 16 telecommunications companies spread across Europe, the Middle East and South-east Asia. This is 18,800 km in length and offers high-speed transmission between the linked countries designed to meet demand in countries with growing economies.

Did you know ??????

Submarine cables can be accidentally damaged. On 26 December 2006, an earthquake damaged SEA-ME-WE3 cables off Taiwan and, on 30 January 2008, a ship's anchor damaged the SEA-ME-WE4 off Egypt.

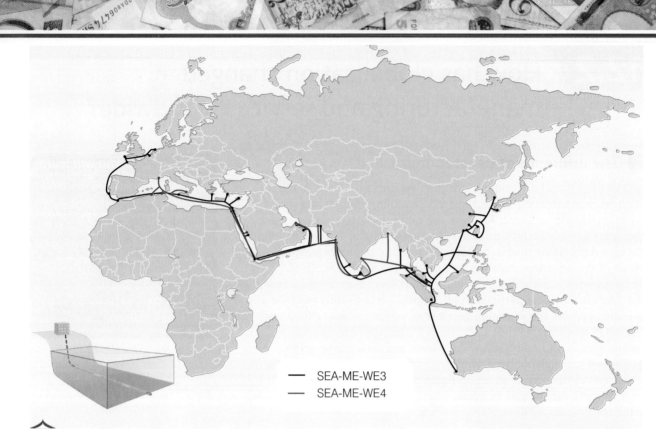

SEA-ME-WE3
SEA-ME-WE4

B *Submarine cables, 2008*

■ Small industrial region with global connections

Developments in ICT have allowed immediate access to people all over the world and fostered developments in small areas, knowing that communication across the world is possible. This has led to a cluster of world-famous companies associated with motor racing in an axis between Northampton and Oxford in the southern part of the British Midlands (map **C**).

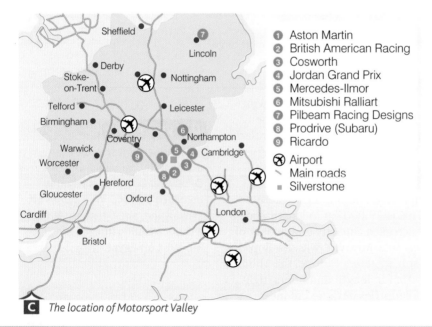

1 Aston Martin
2 British American Racing
3 Cosworth
4 Jordan Grand Prix
5 Mercedes-Ilmor
6 Mitsubishi Ralliart
7 Pilbeam Racing Designs
8 Prodrive (Subaru)
9 Ricardo
✈ Airport
⟍ Main roads
■ Silverstone

C *The location of Motorsport Valley*

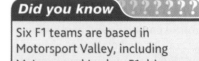

■ The development of call centres abroad

In 2000, an article in *Geography Review* stated 'it seems unlikely that British-based firms would establish **call centres** overseas to reduce their wage bills'. A number of years on, the reverse has been seen to be true. The setting up of call centres is big business and a total of 400,000 people are employed in the UK, often in small towns such as Harrogate, Carlisle, Gateshead and Warrington. Banks and other finance companies such as insurance were among the first to develop centralised call centres and the first to look abroad. Household names such as ASDA, Tesco, BA, BT, Barclays, Lloyds TSB, HSBC and Virgin Media have all set up call centres in India. It is the big cities that house these, such as Mumbai, Delhi, Hyderabad and Bangalore (map **D**). Other important destinations for call centres abroad are South Africa and the Philippines.

AQA *Examiner's tip*

Learn the factors behind the development of call centres in countries such as India.

Key terms

Call centres: offices where groups of people work responding to telephone queries from customers. Employees sit in front of a computer monitor giving them information that they use in their answers to questions.

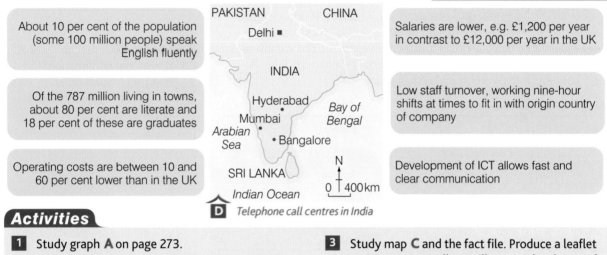

About 10 per cent of the population (some 100 million people) speak English fluently

Of the 787 million living in towns, about 80 per cent are literate and 18 per cent of these are graduates

Operating costs are between 10 and 60 per cent lower than in the UK

Salaries are lower, e.g. £1,200 per year in contrast to £12,000 per year in the UK

Low staff turnover, working nine-hour shifts at times to fit in with origin country of company

Development of ICT allows fast and clear communication

PAKISTAN   CHINA
Delhi ■
INDIA
Hyderabad •   Bay of
Mumbai •   Bengal
Arabian
Sea   • Bangalore
N
SRI LANKA
0 ┬ 400km
Indian Ocean

D *Telephone call centres in India*

Activities

1 Study graph **A** on page 273.

a What is the significance of the 100 value on the *y*-axis?

b What happened to the cost of a transatlantic phone call between 1930 and 1950?

c State two changes in communication costs between 1930 and 2000.

d Give evidence from the graph to support the changes described in c.

2 Study map **B**.

a Describe the pattern of submarine cables shown.

b On a world map outline, draw the routes of SEA-ME-WE3 and SEA-ME-WE4.

c Label the map to show the routes used and the key differences between the two cables.

d Briefly explain the importance of submarine cables to globalisation.

3 Study map **C** and the fact file. Produce a leaflet on Motorsport Valley to illustrate the cluster of motor sports industry and the reasons for it. Your leaflet should include the following:

■ a labelled sketch map to show the advantages of the location

■ a labelled photo of one of the companies present

■ five good reasons for locating an automotive industry here, including one that relates to ICT developments and the need for international communications.

4 Study map **D**. Work in pairs for this activity. Imagine you are director of customer relations for a call centre in India. You are being interviewed about your reasons for choosing to locate abroad for national television news. Present the questions you were asked and your answers to them.

∞links

You can find out more about submarine cables at www.telegeography.com.

More information on Motorsport Valley can be found at www.thebritishmidlands.com.

Visit www.call-centres.com for more on call centres.

What are the advantages and disadvantages of TNCs?

Transnational corporations (TNCs) are large, wealthy organisations. They are companies that have their headquarters in one country, but often have many other branches spread across much of the world. As a result of improvements in transport and communications, TNCs have grown steadily over the last 30 years. Most TNCs have their headquarters in richer areas of the world, especially the USA, UK, France, Germany and Japan. Research and development is usually centred here. Production often occurs in poorer areas where labour costs are lower, laws more lenient or where governments want to seek investment. Production also occurs in richer areas where benefits include a skilled workforce and incentives are often available to attract new investment. The wealth of many TNCs emphasises poverty in certain parts of the world. Table **A** shows the biggest TNCs according to their value.

A *The top ten non-financial TNCs, 2002*

Rank	TNC	Headquarters	Product
1	General Electrics	USA	Aero-engines, engineering
2	Vodafone	UK	Telecommunications
3	Ford	USA	Vehicles
4	British Petroleum	UK	Oil-based activities
5	General Motors	USA	Vehicles
6	Shell	Netherlands/UK	Oil-based activities
7	Toyota	Japan	Vehicles
8	Total, Fina, Elf	France	Oil-based activities
9	France Telecom	France	Telecommunications
10	Exxon	USA	Oil-based activities

TNCs offer many advantages to the countries where they set up branches. They provide jobs in factories making supplies and in services where products are sold. The additional income that people have encourages and benefits local businesses, creating a **multiplier effect**. Training of the workforce leads to the development of skills and companies bring with them new technologies as well as money to invest. Often, the infrastructure is improved as better access, both within and between countries, and communications are needed. Some of the goods may be exported to improve trade or may meet a demand within the country.

However, there are disadvantages. A significant one is that of **leakage**, as well as the fact that in some locations wages are very low and key jobs go to outsiders. If there are problems worldwide economically or within the company, the branch plants are susceptible to closure, especially those that are not performing well. This may mean that incentives to attract companies are not well spent if companies are there only a short time. The government has no say in deciding on the

In this section you will learn

what advantages and disadvantages result from the presence of TNCs in countries

the characteristics of Toyota as a case study of a TNC.

Key terms

Transnational corporation (TNC): a corporation or enterprise that operates in more than one country.

Multiplier effect: where initial investment and jobs lead to a knock-on effect, creating further jobs and providing money to generate services.

Leakage: where profits made by the company are taken out of the country to the country of origin and so do not benefit the host country.

Did you know ??????

The income of Ford and General Motors added together is greater than the GDP of all of sub-Saharan Africa.

AQA **Examiner's tip**

Make sure that you can explain at least three advantages and three disadvantages of TNCs.

future of the TNCs. In some areas, working conditions can be poor and the labour force is expected to work long hours, with little time for breaks. Safety can be compromised where laws are less stringent and workers' health can be jeopardised. The same is true of air and water pollution, where higher levels may be acceptable.

Did you know ??????

On 3 December 1984 a gas leak from a chemical factory in Bhopal, India killed 3,000 people immediately and has since led to the deaths of 15,000 others. Some 100,000 people suffered chronic illness as a result of one of the world's worst industrial accidents.

Case study

Toyota

Toyota began in Toyota, Aichi, Japan in 1937. Seventy years later, it had become the biggest producer of cars in the world, with profits of $11 billion in 2006. Map **B** shows (in red) the countries where Toyota plants are present and table **C** shows production in different regions in 1996 and 2005.

B *The global location of Toyota*

C *Toyota production by region, 1996 and 2005 (1 = 1,000 vehicles)*

Region	1996	2005
North America	782.9	1,535.1
Latin America	3.2	138.5
Europe	150.3	638.2
Africa	85.0	121.1
Asia	257.0	1,029.2
Oceania	67.6	102.2
Overseas total	1,346.0	3,571.3
Japan	3,410.1	3,789.6
Worldwide total	4,756.1	7,360.9

Toyota began to develop overseas in the late 1950s in Brazil. From its headquarters in Tokyo, a huge operation is managed with 250,000 workers employed in 26 countries. The company decided to seek to develop production in the UK in the early 1990s. The reasons for this are given in a statement on the company's website (extract **D**). The location of the Burnaston site where cars were first produced in 1992 is shown in map extract **E** on page 278 and a photo of the site is shown in photo **F**.

◄ ► C + http://   Q-

Home | Contact | Sitemap | News | Forum | Shop

News

Home
Contact
Sitemap
News
Forum
Shop

So, having established the logic behind building in Europe, why the UK? High up on the list of reasons was the strong tradition of vehicle manufacturing in Britain and the large domestic market for our product. In addition, the UK offered us solid industrial transport links to our customers and our 230 British and European supply partners. Another reason was the excellent workforce and favourable working practices.

There was also a supportive positive attitude to inward investment from the British government at both local and national level. Also of course the English language is very much the second language in Japan, making communication and integration so much easier. In December 1989 Toyota Manufacturing UK was established. Another point to remember is that the location of the engine plant alongside the vehicle manufacturing plant at Burnaston would have reduced Toyota's long-term capacity to expand at that site and indeed would have restricted the opportunity for both plants to expand.

D *Reasons for a UK location, taken from www.toyotauk.com*

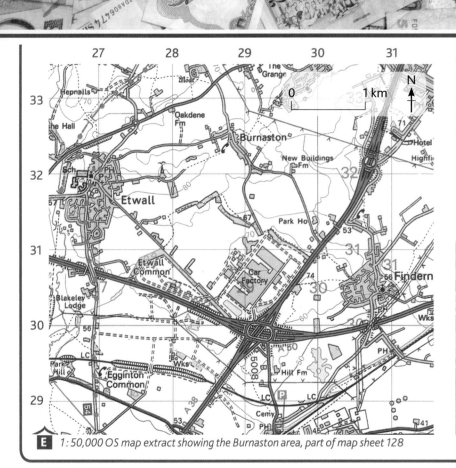

∞ links

You can find out more about Toyota at www.toyotauk.com.

Research the Bhopal gas leak by searching on www.bbc.co.uk.

E 1 : 50,000 OS map extract showing the Burnaston area, part of map sheet 128

F The Toyota plant at Burnaston

Activities

1 Study table **A** on page 276. Work with a partner.

a Find out the identity of a further 10 large TNCs.

b Produce a collage to display the name, logo and line of business in which they are involved. Try to do this in a limited amount of time – say, 20 minutes – to see how you can work under pressure.

2 On a large copy of the table below, summarise the economic, social and environmental advantages and disadvantages of TNCs. One category has been partly done for you to illustrate the types of answers you could include.

3 Study map **B** and table **C** on page 277.

a With the help of an atlas, on an outline world map label 10 countries where Toyota is present. Make sure your labels reflect the global presence of Toyota.

b On your map, draw located bars to show production in each region for 1996 and 2005.

c Describe the pattern shown by your map in 2005.

d Summarise the changes made between 1996 and 2005.

e Where and why do you think Toyota should locate future new branches?

4 Study extract **D** on page 277, map extract **E** and photo **F** above.

a Draw and label a sketch map to describe and explain why the location at Burnaston is a good one.

b What advantages did Toyota believe existed in deciding to locate in UK?

c Suggest some advantages of encouraging major companies such as Toyota to locate factories in the UK.

Category	Advantages	Disadvantages
Economic		
Social	People may feel more secure with a job and regular income.	Working conditions are poor. Workers have to suffer crowded, hot conditions and work for many hours at a stretch.
Environmental		

12.4 How and why is manufacturing in different countries changing?

■ Changes in relative importance of world regions

Manufacturing industry has declined in importance in some regions, while it has become more significant in others. The changes in high-tech manufacturing are shown in graph **A**. In many of the richest areas of the world, manufacturing has declined. For example, in Britain the number of people employed in manufacturing fell from just over 6 million in 1981 to 3.5 million in 2003. This is the result of **de-industrialisation**, which has occurred because of increased mechanisation and the need for industry to be competitive.

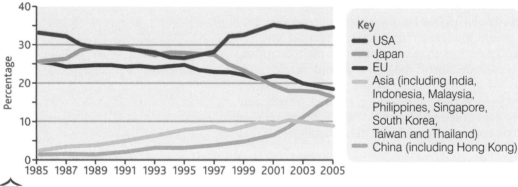

Key
- USA
- Japan
- EU
- Asia (including India, Indonesia, Malaysia, Philippines, Singapore, South Korea, Taiwan and Thailand)
- China (including Hong Kong)

A *World share of high-tech manufacturing, 1985–2005*

Reasons for changes

There are many reasons why some areas have experienced growth in manufacturing industry whereas others have suffered a decline.

Government legislation

This can take many forms, such as:

- setting up areas (**assisted areas/enterprise zones**) where conditions are favourable for new industry
- providing **advanced factories** of various sizes
- offering retraining and removal expenses
- ensuring educational reform is high on the list in areas such as the four Asian 'tigers'
- adding taxes to home-produced goods so that exports are targeted.

Some countries have a minimum wage. In the UK, this is £5.73 for those aged 22 years and older, but in Sri Lanka garment workers are often paid a fraction of a much lower minimum wage.

An EU directive limits the maximum number of hours worked per week to 48 (except in the UK). The average South Korean works 2,390 hours a year in contrast to 1,652 in UK. This has real implications for production. In Sri Lanka, garment workers should not work after 10pm due to International Labour Organization rules, but they are often forced to do so by their employers.

Key terms

De-industrialisation: a process of decline in certain types of manufacturing industry, which continues over a long period of time. It results in fewer people being employed in this sector and falling production.

Assisted areas/enterprise zones: areas that qualify for government help. Enterprise zones are on a smaller scale than assisted areas.

Advanced factories: where buildings for production are built speculatively in the hope that their presence will encourage businesses to buy or rent an existing factory, removing the need to find a site or suitable premises.

Health and safety regulations

Working conditions tend to vary globally. In the UK, adult employees working over six hours are entitled to a 20-minute break. Workers have the right to:

- know how to do their job safely and to be trained to do so
- know how to get first aid
- know what to do in an emergency
- be supplied with protective clothing.

Such regulations often do not exist in some poorer countries or are not enforced. Some workers – unable to travel home after a shift and arrive back on time for the next shift – sleep on the factory floor, although this is illegal.

Prohibition of strikes

In the 1970s, there was much unrest in the UK and various trade unions frequently called strikes. The so-called 'winter of discontent' began with strike action by 15,000 Ford workers in September 1978. There were many groups that joined in and power cuts became the norm. Such disruption had an adverse effect on manufacturing industry and led to the reduction of the power of trade unions. Companies such as Nissan and Toyota only came to the UK on the understanding that strike action would not be allowed. This is also true of many newly industrialising countries (NICs). Trade unions are allowed in Sri Lanka, but workers are likely to be threatened if they join one.

Tax incentives and tax-free zones

Tax incentives take a variety of forms, but all seek to offset costs. For example, One NorthEast (the development agency responsible for the north-east of England) offers job-creation grants, business rate or rent-free periods and help in preparing a business plan while the government gives a grant towards the cost.

Tax-free zones are designated as free from the paying of tax and include zones such as areas of Dubai and Zambia.

■ China: the new industrial giant

China has been dubbed 'the new workshop of the world' (photo **B**), a phrase that was first used with reference to Britain during the 19th century. Chinese steel production in 2003 was thought to be 25 per cent of total world output. China makes 60 per cent of the world's bicycles (photo **D**) and 50 per cent of the world's shoes. One-fifth of all garments exported in 2003 were Chinese and this is expected to rise to 50 per cent in 2010. Between 2000 and 2006, cloth manufacture more than doubled and car production increased by more than six times, while mobile phone ownership increased nine times. The National Bureau of Statistics of China is predicting economic growth of at least 7 per cent to 2018.

B *China: the new workshop of the world*

C Chinese knitwear factory

D 60 per cent of the world's bicycles are made in China

Reasons for China's rapid growth

There are many reasons why China is emerging as the new economic giant.

Government legislation

In 1977, Deng Xiaoping, the successor to Chairman Mao, changed China's policy as he sought to end China's isolation and stimulate Chinese industry. Foreign investment was encouraged but the government maintained overall control over the economy so that China would gain maximum benefit. The setting up of special economic zones (SEZs) between 1980 and 1994 paved the way for foreign investment (map **E**). Here, there were tax incentives to attract interest from outside China. Open cities were also established such as Shanghai and Tianjin. Both these measures were significant as they signalled China's position in openly wanting foreign investment and companies in China and a desire for rapid economic and industrial growth.

The one-child policy, which was introduced in 1979, meant that the population level was much lower than it would have been (estimated at over 2 billion) and people's desires changed. There was an increased demand for electrical household goods, air conditioning, cars and computers.

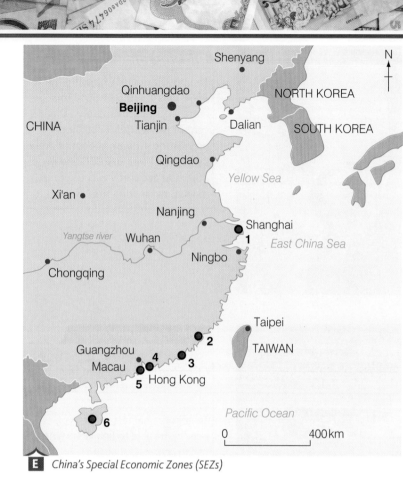

Special Economic Zones (SEZ)
1 Pudong District, Shanghai Muncipality
2 Xiamen, Fujian Province
3 Shantou, Guangdong Province
4 Shenzhen, Guangdong Province
5 Zhuhai, Guangdong Province
6 Hainan Province

E *China's Special Economic Zones (SEZs)*

The home market

China's large and increasingly wealthy population – per capita income in urban households was about £600 in 2001, having risen from just below £200 in 1993 – offers much potential, as do other Asian Pacific areas. China's ambitious urban population numbers some 500 million people.

The Olympics factor

The 2008 Olympics were held in Beijing. This provided China with the perfect opportunity to showcase the nation. The opening ceremony, based on the theme 'One World, One Dream', was important in an attempt to convey China as an open, friendly country and an important, integrated, positive part of the world in the 21st century. The prestige of hosting the games and the image portrayed will be immensely important in stimulating further foreign investment.

The Three Gorges Dam

Industrial development on a large scale demands large resources of energy. China currently generates two-thirds of its electricity at coal-fired power stations. Many new plants are being built. Hydroelectric power (HEP) accounted for 7 per cent of electricity in 2006. China produces more HEP than any other country in the world and is keen to develop new sources of energy. The Three Gorges Dam is the biggest in the world, generating 22,500 mW when fully operational. Together with the development of navigation along the Yangtse, the dam has led to much development.

F *The Olympics in Beijing*

G   *The Three Gorges Dam*

Other factors

Cheap labour is a key reason – wages are 95 per cent lower than in the USA.

Activities

1   Study graph **A** on page 279.

a   What percentage of high-tech manufacturing was produced in the EU in 1985 and 2005?

b   What percentage of high-tech manufacturing was produced in China in 1985 and 2005?

c   Using evidence from the graph, summarise the trends shown.

2   Study map **E**.

a   Draw a sketch map to show the location of some of the main industrial areas and other important items of information that help to explain China's emergence as the new economic giant.

b   Summarise in a series of bullet points the reasons why China has this position. Remember to refer specifically to information about China as evidence.

∞ links

There are many statistics relating to China that can be found in the National Bureau of Statistics of China at **www.stats.gov.cn**.

Research information about China and economic policy, including SEZs, at **www.china.org.cn**.

You can find out more about the Three Gorges Dam at **www. timesonline.co.uk**.

What are the causes and effects of increasing global demand for energy?

There are a number of reasons why there is an increase in demand for energy. Some of these reasons will be considered separately but, in reality, there are links between them. The increasing world population is getting richer and advances in technology increase the availability of products and create an insatiable thirst for energy.

World population growth

Table **A** and graph **B** show the overall global increase in population and the relative importance of different regions.

Increased wealth

A	Overall global increase in population, 1750–2050						
Year	1750	1800	1850	1900	1950	1999	2050
World population (billions)	0.79	0.98	1.26	1.65	2.52	5.98	8.91

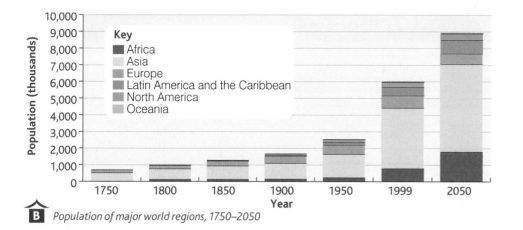

B Population of major world regions, 1750–2050

Consider the following list of things your family might have: a fridge, washing machine, tumble drier, dishwasher, microwave, television, computer, car. The chances are that you will have many, if not all, of these items. This was not always the case. As people have got more money, they look around at what there is to buy that will make their lives more comfortable and desire to own such goods. These goods use energy not only to operate them but also in their manufacture, so increasing personal wealth increases our demand for energy.

Consider the following facts and the impact of these on demand for energy:

- UK average earnings increased to £457 per week in 2007, up by 2.9 per cent from 2006.
- The average wage in China has risen to 1750 yuan a month, four times higher than in 1995.

- Private car ownership in China increased from virtually zero in 1997 to 24 million in 2005. Seven million cars were sold in 2006.

- In the UK, the number of families not owning a car fell from 32 per cent to 27 per cent and the number owning two or more cars rose from 24 per cent (1991) to 29 per cent (2001).

Technological advances

Technological advances have supplied us with increasing amounts of energy and a wide variety of goods that we can purchase. The development of steam using coal led to large-scale production in the UK. It is the use of coal in power stations that has fuelled the Chinese economy. Modern technology allows the development of other sources of energy and a cycle of demand and supply is generated. Research and development is big business. Companies strive to compete with each other and produce smaller mobile phones, faster laptops, in-house entertainment systems that offer best-quality pictures, most innovative games, etc. People see these and want to buy them. This requires energy to develop, make and run the consumer products that are so much a part of our lives today.

C   *Smog in Beijing, home of the 2008 Olympics, before the games*

Social, economic and environmental impacts

The social effects of increased energy use have an impact on people's health. In countries like the UK, the incidence of lung-related diseases such as emphysema and bronchitis was traditionally associated with industrial areas where coal provided the basic source of energy. The incidence of these diseases will increase in areas that are experiencing industrialisation today.

In addition, some cities are shrouded in a haze that blocks out the sun and contains a dangerous mix of chemicals, including those from coal smoke and ozone. Photo **C** shows Beijing before the 2008 Olympic Games, when 1.3 million of the city's 3 million cars were taken off the road and 100 factories were closed. Poor air quality leads to asthma and other respiratory diseases.

How people travel to work may also be affected as people choose to leave cars at home and cycle to work in an attempt to be environmentally friendly and fitter. Where we go on holiday could also be affected if air fares increase in the face of rising fuel costs. More people could choose to holiday at home. Economic effects are clear to see with the cost of petrol soaring in the UK in the early months of 2008.

Environmental effects have had an impact:

- on land, where spoil heaps have built up adjacent to coal mines when unneeded material has been dumped

Did you know ??????

At the opening ceremony of the Beijing Olympics on 8 August 2008, a British smog monitoring team had to close down its website after it recorded an air pollution index (API) of between 101 and 150 micrograms per m^3 (100 is regarded as safe), whereas the official Chinese figures showed 95.

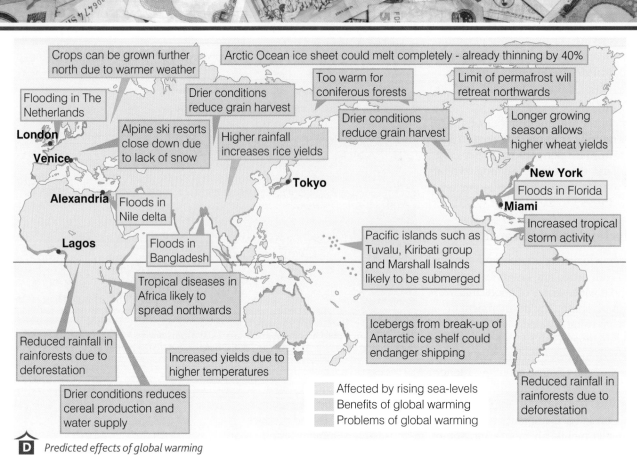

Crops can be grown further north due to warmer weather

Arctic Ocean ice sheet could melt completely - already thinning by 40%

Too warm for coniferous forests

Limit of permafrost will retreat northwards

Flooding in The Netherlands

Drier conditions reduce grain harvest

Drier conditions reduce grain harvest

Longer growing season allows higher wheat yields

London

Alpine ski resorts close down due to lack of snow

Higher rainfall increases rice yields

Venice

Tokyo

New York

Alexandria

Floods in Nile delta

Floods in Florida

Miami

Lagos

Floods in Bangladesh

Pacific islands such as Tuvalu, Kiribati group and Marshall Isalnds likely to be submerged

Increased tropical storm activity

Tropical diseases in Africa likely to spread northwards

Reduced rainfall in rainforests due to deforestation

Increased yields due to higher temperatures

Icebergs from break-up of Antarctic ice shelf could endanger shipping

Reduced rainfall in rainforests due to deforestation

Drier conditions reduces cereal production and water supply

- Affected by rising sea-levels
- Benefits of global warming
- Problems of global warming

D *Predicted effects of global warming*

- on water, where the transportation of oil has led to major pollution incidents such as the *Exxon Valdez* oil spill off Alaska in 1989 and the *Prestige* sinking off the coast of north-west Spain in 2002

- on air, where poor quality is responsible for ill health on a local scale and for substantial effects on a global scale, where global warming is seen to be the main result.

The predicted worldwide impact of global warming is shown in diagram **D**.

∞links

The Environment Agency gives extensive advice at **www. environment-agency.gov.uk**.

You can investigate air quality at **www.airquality.co.uk** and global warming at **www. worldviewofglobalwarming.org**.

Activities

1 Study table **A** on page 284.

a Draw a line graph to show the total world population figures from 1750 to 2050.

b Describe the trends shown by the graph you have drawn.

2 Study graph **B** on page 284.

a Summarise the changes in population in the different world regions.

b Explain how the trends account for an increase in the demand for energy. (You may want to refer to section 12.4 on China here.)

3 a List the products you and your family own that use energy.

b Ask your parents to highlight the items they would not have owned 10 and 20 years ago.

c How will the amount of energy used by you and your family have changed?

4 Explain how increasing levels of wealth and technological advances have resulted in an increasing demand for energy. Try to illustrate your answer using your own experience.

5 Study photo **C** and diagram **D**. Work in pairs. Produce a poster that summarises the social, economic and environmental impacts of increased energy use. You should include at least one map, one photo and evidence of further research. There should be reference to at least two effects in each category. Try to include facts, figures and information as evidence wherever possible.

12.6 How can energy use be sustainable?

■ The use of renewable energy

Renewable energy promises sustainable development. Our reliance on fossil fuels to meet our energy needs is only short term. Supplies of coal, oil and natural gas are limited and renewable alternatives have been used for some time, albeit on a relatively small scale. Hydroelectric power (HEP), solar, tidal and wind power are sustainable options (fact file **A**). They share similar advantages over the use of fossil fuels, but are not without their opponents. The recent surge in the use of **biofuels** also appears at first to offer this possibility, but there is debate about this.

Fact file

Wind power in the UK

- **Aims**: to generate 10 per cent of power by renewable energy sources.

- **Role**: to be responsible for one-third of electricity generated.

- **Number of wind farms and turbines operational in 2008**: 176.

- **Location in 2008**: mostly onshore.

- **The future**: offshore wind farms to become more important, with 8 GW to be generated by 2016.

- **Size of a modern wind turbine**: 100 to 120 m, including the blades. Offshore wind turbines tend to be the largest. For comparison, the London Eye is 135 m high.

- **Offshore operations**: these are usually within 4 to 6 km off the coast, but some planned developments will be over 18 km.

- **Location requirements**: the shallow waters off the coast of the UK are an advantage. Wind farms require an exposed location, whether onshore or offshore, clear of any obstructions (such as buildings onshore). Small differences in distance can mean a real difference in the potential of the site, e.g. a site that is 10 per cent less windy means 20 per cent less energy is generated.

- **Opinions**: critics oppose the development of wind farms on the grounds of noise, their perceived harmful effects on the environment and on birds and wildlife. Noise levels for a distance of 350 m are measured at 35 to 45 dB, while a busy office is measured at 60 dB. Research suggests that turbines are avoided by migrating birds but that, offshore, they can attract fish and have no long-term effects on other sea life. House prices nearby may reduce initially, but generally recover. The cost of generating electricity is more expensive than traditional methods (more than double per kW hour).

In this section you will learn

how the use of renewable energy is sustainable

why some locations are more suited to the development of renewable energy than others and the debate over locations and costs

the importance of international directives and local initiatives in the drive to ensure sustainability.

Key terms

Biofuels: the use of living things such as crops like maize to produce ethanol (an alcohol-based fuel) or biogas from animal waste. It is the use of crops that has become especially important.

Did you know ???????

Biofuels are sometimes known as 'deforestation diesel'. The grain needed to fill the tank of a large Range Rover is enough to feed one person for a year. The EU has a target of 5.75 per cent biofuels for transport by 2010.

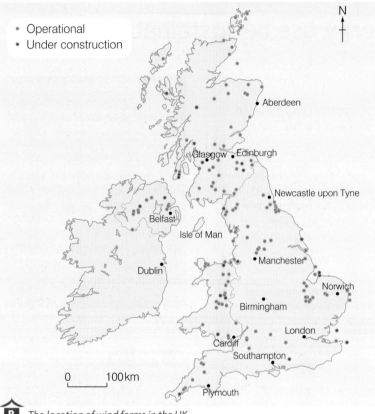

C *Burbo Bank wind farm, Liverpool Bay*

B *The location of wind farms in the UK*

▉ The importance of international directives

As well as ensuring the supplies of energy resources in the long term, there is also a need to care for the environment. Air pollution knows no bounds; it does not stop at international borders and therefore cooperation is needed between countries worldwide if issues relating to air quality and global warming are to be effectively addressed. The Earth Summit in Rio de Janeiro in 1992 marked the first real international attempt to cooperate to reduce emissions. Richer countries agreed there would be no increase in emissions.

The **Kyoto Protocol** in 1997 went further, with an agreement by industrialised countries to reduce greenhouse gas emissions to 5 per cent below 1990 levels between 2008 and 2012. EU countries as a whole should show an 8 per cent reduction, but individual countries have their own targets. The treaty became legally binding in 2005, when enough countries responsible for 55 per cent of the total emissions had signed. The USA has declined to sign the agreement but Australia signed in November 2007, bringing the total number of countries to 181. The poorer nations, including those with many industries, do not have to reduce their emissions. Countries can trade in their carbon credits – the amount of greenhouse gases they are allowed to emit. Therefore, countries putting more pollution into the atmosphere than they should can buy carbon credits from a country below its agreed level.

The Bali Conference in December 2007 sought to establish new targets to replace those agreed at Kyoto. No figure was decided, only a recognition that there would need to be 'deep cuts in global emissions'. The USA agreed to support the Bali 'roadmap', designed to lead the way into the future. Countries have until the 2009 UN climate conference in Denmark to finalise new targets and detail how they will deal with the impacts of global warming.

Local initiatives

The phrase 'think globally, act locally' indicates the need for individuals and groups to seek to reduce pollution and to take responsibility for this. Reducing the use of resources not only increases their life, it also reduces pollution and energy in production. We can seek to seize the initiative by conservation and recycling, and therefore reducing waste and the need for landfill.

Conservation can involve simple things like turning off lights and appliances when they are not being used, filling a kettle with only the water that is needed rather than to the top and buying reusable carrier bags rather than accepting free plastic bags (where these are still available).

Local authorities provide a variety of recycling containers for paper, cans, glass, plastic, cardboard and garden waste, and many encourage composting in an attempt to reduce waste thrown into bins. This in turn reduces the amount that is put into landfill. Figures from the Department for Environment, Food and Rural Affairs (Defra) show that in 2005–06 just over a quarter of household waste was recycled – substantially short of the 40 per cent target for 2010. We are also running out of space for landfills and it is expected that by 2015 some areas will struggle to find suitable sites.

AQA *Examiner's tip*
Be able to give examples of attempts at achieving sustainable development operating at international and local scales.

∞ links

You can investigate biofuels by searching www.bbc.co.uk. This website is also useful for information on offshore wind farms.

The British Wind and Energy Association website at www.bwea.com gives lots of information on wind farms in the UK.

Local authority websites provide information on recycling and waste disposal.

Activities

1 Study map **B**.
a Describe the distribution of wind farms shown on the map.
b Explain reasons for the distribution you have described.

2 Study photo **C**. Draw a labelled sketch to show the characteristics of an offshore wind farm and the advantages of its location.

3 Different people have contrasting views on the development of wind farms such as Burbo Bank (photo **C**) and onshore wind farms.
a Choose one of the following who you think would support wind farms and one who you think would oppose them.
  i a resident overlooking the new development in favour of renewable energy
  ii a NIMBY ('Not In My Back Yard') supporter of wind energy
  iii an electricity-generating company director
  iv a member of the Royal Society for the Protection of Birds (RSPB)
  v a member of Greenpeace for action against climate change
  vi a resident who wants cheap electricity.
b From the point of view of each person, give your views explaining fully why you are for or against the development of wind farms.

4 a Explain the importance of an international approach to reduce air pollution.
b In your view, how successful have international initiatives been? Provide evidence to back up your answer.

5 Use your own experience of where you live and research what your local authority does to encourage energy conservation. Work in pairs. You have been commissioned by your local council to produce a leaflet encouraging people to conserve energy. Design and present your leaflet so that it is informative, clear, well illustrated and eye-catching.

Food production has increased significantly throughout the world in the last 50 years (graph **A**). The quest to increase food production worldwide – whether out of necessity for survival or to sell to areas where food items are desired, but otherwise unobtainable all year round – has had significant impacts. These can be categorised as environmental, political, social and economic.

Environmental impacts

Transporting food longer distances – the idea of **food miles** – increases our **carbon footprint**. People in the UK demand out-of-season produce and this demand is met by importing food; indeed, our overall demand for food is largely met via imports. One half of vegetables and 95 per cent of fruit comes into this category. It is produce which is imported by air that attracts most attention and concern. We import 1 per cent of food by air, but this accounts for 11 per cent of carbon emissions resulting from transportation of UK food. Therefore, the more we rely on imported food – especially air-freighted food – the more we are contributing to air pollution and global warming.

Beans in Kenya are produced in a highly environmentally friendly manner. 'Beans there are grown using manual labour – nothing is mechanised,' says Professor Gareth Edwards-Jones of Bangor University, an expert on African agriculture. 'They don't use tractors, they use cow muck as fertiliser; and they have low-tech irrigation systems in Kenya. They also provide employment to many people in the developing world. So you have to weigh that against the air miles used to get them to the supermarket.'

When you do that you make the discovery that air-transported green beans from Kenya could actually account for the emission of less carbon dioxide than British beans. British beans are grown in fields on which oil-based fertilisers have been sprayed and which are ploughed by tractors that burn diesel. According to Gareth Thomas, Minister for Trade and Development: 'Driving six and a half miles to buy your shopping emits more carbon than flying a pack of Kenyan green beans to the UK.'

Apples are harvested in September and October. Some are sold fresh while the rest are chill stored. For most of the following year, they still represent good value (in terms of carbon emissions) for British shoppers. However, by August those Cox's orange pippins and Braeburns will have been in store for 10 months. The amount of energy used to keep them fresh for that length of time will then overtake the carbon cost of shipping them from New Zealand. It is therefore better for the environment if UK shoppers buy apples from New Zealand in July and August rather than those of British origin. In Britain, lettuces grown in winter are produced in greenhouses or polytunnels that require heating. At those times it is better – in terms of carbon emissions – to buy field-grown lettuces from Spain. However, in summer, when no heating is required, British is best.

In this section you will learn

how efforts to meet the global demand for food can have both positive and negative effects

how these impacts can be classified as environmental, political, social and economic, and that at times the categories overlap.

Key terms

Food miles: the distance that food items travel from where they are grown to where they are eaten.

Carbon footprint: the amount of carbon generated by things people do, including creating a demand for out-of-season food.

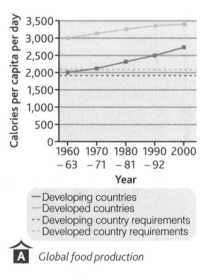

A Global food production

Did you know ?????

A quarter of Europe's cut flowers are supplied by Kenya in an industry that was worth £77 million in 2003 and 8 per cent of Kenya's export earnings.

As populations increase in certain areas of the world, there is pressure to increase food production. This may involve farming on land that is not really suitable, but will just suffice, providing minimal returns. Photo **B** illustrates the difficult conditions in which people try to grow food in order to survive. Here, the already poor-quality land is likely to become even poorer. As the meagre crops are harvested, no goodness is returned to the soil and so it becomes exhausted. The lack of vegetation cover makes the area prone to soil erosion, where it is easily washed or blown away (diagram **C**).

Political impacts

B Marginal land in Darfur

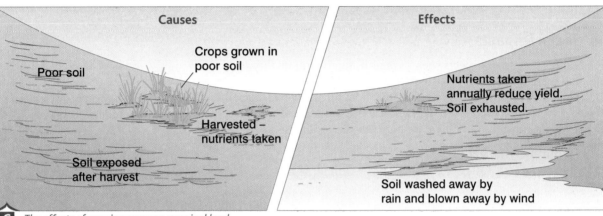

C The effects of growing crops on marginal land

Water is essential for farming and food production. In areas where water is limited or available only seasonally, irrigation is essential. This means having access to a permanent water supply – usually a river – controlling its flow and distributing water to where it is needed.

The River Indus is a river that has this role in parts of northern India and Pakistan. The flow of the Indus is seasonal: a huge amount of water results in flooding in summer, whereas flows are much less in winter due to seasonal variation in rainfall. The Indus supplies the fertile Punjab, which crosses the India–Pakistan border.

Did you know ??????

The River Indus is 3,200 km in length, making it the 24th longest river in the world.

After the Independence Act of 1947, it took 13 years of negotiation before the Indus Waters Treaty was signed by the two countries (map **D** on page 292). The disputed region of Kashmir lies within the drainage basin, where territory is not the only issue. There had been concerns that during times of conflict India could build dams and cut off water to Pakistan, perhaps even diverting rivers. The signing of the Indus Waters Treaty in 1960 meant that Pakistan had the westward-flowing rivers and India had the eastward-flowing rivers. The construction of two dams on the Jhelum and Indus Rivers gave Pakistan water independent of upstream control by India. However, there is resentment in that part of Kashmir in India as people believe that farming and irrigation has been limited due to them being deprived of water that should have been theirs. In June 2006, talks about the Wuller Barrage that India wants to build on the Jhelum River for navigation raised fears about India controlling Pakistan's water.

D The Indus Waters Treaty

India:
1 Sutlej
2 Beas
3 Ravi

Pakistan:
4 Indus
5 Chenab
6 Jhelum

E The Indus River

Social impacts

Social issues relate to aspects such as health, safety and quality of life, and include the right to a clean water supply. Growing cash crops for a source of income can offer economic benefits. However, socially there are potentially more problems. The areas around Lake Naivasha in Kenya and north of Mt Kenya are home to the profitable flower industry (diagram **F**).

∞links

See spread 10.7 on p.240 for more on flower production in Kenya

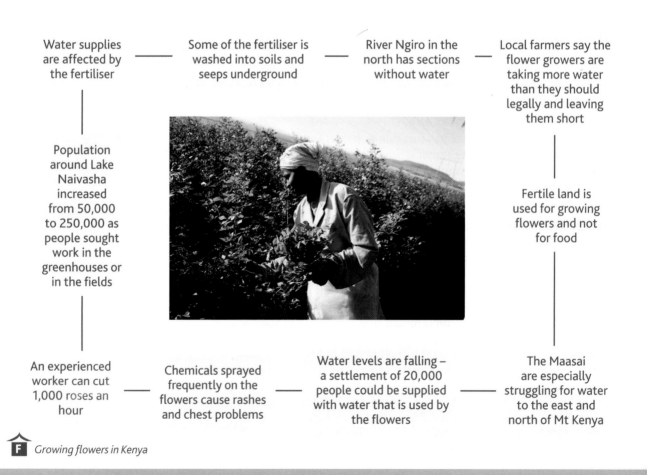

Water supplies are affected by the fertiliser

Some of the fertiliser is washed into soils and seeps underground

River Ngiro in the north has sections without water

Local farmers say the flower growers are taking more water than they should legally and leaving them short

Population around Lake Naivasha increased from 50,000 to 250,000 as people sought work in the greenhouses or in the fields

Fertile land is used for growing flowers and not for food

An experienced worker can cut 1,000 roses an hour

Chemicals sprayed frequently on the flowers cause rashes and chest problems

Water levels are falling – a settlement of 20,000 people could be supplied with water that is used by the flowers

The Maasai are especially struggling for water to the east and north of Mt Kenya

F Growing flowers in Kenya

Economic impacts

Growing cash crops as well as food crops is often the way forward for many small-scale farmers – some even seek to sell any surplus of food they produce. Additional cash allows investment in the farm and other items to be bought. However, there are problems. There is often a need to intensify production, which means increasing the use of fertilisers and pesticides. These cost money and a vicious circle can be set up (diagram **G**).

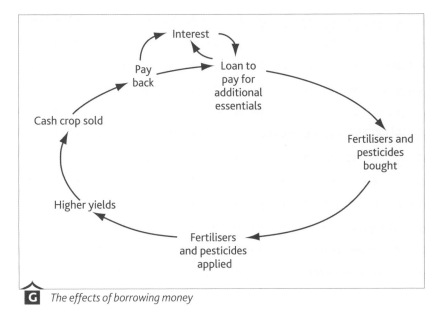

G   *The effects of borrowing money*

Buying locally produced food

In the UK, farmers benefit from local people buying food they produce and there are arguably more widespread economic, social, political and environmental benefits. Buying from local, or indeed regional or national sources, should benefit the domestic farming industry and address environmental concerns. Others might argue from a different perspective. What would the effect on Kenya be if the UK stopped importing strawberries, for example?

In the UK, we can ensure that we support local produce by:

- looking at labels in supermarkets, which increasingly give the specific origin of foods
- visiting specialist local shops
- buying online from 'local' producers
- supporting local farmers' markets
- attending regional agricultural shows, which celebrate and sell local produce
- heeding the advice of celebrity chefs such as Gordon Ramsay who joined the debate in May 2008 by saying 'There should be stringent licensing laws to make sure produce is only used in season. I don't want to see strawberries from Kenya in the middle of March. When we haven't got it, we take it off the menu.'

Activities

1 Working in small groups, choose at least two different types of non-seasonal fruit and vegetables each. Visit one (or more) supermarkets, making sure in advance that they stock the produce you have decided on. If not, you will need to change your choice of items.

a Write down the origin of the items you have selected and whether they have been brought in by air. Many supermarkets now give this information.

b Display the origins of the fruit and vegetables on a world map.

c Summarise what you have discovered about the origin of fruit and vegetables eaten in the UK.

d Describe the environmental impact of this.

2 Study photo **B** and diagram **C** on page 291.

a Give the meaning of 'marginal land'.

b Use the photo to illustrate the difficulties of farming on marginal land.

c Use the diagram to explain the negative effects of farming on marginal land.

3 Study map **D** (and photo **E**) on page 292.

a Draw a sketch map to show the Indus and its five tributaries, international boundaries, main regions and places.

b Label the map to show the cause of conflict linked to water.

c Draw a timeline to show the sequence of events described that contributed to the conflict.

4 Working in pairs, produce an advert for locally produced food in your area.

a Decide whether your advert is to be printed in a newspaper or spoken on the radio.

b Carry out some research to discover which foods are grown locally. Your local newspaper, library, farm shops, farmers' markets and the internet may be useful.

c Produce your advert. It needs to be informative but hard-hitting if it is to be successful.

d Present your advert to the rest of the class.

∞links

The case study on 'Flower growing in Kenya' on www.learningafrica.org.uk is particularly useful, as is www.farmersmarkets.net.

13 Tourism

13.1   Why has global tourism grown?

■ Growth in tourism

Tourism is the world's largest industry, worth $500 billion dollars in 2007. Leisure accounts for 75 per cent of all international travel. The World Tourism Organization (WTO) measured a rise in international tourism arrivals (i.e. people arriving in all countries from abroad for a holiday) of 6.1 per cent between 2006 and 2007 (table **A**). There were nearly 900 million tourist travellers in 2007 and this is set to rise to a massive 1.6 billion by 2020. In most countries, domestic tourism (people going on holiday in their own country) is between four and five times greater than international tourism.

> **In this section you will learn**
> the many types of landscapes and holidays that attract people and why.

A   *International tourist arrivals (millions)*

Region	2006	2007	Percentage change 2006–07	Percentage of world tourism
Africa	40.9	44.2	+7.9	4.8
Americas	135.7	142.1	+4.7	16.1
Asia and Pacific	167.8	184.9	+10.2	19.8
Europe	460.8	480.1	+4.2	54.4
Middle East	41.0	46.4	+13.4	4.9
World	846[1]	898[1]	+6.1	100.0

[1] To the nearest million.

The tourism industry is therefore one of the greatest providers of jobs and income in countries at different stages of development. The reliance of different parts of the world on tourism varies. For 83 per cent of countries, tourism is one of the top five sources of foreign exchange. Caribbean countries get half their GDP from tourism. The top six tourist destination countries are France, Spain, the USA, China, Italy and the UK. Germans spend more per person than any other nation on holiday, followed by Americans, British, French and Japanese.

Factors affecting tourism's growth

Growth in tourism is explained by three sets of factors.

1 Social and economic factors

Since the 1950s people have become wealthier. Incomes are larger and so is disposable income (the amount left to spend as you wish after essentials such as housing, food and bills are paid). Most families have two working parents whereas in the past it was usually one.

> **AQA Examiner's tip**
> You will be asked to interpret tables of statistics. Usually, we look for patterns or trends and drawing a graph can be the easiest way to spot these. Sometimes there are few differences in the data. Nevertheless, tiny differences can still be important and may show a trend, and this is what you need to pick out.

People have fewer children; it is less expensive to take a small family away than a large one. Car ownership has also grown rapidly.

People have more leisure time. Holiday leave time has increased from two weeks per year in the 1950s to between four and six today. Life expectancy has risen so more people are retired. Many have good pensions and can afford several trips a year. They also have more time to travel.

2 Improvements in technology

Travel today is quick and easy – motorways, airport expansion and faster jet aircraft have all contributed to this. Flying has become cheaper and booking online is quick and easy. In 2008 the rapid rise in oil prices had an impact on the cost of flights and more people took domestic holidays to save money.

3 Expansion of holiday choice

During the 1950s and 1960s coastal resorts were popular and in the UK the National Parks were opening and offering new opportunities. The 1970s saw a decline in seaside holidays due to competition from cheap package holidays to mainland Europe, especially Spain. Packages are now available to destinations all over the world that offer a huge variety of sights and activities. **Ecotourism** and unusual destinations such as Alaska are expanding rapidly.

Key term

Ecotourism: tourism that focuses on protecting the environment and the local way of life. Also known as green tourism.

B *Budget airlines have made international travel quicker and cheaper*

◼ Tourist attractions

Many people choose to visit cities to enjoy the culture associated with museums, art galleries, architecture or shops and restaurants. Cities such as London, Rome and Paris have a huge amount to offer tourists of every age. The natural landscape is also a major 'pull' for tourists, particularly mountains such as the Alps in Europe or the beautiful stretches of coastline found in the Mediterranean or the Caribbean.

Activities

1   Study table **A** on page 295.

a   Which continent had the greatest tourism business in 2007?

b   Why do you think this continent has such a large tourist industry?

c   Which continent had the smallest tourism business in 2007?

d   Why do you think this continent has not yet developed its tourism as much as other regions?

e   Which continent's tourism grew by the largest percentage between 2006 and 2007?

f   Which continent's tourism grew the least?

g   Is there a relationship between scale of tourism and rate of growth? Explain your answer.

2   Study photos **C, D** and **E** opposite.

a   Draw a field sketch of one of the photos. Label all the pull factors that might attract people to come to that region of Italy.

b   What types of people are most likely to be attracted to each area? Remember to consider the seasons. Some landscapes are more attractive in winter than in summer.

c   Italy is a base for holidays all year round. Explain why this is so.

d   For the three Italian holiday types shown, assess their tourist potential for members of your family and/or friends. Give your reasons. Who would be most attracted to what and why?

C   *The Italian Alps*

D   *Venice is a popular tourist destination*

E   *Coastal tourism in Vernazza*

Italy

Italy is a country with a great variety of landscapes. Photos **C**, **D** and **E** show the three key types: mountains, cities and coastline. All the places shown have a busy tourist business, making an important contribution to the national economy. Venice is well known for its canals and Renaissance architecture; Florence for its art galleries. Skiing in the Alps and sunbathing on the coast are popular with both Italians and visitors from other countries.

The example of Italy illustrates point 3 above – expansion of holiday choice. People want to visit a greater variety of places and the tourist industry grows and adapts to supply what the market demands.

AQA Examiner's tip

Photos can be labelled directly with arrows to point out the features. There needs to be space around each photo because the labels need to be detailed. Keep what you write clear, straightforward and to the point.

Tourism is an important part of the economies of many richer countries, especially those in Western Europe and North America. Today, it is increasingly seen by developing countries as one of the best ways to earn foreign income, provide jobs and improve standards of living. Countries want to take advantage of the growing numbers of tourists and the money they have to spend.

> **In this section you will learn**
>
> the economic importance of tourism to countries in contrasting parts of the world
>
> how to compare and contrast tourist regions.

The economic importance of tourism

Tables **A** and **B** show the top 20 countries for tourist receipts (national income from tourism) and most tourist arrivals (number of tourists every year).

- France has had more tourists than any other country for many years. In 2007 it earned the largest amount of any nation from this source. French tourism includes every type of holiday such as city breaks, holiday cottages, camping and skiing.

- The USA earns more than any other country from tourism, yet has the third largest number of visitors. Europeans consider a trip to the USA as more special than staying in Europe, so they are likely to stay longer and spend more.

- China is high in both tables **A** and **B**. For many people, distance makes it too expensive a place to visit, but its variety of unusual landscapes and unique culture attracts increasing numbers with both time and money. This trend is likely to continue.

- In the Caribbean almost 50 per cent of visitors come from the nearby USA, with France, Canada and the UK also important sources of business. Expenditure per tourist ranges between $324 per holiday in Belize to $2,117 in the Virgin Islands, which attracts the wealthiest visitors.

Essential jobs are created in all countries from tourism, but the contribution this industry makes to GDP varies greatly between wealthier and poorer countries. Rich countries have a broadly balanced economy, of which tourism is one part. On the other hand, in less well-off countries tourism can be essential. In the Caribbean, for example, several small island countries rely heavily on tourism to provide national income and employment. Around 80 per cent of Barbados's national income comes from this business.

A *Countries with the largest tourist receipts, 2005*

Country	Annual tourist income ($ millions)
USA	66,547
Spain	33,609
France	32,329
Italy	26,915
China	20,385
Germany	19,158
UK	17,591
Austria	11,237
Hong Kong	10,117
Greece	9,741

B *Countries with the most tourist arrivals, 2005*

Country	Number of tourists (millions)
France	76.0
Spain	55.6
USA	46.1
China	41.8
Italy	36.5
UK	30.0
Germany	21.5
Mexico	20.6
Turkey	20.2
Austria	20.0

Dubai

Dubai, a tiny state, is one of the United Arab Emirates (UAE). It is located on the Arabian Gulf coast neighbouring Saudi Arabia (map **C**). Because it is easily accessible from Europe, Asia and Africa (120 airlines fly there), tourism in Dubai is growing quickly. Hotel revenue was up by 22 per cent in the first quarter of 2008 compared with the same period the previous year and number of hotel and apartment bed nights increases by 2 to 3 per cent annually. Around 2.8 million people visited the principality in 2000, 4.9 million in 2003 and 5.4 million in 2004. These numbers are predicted to grow further to 10 million by 2010, which would make Dubai one of the world's top tourist destinations.

The state is famous for its duty-free shopping malls with huge department stores and its markets (photo **D**). Prices are reasonable and there is huge variety. Emirates Airlines, which is based in Dubai, carries millions of long-haul passengers to hundreds of destinations and many stop over in Dubai as part of their trip.

Sightseeing is popular – the markets, the zoo, the dhow-building yards (traditional boats). Watersports, and especially diving, are growing in popularity. Excursions out from the city allow the visitor to see the desert and its wildlife (photo **E**). Bird-watching trips take visitors to the wetland mudflat areas, where there are 400 species.

C *Location of Dubai*

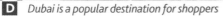

D *Dubai is a popular destination for shoppers*

E *Desert trekking in Dubai*

Benefits of tourism in poorer countries

- Many people are employed to serve tourists such as waiters (photo **F**), souvenir shop assistants and tour guides. In Antigua and Barbuda 30 per cent of the population work in these jobs, but in Jamaica only 8 per cent.

- Tourists spend their holiday money in pounds sterling, US dollars or euros. This foreign exchange is essential to poorer countries. It can be used to buy goods and services from abroad.

- Many governments tax visitors to help pay for the extra services they use such as water supply, drainage, electricity and roads.

- Extra jobs are created indirectly. Hotels buy some produce from local suppliers to feed the visitors.

- Many small businesses have been started up to serve the tourists themselves and supply the services they demand. These include taxis, bars and restaurants, builders and maintenance workers.

Activities

1 Study tables **A** and **B** on page 298.

a Draw a pie chart using the data in table **A**. Calculate the percentage that each country/region is out of the total figure. 100 per cent represents 360 degrees, so multiply each percentage you have calculated by 3.6 to obtain the number of degrees for each sector in your pie chart. Label each sector with the appropriate country/region or add a key to explain your shading. The data in the table is in size order. If you follow this in your chart, starting at the top, it will be easier to read and interpret.

b Which are the top three countries for tourist receipts?

c Which are the top three countries for tourist arrivals?

d Comment on any similarities and differences in your answers to b and c.

e Compare the balance of richer and poorer countries in the two tables.

2 Study photos **D** and **E** on page 299.

a What are the attractions that Dubai has to offer to tourists?

b Conduct some Internet research to discover what is available for tourists outside Dubai and its shopping malls.

c Which groups of people do you think benefit the most from the growth of tourism in Dubai?

d Would you like to go to Dubai? Why?

F Working as a waiter

links

Find out more about Dubai's ambitions at **www.nakheel.com**.

13.3   How do we manage tourism in the UK?

The growth of tourism in the UK

Almost all UK tourism used to be domestic – British people holidaying in the UK. Only the wealthy and privileged were able to go abroad, in other words, were international tourists. Domestic holidays can be cheap or expensive. Camping costs little, but good-quality hotels cost more than similar ones in other countries.

Domestic tourism grew quickly in the 1950s and 1960s as the growing UK economy provided higher pay and more time off work. Having an annual holiday became common. UK seaside holidays peaked in the early to mid-1970s, with 40 million visitors annually (graph **A**). After that, Britain's seaside resorts declined as package holidays abroad grew in number and affordability.

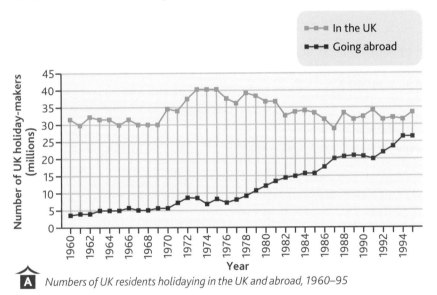

Legend: In the UK / Going abroad

A   Numbers of UK residents holidaying in the UK and abroad, 1960–95

Between the world wars, wealthier people began to go abroad. Cheap package holidays with guaranteed hot weather attracted people of all incomes. It was often cheaper to go to a Spanish resort like Benidorm than to holiday in the UK. British weather was seen as too unreliable. Many small coastal hotels were forced to survive by housing the homeless during the 1980s, decreasing their reputation even more.

The contribution of tourism to the UK economy

The UK economy earns over £80 billion every year from tourism and leisure. This amount usually grows slightly annually. Around 27.7 million overseas visitors spend over £13 billion of this sum. Restaurants and hotels make up a large proportion of these earnings, at £20 billion and £16 billion generated respectively. More than 100 new hotels opened in the UK between September 2004 and December 2005, creating more jobs and income. The London Eye is the most visited paying attraction in the UK, with 3.7 million visitors each year.

Britain has been a key tourist destination for many years – one of the most popular in the world. Almost 26 million people arrive here every year to see what Britain is about. London is one of the world's favourite cities.

But the challenge facing us now is to create a competitive, world-class tourism industry in Britain. We must have a tourism industry which provides affordable quality, which is open to all and which makes the best use of Britain's resources. And a tourism industry which concentrates on our key resource – people.

UK Prime Minister Tony Blair, speaking in 1999

B *Britain as a tourist destination*

The Butler tourist resort life-cycle model

This model says that any tourist resort starts on a small scale, develops into something more significant, then either goes into decline or makes changes to maintain its attractions (graph **C**). There are six stages.

1 Exploration

Small numbers of visitors are attracted by something particular: good beaches, attractive landscape, historical or cultural features. Local people have not yet developed many tourist services.

2 Involvement

The local population sees the opportunities and starts to provide accommodation, food, transport, guides and other services for the visitors.

3 Development

Large companies build hotels and leisure complexes and advertise package holidays. Numbers of tourists rise dramatically. Job opportunities for local people grow rapidly, but this brings both advantages and disadvantages.

4 Consolidation

Tourism is now a major part of the local economy, but perhaps at the expense of other types of development. Numbers of visitors are steady making employment more secure. However, some hotels and other facilities are becoming older and unattractive, so the type of customers attracted goes downmarket. Rowdiness becomes a problem.

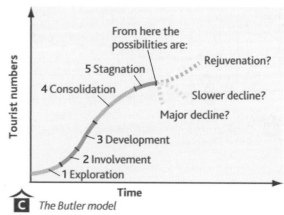

C *The Butler model*

5 Stagnation

The resort becomes unfashionable and numbers of visitors start to decline. Businesses change hands and often fail.

6 Decline or rejuvenation

Decline: visitors prefer other resorts. Day trippers and weekenders become the main source of income.

Rejuvenation: attempts are made to modernise the resort and attract different people to enjoy new activities.

Blackpool is a good example of a resort reinventing itself. Day trippers and weekenders now bring in most of the income, although websites and brochures make a huge effort to attract people for longer periods.

Blackpool, a UK coastal tourist resort

Blackpool's growth and stagnation

Located on the Lancashire coast in the north-west of England, Blackpool became a major tourist centre during the 19th century to serve the inhabitants of northern industrial towns. Factory workers could increasingly afford a holiday, travelling by train to a nearby coast. Blackpool boomed between 1900 and 1950. However, as people's disposable income increased, they preferred to try somewhere different and soon package holidays created huge competition for Britain's seaside resorts. Blackpool's summer weather can be unreliable, which proved a disadvantage.

D   *Factory workers on holiday, 1951*

As one of the larger resorts, Blackpool did attract some private investment and local authority grants to upgrade hotels, turn outdoor pools into indoor leisure centres and increase car-parking provision. Many smaller failing hotels were converted into self-catering holiday flats. Decline continued, but more slowly. Blackpool's attractions still made it a little different to other resorts. The famous Blackpool Tower, modelled on the Eiffel Tower in Paris, gives fantastic views up and down the Lancashire coast and inland towards the Pennines. The complex there includes the Tower Ballroom, famous for national ballroom dancing competitions, and the Tower Circus. The town upgraded its zoo and a Sealife Centre was built. The Blackpool Illuminations – a light show stretching along the Golden Mile (the central section of the sea front) – began in 1879 and has been upgraded several times with advancing technology.

E   *Blackpool Tower and the Illuminations*

Blackpool should have been quicker to fight the competition from package holidays. Eventually it lost much of its family holiday business and came to rely on day trippers and stag and hen party business – not popular with residents and bad for the town's image.

The supercasino

One approach taken by Blackpool to get out of recession was to apply to the government to be the home of the UK's first supercasino, a huge leisure and entertainment complex based on those in Las Vegas and Atlantic City in the USA. Blackpool was the favourite applicant on the shortlist. The proposed sea-front site included a conference centre and a wide variety of entertainments. Some 20,000 jobs and £2 billion of investment would have been generated. The town's high 3.6 per cent unemployment rate would have been greatly improved. Nevertheless, not everyone was in favour, but the final decision closed the debate – Blackpool lost the supercasino development to Manchester, coming only third in the vote. The town council was shocked.

Despite the supercasino setback, Blackpool is aiming for regeneration. The Blackpool Masterplan spent millions improving the town for the casino bid, so the town now looks less run-down. A new department store opened in 2008.

High stakes over Blackpool supercasino

Blackpool is suffering from a very slow, long and terminal decline. We need shock and awe. If not, why would people come? We have always had wacky buildings and we have always been the people's playground. It is Blackpool's job to be at the cutting edge of public taste.

F   *From the* Guardian

Alan Cavill, Head of Corporate Policy and Development at Blackpool City Council, quoted in the *Guardian*, 8 September 2006

External factors affecting UK tourism in the early 21st century

Tourism can be limited by political and economic situations. Two key issues have caused difficulties in the early 21st century.

Terrorism

The destruction of the World Trade Center in New York on 11 September 2001 had a huge impact on travel. The USA stepped up its security overnight, as did the UK and the EU. Airport security checks have multiplied and check-in times increased. London is a terrorist target: the Underground bombing of 7 July 2005 is an example. In the aftermath of such events, visitor numbers decline sharply.

Exchange rates and the banking crisis

Currency exchange rates control value for money for tourists on holiday. In 2008 the euro was high against the pound, valued at around 79p (compared with 68p previously), so holidaying in France and other Eurozone countries became more expensive. At the same time, the US dollar was valued at almost two to the pound, making the USA a much more attractive holiday destination. In 2009 exchange rates make the UK an attractive destination for visitors from abroad.

The banking crisis of autumn 2008 may mean people have less money to spend. People may reduce the number of holidays they take or even manage without one until the economic situation improves. On the other hand, the weaker pound should attract more foreign visitors in 2009.

Activities

1 Study graph **A** on page 301.
  a In 1960 how many people holidayed within the UK and how many went abroad?
  b Which year had the largest gap between the sets of statistics?
  c Which year had the smallest gap?
  d Describe the trend of holidays taken in the UK.

2 Study graph **C** on page 302. To what extent has Blackpool's development as a tourist resort followed the Butler model? Write at least two paragraphs.

3 Work with a friend to put together some suggestions for making Blackpool attractive to tourists in the future. Use the Internet to give you some ideas. Put together a 'Tourism Plan for Blackpool' using ICT if you wish. Don't forget to justify your plan.

⊂⊃ **links**

Find out more about Blackpool tourism at **www.visitblackpool.com**.

13.4 What is the importance of National Parks for UK tourism?

National Parks are large areas of mainly rural land. England's Parks cover 7 per cent of its land area and Scotland's two take up 7.3 per cent. In Wales, a huge 20 per cent makes up its three National Parks (map **A**). Globally, there are 6,000 National Parks, protecting 12 per cent of the world's land surface. The UK's first was the Peak District of Derbyshire, created in 1951 by an Act of Parliament. The post-war period saw many efforts to improve people's quality of life. National Parks aim to conserve natural and cultural landscapes while allowing access for visitors to enjoy them. The first Scottish Park was designated in 2002. In England the most recent is the New Forest (2004); the South Downs National Park should be in place by 2010. It is likely to become the busiest as it is in such a densely populated region.

In this section you will learn

- the contribution of tourism to the UK economy

- the importance of National Parks for UK tourism

- the conflicts caused by tourism in the Lake District

- how to devise management strategies for tourism in the Lake District.

1   Loch Lomond and the Trossachs
2   Lake District
3   Peak District
4   Snowdonia
5   Pembrokeshire
6   Brecon Beacons
7   Exmoor
8   Dartmoor
9   New Forest
10   South Downs
11   Norfolk Broads
12   North Yorkshire Moors
13   Yorkshire Dales
14   Northumberland
15   Cairngorms

- National Park
- Motorway
- Major urban areas

A   The location of National Parks in Britain

Many National Parks are uplands such as Snowdonia (photo **B** on page 306) and the Lake District; a few are lowlands (the Norfolk Broads) and coastal (Pembrokeshire). Land remains privately owned (81 per cent), mostly by farmers, but the Forestry Commission, the National Trust, the Ministry of Defence and the water authorities also own some areas. The National Park Authorities only directly control 1 per cent. Local people make their living from the land and local businesses. Tourism provides much-needed jobs.

Key terms

National Park: an area usually designated by law where development is limited and planning controlled. The landscape is regarded as unusual and valuable and therefore worth preserving.

Honeypot site: a location attracting a large number of tourists who, due to their numbers, place pressure on the environment and people.

Activities

1. Study map **A**. One of the aims of National Parks is to serve the population with recreation facilities.

   a. Describe the locations of the National Parks in relation to the major urban areas.

   b. Based on your answer to a, how accessible do you think the National Parks are for most of Britain's population? Make sure you name Parks and cities in your answer.

   c. Which National Parks are the most accessible? Explain why.

2. Study photo **B**. What types of activities might people undertake in this landscape?

B Snowdonia National Park

The Lake District

Attractions and opportunities for tourism

The English Lake District is a glaciated upland area in Cumbria, north-west England. It stretches 64 km from north to south and 53 km east to west. It became a National Park in 1951. Famous for its stunning scenery, abundant wildlife and cultural heritage, it is considered to be England's finest landscape. What makes it so special for tourists? Landscape, culture and activities all play a major role.

The ribbon lakes and tarns are part of a unique and hugely varied landscape, as well as being a major recreational resource. Lake Windermere specialises in ferry cruises. Most people sail between the main centres of Windermere town (photo **C**) and Ambleside. Small boats are allowed on many lakes. Areas are set aside for windsailing and power-boating so the activities do not clash and quiet areas are left for people seeking peace and quiet. Fishing from the shore or boats is increasing in popularity.

Walking is one of the most popular reasons why people visit the Lake District, whether for a day or longer. Routes vary from short and relatively flat to extremely long and tough. Known as the 'birthplace of mountaineering', even the most experienced climbers find plenty of challenges in scaling back walls of corries and sides of U-shaped valleys (photo **D**). Public access to the fells (open uplands) is unrestricted. Many guides have been written for walkers and climbers. The poet Wordsworth wrote one in 1810 but the most famous series of guides is by Wainwright.

Historical and cultural sites also attract tourists. The Lake District has been occupied since the end of the ice age 10,000 years ago and evidence of early settlement remains in the landscape. The land has been farmed for centuries, leaving a distinctive field pattern with drystone walls. Many 19th-century writers and artists, such as John Ruskin, loved the area. Beatrix Potter's family, residents of London, had a summer home there, which is why she settled there later at Hill Top beside Lake Windermere (map **E** opposite).

C Windermere town centre

D Striding Edge, Helvellyn

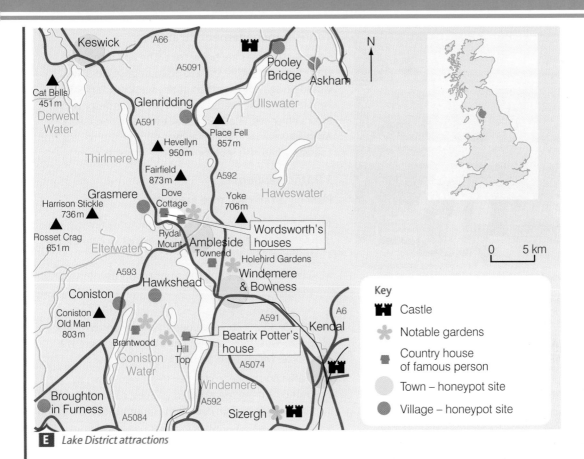

E *Lake District attractions*

Impacts of tourism

There are many times the number of visitors in the Lake District National Park as local population: 12 million tourists (22 million visitor days) to 42,239 residents. The impacts are potentially huge and must be managed as well as possible. There are both negative and positive impacts.

1 Traffic problems

Over 89 per cent of visitors come by car, often just for the day. Many roads, including A roads, are narrow and winding. Buses and large delivery vehicles have to use these to service both locals and tourists. Queues are a common problem, especially towards the end of the day when day trippers are heading home. Towns like Bowness-on-Windermere were not built originally for the huge volumes of traffic that arrive daily in the summer, especially at weekends. Congestion and parking are serious problems. Bowness has built a new car park at Braithwaite Fold on the edge of town and has extended another, but capacity is still inadequate. In desperation, in the countryside people park on grass verges, causing serious damage.

2 Honeypot sites

The Lake District has both physical and cultural **honeypot sites**. Beauty spots, small shopping centres and historic houses all attract hundreds of visitors daily. Cat Bells is quite an easy climb, so many people walk up this smaller mountain. It therefore suffers from serious footpath erosion (photo **F**). Across the Lake District, 4 million people walk at least 6 km every year. Several areas have scarred landscapes. Bowness is an extremely busy shopping and recreation centre in summer. Hill Top, Beatrix Potter's house, attracts families (photo **G** on page 308). Honeypot sites need to provide access and facilities while remaining as unspoilt as possible.

F *Footpath erosion in the Lake District*

G *Hill Top, Beatrix Potter's house*

3 *Pressure on property*

Almost 20 per cent of property in the Lake District National Park is either second homes or holiday let accommodation (15 per cent of all housing in 2007). Some local people make a good income from owning and letting such property, and this is often forgotten by those who are more critical of second homes. The main issues include the following:

- Holiday cottages and flats are not occupied all year.
- The same is true of second homes, so their owners are not part of the community full time.
- Holidaymakers do not always support local businesses, often doing a supermarket shop at home before their trip. On the other hand, the main supermarket in Windermere is often full of visitors buying a great deal of food and drink for their stay.
- Demand for property from outsiders increases property prices in the Lake District, causing problems for local people who are forced out to find affordable homes on the edge of the region in Kendal or Penrith. This is the most serious tourist problem affecting local communities.

4 *Environmental issues*

Water sports are not allowed on some of the lakes, but Windermere, the largest lake, has ferries and allows power-boating, windsurfing and other faster and more damaging activities (photo **H**). The main issue is the wash from faster vehicles eroding the shore. Fuel spills are not uncommon, causing pollution.

Tourism management strategies

Several strategies are being tried. The aim is to limit tourist impact rather than to discourage visitors, which would be against the ethos of any National Park. People remain attracted to honeypot sites despite their inconveniences. The 1997 Lake District National Park Authority management plan was put in place to cater for tourists while also looking after the interests of residents.

1 *Traffic solutions*

Planning an efficient road network:

- County strategic roads, often dual carriageways, are built on the edges of the Lake District to help move traffic in and out as efficiently as possible.
- Distributor roads link the small towns and key tourist villages.
- Access roads are small and take less traffic. Many people do not drive beyond the larger settlements. Some routes are 'scenic' and sometimes there is a choice, which splits traffic between routes.
- Traffic on smaller roads can be slowed down by traffic-calming measures in villages, cattle grids in the countryside and an overall maximum speed limit.
- Heavy lorries should be kept off scenic roads.

Planning public transport:

- Where possible bus lanes operate in towns, although narrow streets limit this.
- Park-and-ride schemes encourage people to leave their cars at the edge of the National Park and go by bus. Costs are lower than town car parks.
- Buses in the most rural areas remain a difficulty as roads are so narrow.

H *Overcrowding on Lake Windermere*

2 Honeypot management

Footpaths:

- Repairing footpaths improves appearance and encourages people to stay on the path.
- Reinforcing path surfaces reduces future damage.
- Signposting routes limits the number of paths.

Parking:

- Fence off roadsides so people cannot damage verges.
- Develop several new small car parks and hide them by landscaping using tree planting.
- Reinforce car-park surfaces to prevent damage. 'Waffles' are large concrete slabs with holes in them, like an edible waffle. Soil fills the holes and grass grows, giving a hard green surface.

Litter:

- Bins should be provided at key points and emptied regularly. Overflowing bins encourage more litter.
- Designated picnic areas mean litter has to be dealt with in fewer places.
- Signs encouraging people to be responsible reduce litter.

3 Property prices

This is the most difficult issue. Management strategies cannot control house prices. Local authorities could build more homes for rent and developers could erect more low-cost homes for sale. Little has yet been achieved.

4 Environmental issues

Speed limits for boats can limit the amount of wash caused, but to prevent erosion speeds would have to be very low, which clashes with the main pleasure of the sport – going fast! The speed limit on Windermere is 18 kph. Limiting the noisiest and most damaging sports to certain parts of the lake can restrict the amount of damage done.

Tourism conflicts and opportunities

1 Farming

Tourism and farming are often thought to be in conflict, which can be true. Visitors can trample crops and disturb livestock, but signs and education have limited these problems. Tourists have offered hill farmers new opportunities for diversification in difficult economic times. Income can be made from B&B accommodation, holiday cottages converted from farm buildings, camping and caravan sites. Activities such as pony trekking and paintballing can be offered.

2 Employment

The impact of tourism on employment must be positive as so many jobs are created. Many businesses thrive and make a profit. Conversely, seasonality is a problem, as well as low pay. Visitor numbers can be unpredictable.

I *Elterwater village*

∞ links

Find out more about the Lake District at **www.lake-district.gov.uk**.

Why do so many countries want mass tourism?

Advantages and disadvantages of mass tourism

Mass tourism involves large numbers of tourists coming to one destination (photos **A** and **B**). There is usually a particular purpose and a particular type of location, such as skiing in a mountain resort or sunbathing at a beach location. Many countries and regions want to develop mass tourism because they believe it will bring many advantages (table **C**).

A The Alhambra in Granada, Spain, is one of Europe's most popular tourist attractions

B The Parthenon in Athens, Greece, is the most visited ancient monument in Europe

Did you know ??????

In the USA the first mass tourist resorts were Atlantic City in New Jersey and Long Island in New York State. In Europe, Ostend developed to serve the citizens of Brussels, Deauville attracted Parisians and Boulogne, people from Normandy.

Key terms

Mass tourism: tourism on a large scale to one country or region. This equates to the Development and Consolidation phases of the Butler tourist resort life-cycle model.

C *Advantages and disadvantages of mass tourism*

Advantages	Disadvantages
• Tourism brings jobs. People who previously survived on subsistence agriculture or day labouring gain regular work with a more reliable wage.	• The activity may be seasonal – skiing only happens in winter. Local people may find themselves out of work for the rest of the year.
• New infrastructure must be put in place for tourists – airports, hotels, power supplies, roads and telecommunications. These also benefit the local population, although they may bring with them pollution and over-development.	• The industry is dominated by large travel companies who sell package holidays by brochure or on the internet.
• Construction jobs often go to local people, but they are temporary.	• Lower- and middle-income customers are the target market – this type of tourism does not appeal to wealthier groups of people.
• New leisure facilities may be open to local people.	• Few local employees are well paid. The higher level jobs are often taken by people from the companies involved in developing the resort, who are not locals.
	• Investing companies are usually based in countries at further stages of development. Mass tourist resorts are increasingly in countries at lesser stages of development. Profits therefore go outside the tourist country – they do not benefit the host country.
	• New building developments need land. Local farmers may be tempted to sell their land to developers or development around them makes farming their land almost impossible. Local food production decreases at a time it needs to be increasing to fulfil additional demands.
	• Tourists can be narrow-minded and often prefer familiar food, so much is imported rather than produced locally.
	• Local people may not be able to afford the new facilities put in place for tourists.

Tourism in Jamaica

Jamaica is one of the Caribbean's main tourist destinations, with 1.3 million visitors in 2001. After this, competition from other islands began to be something of a problem. Tourism is the country's second biggest earner and 220,000 Jamaicans work directly in this sector. Other local businesses also depend on tourism, such as food production for visitors and other hotel suppliers. Jamaica has much to offer the tourist (map **D** and fact file **E** on page 312) including watersports, for which it is famous, wildlife sanctuaries and, increasingly, golf.

Case study

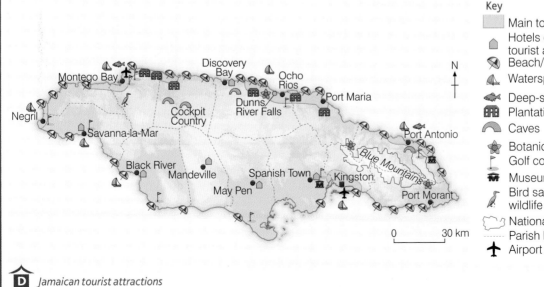

Key
- Main tourist areas
- Hotels outside main tourist areas
- Beach/bathing areas
- Watersports
- Deep-sea fishing
- Plantation house
- Caves
- Botanic garden
- Golf course
- Museum
- Bird sanctuary/wildlife reserve
- National Park
- Parish boundary
- Airport

D *Jamaican tourist attractions*

E *Jamaican tourism fact file, 2001*

Jamaican tourism facts and figures	
Total number of tourist arrivals	1,322,690
Population employed in tourism	8%
Tourist expenditure (per person)	US$931
Total foreign exchange earnings	US$1.3 billion
Contribution to GDP	20%
Cruise ship arrivals (2002)	865
Rooms available to tourists (2002)	14,388
Visitors from USA and UK (% of total)	70% and 10%

⊙⊙ links

Find out more about Jamaican tourism at **www.visitjamaica.com**.

A recent trend has been the growth in community tourism. This involves local people directly because visitors become a part of their home and village during their stay. Families provide bed and breakfast accommodation and other local businesses, such as restaurants and bars, supply their other needs. This style of holiday provides greater interaction between tourists and local people, gives the visitors a clearer idea of local life, supports local businesses and uses fewer resources. Money goes directly to the people rather than to large international businesses.

Jamaica needs to maintain its tourist resources into the future and some companies offer ecotourism, which is becoming more popular. This utilises the inland area of the island such as the Blue Mountains as well as parts of the coast, spreading tourists further around the island. Nature reserves are increasing and eco-lodges are being built. Tourist densities are kept low in these areas, which keeps pressure off the environment.

Activities

1   Study map **D** on page 311 and fact file **E**.

 a   Name two main tourist areas in Jamaica.

 b   Which tourist activities are found in these areas?

 c   Make two separate lists of the advantages and disadvantages to Jamaica of the holiday industry.

 d   Research on the internet or in holiday brochures to discover any other advantages or disadvantages you could add to your lists.

 e   Write a minimum of three paragraphs to discuss whether tourism is beneficial to Jamaica's people and economy. Add a clear conclusion at the end of your discussion.

2   This is a group exercise. At least three members of the class should be in each group. Imagine a new tourist enterprise is being set up in Jamaica, aiming to be as eco-friendly and sustainable as possible. (Read pages 316–317 later in this chapter to understand sustainable tourism better.) You are members of the travel company (you may be from Jamaica or from a country that tourists come from – you choose) and you are preparing your proposals for the government of Jamaica and a local environmental interest group.

 a   Plan an exciting programme for the clients. Name locations and the activities on offer there.

 b   What type of accommodation is provided? How is it serviced?

 c   How would the workforce be recruited?

 d   Emphasise the advantages of your proposals and how you would aim to overcome any difficulties.

 e   Compare your plans with those of other groups.

13.6 / What attracts people to extreme environments?

Extreme environments and activities

Extreme environment tourism involves dangerous landscapes often with a difficult climate, and places that are sparsely settled (or not occupied at all), access to which is limited. Increasing numbers of tourists are attracted to extreme environments. Others prefer even more of a thrill, pursuing activities such as rock climbing, paragliding and white-water rafting. Some of these activities have to be undertaken in extreme environments: ice-walking is one example because you need an ice cap or glacier to walk across. Many extreme activities can be done in a variety of places, for example paragliding and microlighting are done on the South Downs in Sussex.

Extreme environments are spread across the globe and cover a wide variety of locations: mountains, deserts, rainforests, caves and ice-covered terrain (photos **A** to **D**). Also known as shock or adventure tourism, it caters for a niche market. It involves an element of risk and people often choose such a trip for the adrenaline rush they get from the dangerous activities and sports involved. Examples include ice-diving in the White Sea, north Russia, with almost freezing temperatures, and travelling across the Chernobyl Zone of Alienation in Ukraine – the area devastated by nuclear contamination in 1986. In Jamaica such activities include climbing waterfalls and cliff-diving. Adventure tourism is one of the fastest-growing types of tourism in the world.

In this section you will learn

what we mean by 'extreme' tourism

which environments are classified as extreme

what is happening in Antarctica in tourism development and the likely consequences of this

how to assess whether particular remote landscapes should be developed for tourism on any scale.

Key terms

Extreme environments: locations with particularly difficult environments where the development of tourism has only recently occured due to a niche market demand for somewhere different with physical challenges.

A *Extreme tourism in Greenland*

B *Extreme tourism in the desert*

C *Extreme tourism in the Amazonian rainforest*

D *Extreme tourism in Nepal*

The target market

Adventure tourists look for physical challenge and risks. They are often around 30 years old, unmarried and without children, have high-powered jobs and a good income – these trips are expensive. Groups are small and distances great. However, there are enough wealthy individuals with a taste for something completely different to allow this sector to grow. It will never be large but in some areas it is increasing in significance. Most companies advertise on the internet rather than by brochure.

Little investment is needed to set up such trips. The usual costly expenses of building hotels and roads are irrelevant. Part of the experience is to sleep 'rough' and travel over untouched landscapes. This tourism sector is growing rapidly in Peru, Chile, Argentina, Azerbaijan and Pakistan. Northern Pakistan is one of the most mountainous and difficult landscapes in the world and even its risky political situation as the base of Al Qaeda terrorists adds a thrill for some.

Case study

Antarctica

Small-scale tourism began in Antarctica in the 1950s when commercial shipping began to take a few passengers. The first specially designed cruise ship made its first voyage in 1969. Some 9,000 tourists in 1992–93 have now grown to 37,000 in 2006–07 and to 46,000 in 2007–08 (table **E**). This is thousands more than the scientific workers and their support staff who are there temporarily for research purposes. Over 100 tourist companies are involved. In 2006, 38.9 per cent of visitors were American, 15.4 per cent British, 10.3 per cent German and 8.4 per cent Australian.

Tourists from the northern hemisphere usually fly to New Zealand or Argentina, taking their cruise ship onwards for one to two weeks. Smaller boats take them ashore at key locations for short visits, mainly to the peninsula or nearby islands (map **F**).

E *Antarctic tourist numbers, 2007–08*

Country of origin	Numbers	Percentage of total
USA	16,533	35.9
UK	7,372	16.0
Germany	5,090	11.0
Australia	3,338	7.2
Canada	2,809	6.1
Japan	1,720	3.7
Switzerland	1,296	2.8
Netherlands	1,213	2.6
Others	6,698	14.5
Total	46,069	100.0

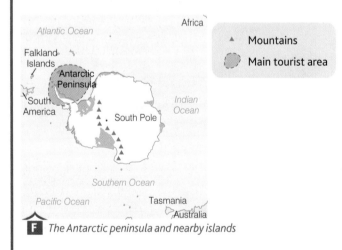

F *The Antarctic peninsula and nearby islands*

The environmental impact of an individual tourist is much greater than that of a researcher. Landing sites are chosen for a special feature, so they quickly become honeypots. More than 99 per cent of Antarctica is covered with ice, so little is left for tourist activity. Few visitors go on the ice. Walking, kayaking, skiing, climbing, scuba diving and helicopter/small aircraft flights are some of the activities offered (chart **G** opposite).

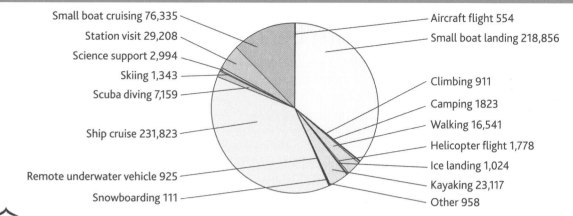

Small boat cruising 76,335
Station visit 29,208
Science support 2,994
Skiing 1,343
Scuba diving 7,159
Ship cruise 231,823
Remote underwater vehicle 925
Snowboarding 111

Aircraft flight 554
Small boat landing 218,856
Climbing 911
Camping 1823
Walking 16,541
Helicopter flight 1,778
Ice landing 1,024
Kayaking 23,117
Other 958

G *Tourist numbers in Antarctica, 2007–08*

Tourists only spend a short time ashore, but the impacts do not always reflect this. They want to visit the most picturesque and wildlife-rich areas. The impact is uneven but in places too great. Animals, especially penguins and seals, are disturbed by more than a few people (photo **H**). Not used to humans, they do not like to be touched. If they leave as a result, they may abandon eggs and young.

There have been accidents when ships have struck uncharted rocks or ice floes. The great majority of shipping in Antarctic waters is tourist-based. Oil spills are becoming an increasing hazard for wildlife. Tourist ships must discharge all waste materials well away from the shore of Antarctica.

Coping with tourism in Antarctica

All tour operators are members of IAATO, which directs tourism to be safe and environmentally friendly. Around 100 companies are involved. In line with the Antarctic Treaty, tourism is an acceptable activity in Antarctica – it is the scale that has to be controlled. Visitors are not allowed to visit Sites of Special Scientific Interest (SSSIs) in order to conserve precious wildlife and landscapes. Bird Island on South Georgia is one example.

Although tourist numbers have increased rapidly in Antarctica, protection remains a high priority. A permit must be gained for any activities on the continent. No ship carrying over 500 passengers can land in Antarctica. Nevertheless, there is concern that larger ships will eventually be allowed to land and that the volume of tourists will be beyond sustainable limits.

H   *The impact of tourists can be great*

Activity

Use the information about Antarctica to help you make a study of tourism in this extreme environment. Use the Internet to help you investigate the following themes:

a   What are the attractions to tourists?

b   Where do tourists come from? (Here you could use table **E** opposite to make a flow map.)

c   What are the issues associated with tourism?

d   How should tourism be controlled in the future?

Did you know ??????

The ice in an Antarctic iceberg is around 100,000 years old. The continent is actually a desert, with only 254 mm of precipitation each year.

links

Find out more about tourism in Antarctica at **www.coolantarctic.com**.

Stewardship and conservation

In the Middle Ages a steward was in charge of his master's estate and business. Today, the term **stewardship** is used to mean careful management of the environment on a large scale: regionally, nationally and globally. All types of development and resource exploitation are planned sustainably. Development priorities are set and strategies created to achieve these. It is the ideal way to develop any country and any economy but, despite all the moves towards globalisation, it is not yet the way that all governments think.

Conservation is stewardship on a smaller and more manageable scale. People feel much more involved at this level. An individual building can be conserved and protected because of its historical importance. Habitats and landscapes in rural areas can be protected. Careful use of non-renewable resources is also a form of conservation; planned use allows them to last as long as possible. Conservation can involve improving energy efficiency and recycling waste.

The aims of ecotourism

Ecotourism is environmentally friendly tourism. Tourism is being increasingly blamed not only for environmental but also for social and cultural damage. Beaches become polluted, coral reefs degraded and economies too dependent on tourism at the expense of local food production and other essential industries and services. Ecotourism caters for a small but growing niche market of environmentally aware tourists – it is the fastest expanding tourism sector. Inevitably, these holidays costs more, which means they attract only people with enough money. Less well-off people may be aware of the need for sustainable tourism but cannot afford it. Should they go without a holiday as a result?

Ecotourists want to experience the natural environment directly, undertaking activities such as trekking and bird-watching. They want their holiday to have as little impact on the environment as possible. Energy use should be sustainable and no waste should be generated that cannot be dealt with efficiently. Ecotourists prefer small-scale accommodation in lodges that may not even have electricity, not large hotels. They eat local food. Local people are their guides as their knowledge and experience is seen to be more valuable. The impact on the environment is low but, because ecotourism is small in scale, the price paid by each tourist is high. The market for such tourism is therefore limited.

In this section you will learn

the concept of sustainable tourism

different types of sustainable tourism and be able to compare them.

Key terms

Stewardship: the personal responsibility for looking after things, in this case the environment. No one should damage the present or future environment.

Conservation: the careful and planned use of resources in order to manage and maintain the natural environment for future generations.

AQA Examiner's tip

Learn the difference between 'stewardship' and 'conservation'.

AQA Examiner's tip

Ensure you can use an example to explain how ecotourism can contribute to sustainable development.

The Galapagos Islands

The 50 volcanic Galapagos Islands lie 1,000 km off the west coast of South America in the Pacific Ocean. They belong to Ecuador. Here, Charles Darwin formulated his theory of evolution. Around 90 per cent of these islands are designated as National Park or marine reserve. Protection began in the 1930s. The islands are among the most fragile and precious ecosystems in the world, becoming the first Unesco World Heritage Site in 1979. The area is also a biosphere reserve and whale sanctuary.

Today, tourists visit under strict rules. They arrive mainly by small ships that tour the islands and allow people onshore only at specific locations in limited numbers. An eight-day cruise costs around £800 without flights. The Galapagos Conservation Trust receives £25 from every visitor out of the price of their holiday. This pays for conservation work on the islands. The tour boats are owned by locals and take 10 to 16 tourists each, many accompanied by professional guides. Visitors are given accurate information and prevented from causing damage.

Galapagos tourism has brought great benefits to Ecuador because it supports the National Park and generates income. However, some sites are over-used, oil from boats can pollute the area and the islands' water supply is put under pressure. Nevertheless, in general, the benefits outweigh the difficulties.

The benefits of ecotourism in the Galapagos Islands include the following:

- Environmental benefits – local people have to make a living. The Galapagos Islands have such a fragile environment with so many limitations for agriculture and other basic industries that opportunities for development without environmental damage are few. Carefully planned ecotourism should offer opportunities for the present inhabitants without compromising those of future ones.

- Economic benefits to the local economy – local businesses have been started to provide the needs of tourists. Tourists usually stay in small guest houses, often run as family businesses. Companies have grown up to provide boat trips around islands and between the islands.

- Economic benefits to the lives of individuals – people are employed in guest houses, on boats and as guides. The income is enough to make a difference to a household. Many visitors give tips, which go directly to local people.

A *Ecotourism on the Galapagos Islands*

Activities

1
a Write a definition of ecotourism.

b What are the main characteristics of ecotourism?

c Do you think ecotourism is a good thing? Why?

2 Choose one of this chapter's case studies – Jamaica or the Galapagos Islands. Write an extract from a travel company brochure to sell this holiday to potential customers. Think of your target market and make sure you use language that would appeal to them. For your chosen tourist destination, find a photo from the internet to illustrate the holiday you are advertising.

3 The Geography and Biology departments at your school want to run a trip to the Galapagos Islands. Your class has been asked to help promote the trip by producing posters and PowerPoint presentations. The trip is to be as environmentally sensitive as possible so you need to stress the principles of ecotourism. Work in pairs or small groups and use the Internet to help you.

Examination-style questions

It is important to note that the examination-style questions in the following section are practice questions in the style you would expect to find in the examination paper, but are not representative of the full range of questions and mark schemes you will find in the actual examination.

These questions have been written based on existing content within the student book; in the examination you will not necessarily be tied to one specific case study.

1 The restless earth

Foundation Tier

(a) Study Figure B on page 9.
Are the following statements true or false?
Tick the correct box.

	True	False
Oceanic crust is denser than continental crust.		
Oceanic crust is 1500 million years older than continental crust.		
Oceanic crust can be renewed and destroyed.		

(3 marks)

(b) Use the case study of the Indian Ocean tsunami (pages 32–33) to describe its effects. *(6 marks)*

Higher Tier

(a) Study Figure B on page 9.
Outline differences between oceanic and continental crust. *(4 marks)*

(b) Use the case of the Indian Ocean tsunami (pages 32–33) to explain its cause. *(8 marks)*

Common question

Study Figure C on page 13 and Figure A on page 22.
Describe how a composite volcano is different from a supervolcano. *(6 marks)*

2 Rocks, resources and scenery

Foundation Tier

(a) Study Figure A on page 36.
Complete the sentences below to describe the rock cycle. Choose the correct words from this list.
Weathered, eroded, deposited, igneous, sedimentary, metamorphic
Cooling magma leads to the formation of _____ rock.
Changing temperatures on the earth's surface cause rocks to be _____.
The transportation of material to the seabed is the beginning of the formation of _____ rock.
Pressure or heating affecting rocks causes _____ rock to form. *(4 marks)*

(b) Study Figure B on page 49 and Figure C on page 46. Explain the formation of limestone pavement. *(4 marks)*

Higher Tier

(a)   Study Figure A on page 36.
      Explain what is meant by the 'rock cycle'.   *(4 marks)*

(b)   Study Figure C on page 46. Explain the formation of the underground features
      in a limestone area.   *(8 marks)*

Common question

Study Figure B on page 37.
Describe the process of freeze–thaw weathering.   *(4 marks)*

3   Challenge of weather and climate

Foundation Tier

(a)   Study Figure C on page 55.

      Maximum temperatures and sunshine hours vary throughout the UK.
      Circle the correct word(s) to complete the sentences below that describe this.

      The highest temperatures (of 16.4 °C or more) are in the south-east/north-west.

      The least sunshine (711–1140 hours) occurs mainly in Scotland/Wales.

      The number of sunshine hours decreases from south-east to north-west/north-west
      to south-east.

      There is a weak/strong link between the two maps.   *(4 marks)*

(b)   Study Figures D and E on page 59. Describe what is likely to happen to temperature,
      clouds and precipitation as a cold front passes overhead.   *(3 marks)*

Higher Tier

(a)   Study Figure C (rainfall map) on page 55.
      Describe the pattern of rainfall shown in Figure C.   *(3 marks)*

(b)   Study Figures D and E on page 59.
      Explain the change in precipitation as a cold front passes overhead.   *(4 marks)*

Common question

Use the case study of Boscastle to describe issues resulting
from the flash flooding.   *(F Tier 6 marks/H Tier 8 marks)*

4   Living world

Foundation Tier

(a)   Study Figure B on page 79.

      There are different parts of an ecosystem. Draw a line to each of the statements
      to complete the sentences correctly.

      Consumers   'make' their own food

      Decomposers   eat other living things

      Producers   break down dead plants and animals

                                                                                   (2 marks)

(b)   Refer to the case study of Epping Forest on pages 86–87 to describe how
      the deciduous woodland is used.   *(6 marks)*

Higher Tier

(a)   Study Figure B on page 79.
        Explain how a producer is different from a consumer.   *(3 marks)*
(b)   Use the case study of Epping Forest on pages 86–87 to explain why the
        deciduous woodland is currently managed.   *(4 marks)*

Common question

        Study Figure C on page 87.
(a)   Give the 6-figure grid reference for the church on the map.   *(2 marks)*
(b)   Describe the map evidence that suggests that people visit Epping Forest for recreation.   *(4 marks)*

5   Water on the land

Foundation Tier

(a)   Study Figures D and E on pages 109 and 110.
        Use Figures D and E to complete the following fact file.

Month of maximum rainfall	_____
Temperature range	_____
The amount by which rainfall on 24 and 25 June 2007 exceeded the June average	_____

                                                                                (3 marks)
(b)   Use the case study of Three Gorges Dam on pages 117–118 to describe the costs
        and benefits of the scheme.   *(6 marks)*

Higher Tier

(a)   Study Figure E on page 110.
        Explain how the rainfall figures on 24 and 25 June help to account for the flooding
        in Sheffield on 26 June 2007.   *(3 marks)*
(b)   Refer to the case study of Three Gorges Dam on pages 117–118. Do the benefits of
        building the dam outweigh the costs? Explain your decision.   *(8 marks)*

Common question

        Study Figure G on page 106.
        Draw a labelled sketch to show the main features of the meander in the photograph.   *(4 marks)*

6   Ice on the land

Foundation Tier

(a)   Study Figure B on page 126.
        Are the following statements true or false?
        Tick the correct box.

	True	False
In summer, there is an overall loss of ice in the glacier.		
There are four times when ice melting is equal to snow being added.		
In winter, melting is higher than snow being added.		

                                                                                (3 marks)

(b)   Using the case study of French Alps on pages 135–137, describe the attractions
for tourists.   *(4 marks)*

Higher Tier

(a)   Study Figure B on page 126.
Describe the annual glacial budget.   *(4 marks)*

(b)   Use the case study of French Alps on pages 135–137 to explain how conflicts
can be managed.   *(6 marks)*

Common question

Study Figures B and C on page 131.
With the help of a diagram(s), explain the formation of a corrie.   *(F Tier 6 marks/H Tier 8 marks)*

7 The coastal zone

Foundation Tier

(a)   Study Figures D and E on page 145.
Constructive waves are different from destructive waves.
Complete the sentences below to describe this. Circle the correct word.
Constructive waves are smaller/larger in height than destructive waves.
Constructive waves have a weaker/stronger backwash than swash.
The crest on constructive waves plunges less/more than on destructive waves.   *(3 marks)*

(b)   Study Figure A on page 158. Describe the economic effects of rising sea levels.   *(4 marks)*

Higher Tier

(a)   Study Figures D and E on page 145.
Draw a labelled diagram to show two differences between destructive
and constructive waves.   *(4 marks)*

(b)   Use the case study of Keyhaven Marshes on page 168 to explain how the area
is managed and is used.   *(8 marks)*

Common question

Study Figure B on page 146.
Describe the effect of mass movement on the landscape and people.   *(6 marks)*

8 Population change

Foundation Tier

(a)   Use the information on pages 170–171 about features of population change.
Draw a line to each of the statements to complete the following definitions:

Birth rate	the rise in the number of people at an ever-increasing rate
Natural population change	the number of babies born per 1,000 per year
Exponential population growth	the difference between the birth and death rate

(2 marks)

(b)   Using the case study of China on pages 181–183, describe how
China tried to control population.   *(6 marks)*

Higher Tier

(a) Study Figure A on page 170.
Explain why global population growth up to 2050 can be described as exponential. *(4 marks)*

(b) Use the case study of China on pages 181–183 to explain the positive and negative effects of the one-child policy. *(8 marks)*

Common question

Study Figure C on page 187.
Describe how the UK population over 50 changed between 1951 and 2003. *(4 marks)*

9 Changing urban environments

Foundation Tier

(a) Study Figure A on page 194.
The urban population varies throughout the world.
Complete the sentences below to describe this. Circle the correct word or number.
Urban population increased/decreased in all areas of the world between 1990 and 2000.
Least/most people live in urban areas in more developed regions.
The rate of predicted increase from 2030 to 2050 is fastest/slowest in more developed regions.
Growth in least developed regions is predicted to be 13%/15% between 2030 and 2050. *(4 marks)*

(b) Study Figure C (photos c and d) on page 199.
Describe the advantages of building homes on the edge of cities. *(4 marks)*

Higher Tier

(a) Study Figure A on page 194.
Describe the trends shown in Figure A. *(4 marks)*

(b) Study Figure C on page 199.
Explain how the increase in demand for housing is being met in different parts of cities. *(8 marks)*

Common question

Study Figure I on page 204.
Explain how the image of the CBD has been renewed. *(6 marks)*

10 Changing rural environments

Foundation Tier

(a) Study Figure A on page 220.
Identify two pressures on the rural–urban fringe. For each one, describe why the rural–urban fringe is an attractive location. *(3 marks)*

(b) Study Figure C on page 245.
Outline the effects of soil erosion. *(4 marks)*

Higher Tier

(a) Study Figure A on page 220.
Describe the changes in pressure on the rural–urban fringe shown in the figure. *(4 marks)*

(b) Study Figure C on page 245.
Explain how the soil erosion can be reduced. *(6 marks)*

Common question

Study page 227.
Describe advantages and disadvantages of the growth in second home ownership.   *(6 marks)*

 The development gap

Foundation Tier

(a)   Study the information on pages 261–262.
Describe how loans are different to aid.   *(3 marks)*

(b)   Study Figure D on page 264.
Outline advantages of long-term aid for recipient countries.   *(4 marks)*

Higher Tier

(a)   Study the information on page 262.
Explain how conservation swaps can reduce the debt of poorer countries.   *(4 marks)*

(b)   Study Figure D on page 264.
Explain the disadvantages of aid for recipient countries.   *(6 marks)*

Common question

Study Figure B on page 255.
Use Figure B to complete the following fact file.

Country with the highest HDI	_____
Country with the lowest percentage of population with access to clean water	_____
Country with the highest infant mortality and lowest number of doctors per thousand	_____
Country with the lowest death rate, but a GNP of under US $5000	_____

(4 marks)

12 Globalisation

Foundation Tier

(a)   Study Figure B on page 277.
Are the following statements in this table true or false?
Tick the correct box.

	True	False
Toyota plants are found in all inhabited continents		
There are least Toyota plants in Asia		
There are two countries in Europe that have Toyota plants		

(3 marks)

(b)   Study the information on pages 280–283.
Explain how government legislation led to China becoming the new economic giant.   *(4 marks)*

Higher Tier

(a)   Study Figure B on page 277.
Describe the pattern shown in Figure B. *(3 marks)*

(b)   Study the information on pages 280–283.
Explain why China has emerged as the new economic giant. *(8 marks)*

Common question

Study Figure D on page 275.
Explain why call centres have developed abroad. *(6 marks)*

13 Tourism

Foundation Tier

(a)   Study Figure C on page 302.
Draw a line to each of the statements to complete the sentences correctly.

Development   small number of initial visitors to an area

Exploration   falling numbers of visitors as people go elsewhere

Stagnation   investment by companies building facilities as rapid expansion occurs

(2 marks)

(b)   Use the case study of Jamaica on pages 311–312 to describe one advantage and one disadvantage of tourism. *(4 marks)*

Higher Tier

(a)   Study Figure C on page 302.
Describe the resort life cycle model. *(4 marks)*

(b)   Use the case study of Jamaica on pages 311–312 to describe the economic advantages and disadvantages of tourism. *(8 marks)*

Common question

Using the case study of Antarctica on pages 314–315, describe the impact of tourism in this extreme environment. *(6 marks)*

Glossary

A

Ablation: outputs from the glacier budget, such as melting.

Abrasion: a process of erosion involving the wearing away of the valley floor and sides (glaciers) and the shoreline (coastal zones).

Abrasion (*rivers*): occurs when larger load carried by the river hits the bed and banks, causing bits to break off.

Accumulation: inputs to the glacier budget, such as snowfall and avalanches.

Adaptations: the ways that plants evolve to cope with certain environmental conditions such as excessive rainfall.

Advanced factories: where buildings for production are built speculatively in the hope that their presence will encourage businesses to buy or rent an existing factory, removing the need to find a site or suitable premises.

Age structure: the proportions of each age group in a population. This links closely to the stage a country has reached in the demographic transition model.

Aggregate: crushed stone made from tough rocks such as limestone, used in the construction industry and in road building.

Agribusiness: running an agricultural operation like an industry. Inputs and outputs are both high.

Aid: money, food, training and technology given by richer countries to poorer ones, either to help with an emergency or to encourage long-term development.

Air pollution: putting harmful substances into the atmosphere such as carbon dioxide.

Anticyclone: an area of high atmospheric pressure.

Aquifer: an underground reservoir of water stored in pores and/or joints in a rock, e.g. chalk.

Arable farming: growing crops.

Arch: a headland that has been partly broken through by the sea to form a thin-roofed arch.

Areas of water deficit: locations where the rain that falls does not provide enough water on a permanent basis. Shortages may occur under certain conditions, e.g. long periods without rain.

Areas of water surplus: areas that have more water than is needed – often such areas receive a high rainfall total, but have a relatively small population.

Arête: a knife-edged ridge, often formed between two corries.

Arid: dry conditions typically associated with deserts.

Asian 'tiger': one of the four east Asian countries of Hong Kong, South Korea, Singapore and Taiwan, where manufacturing industry grew rapidly from the 1960s to the 1990s.

Assisted areas/enterprise zones: areas that qualify for government help. Enterprise zones are on a smaller scale than assisted areas.

Attrition: load carried by the river knocks into other parts of the load, so bits break off and make the material smaller.

Attrition: the knocking together of pebbles, making them gradually smaller and smoother.

Avalanche: a rapid downhill movement of a mass of snow, ice and rocks, usually in a mountainous environment.

B

Backwash: the backward movement of water down a beach when a wave has broken.

Bar: a spit that has grown across a bay.

Batholith: a huge irregular-shaped mass of intrusive igneous rock that only reaches the ground surface when the overlying rocks are removed.

Bay: a broad coastal inlet often with a beach.

Beach: a deposit of sand or shingle at the coast, often found at the head of a bay.

Bilateral aid: aid given by one government to another. It may include trade and business agreements tied to the aid.

Biofuels: the use of living things such as crops like maize to produce ethanol (an alcohol-based fuel) or biogas from animal waste. It is the use of crops that has become especially important.

Biological weathering: weathering caused by living organisms such as tree roots or burrowing animals.

Biomes: global-scale ecosystems.

Birth rate (BR): the number of babies born per 1,000 people per year.

Bottom-up aid: aid used to provide basic health care for communities, clean drinking water and money for education.

Brownfield sites: land that has been built on before and is to be cleared and reused. These sites are often in the inner city.

Bulldozing: the pushing of deposited sediment at the snout by a glacier as it advances.

C

Caldera: the depression of a supervolcano marking the collapsed magma chamber.

Call centres: offices where groups of people work responding to telephone queries from customers. Employees sit in front of a computer monitor giving them information that they use in their answers to questions.

Capital intensive: farming to achieve maximum production via inputs of money to allow purchase of fuel, fertilisers and buildings that will allow maximum output.

Carbon credits: a means of trading carbon between organisations or countries in order to meet an overall target.

Carbon footprint: the amount of carbon generated by things people do, including creating a demand for out-of-season food.

Carbon sink: forests are carbon sinks because trees absorb carbon dioxide from the atmosphere. They help to address the problem of global carbon emissions.

Carbonation: weathering of limestone and chalk by acidic rainwater.

Cave: a hollowed-out feature at the base of an eroding cliff.

Cavern: a large underground cave.

Cement: mortar used in building, made from crushed limestone and shale.

Central business district (CBD): the main shopping and service area in a city. The CBD is usually found in the middle of the city so that it is easily accessible.

Channel: the part of the river valley occupied by the water itself.

Chemical weathering: weathering that involves a chemical change taking place.

Child mortality: the number of children that die under five years of age, per 1,000 live births.

Choropleth map: a map where areas are shaded to show a range of figures. The higher categories are shown in darker colours and the colours get lighter as the figures reduce.

City Challenge: a strategy in which local authorities had to design a scheme and submit a bid for funding, competing against other councils. They also had to become part of a partnership involving the local community and private companies who would fund part of the development.

Clear felling: absolute clearance of all trees from an area.

Cliff: a steep or vertical face of rock at the coast.

Climate: the average weather conditions recorded over a period of at least 30 years.

Cold front: a boundary with warm air ahead of cold air.

Collision: the meeting of two plates of continental crust. They are both the same type so they meet 'head on' and buckle.

Commercial farming: farming with the intention of making a profit by selling crops and/or livestock.

Common Agricultural Policy (CAP): a policy to support and control farming in the EU.

Commuter village: a village located in the rural–urban fringe, many of whose inhabitants commute to work in surrounding towns or cities.

Commuting: The daily movement of people travelling between home and work and back again.

Composite volcano: a steep-sided volcano that is made up of a variety of materials, such as lava and ash.

Congestion charging: charging vehicles to enter cities, with the aim of reducing the use of vehicles.

Conservation: The careful and planned use of resources in order to manage and maintain the natural environment for future generations.

Conservation: the thoughtful use of resources; managing the landscape in order to protect existing ecosystems and cultural features.

Constructive wave: a powerful wave with a strong swash that surges up a beach.

Consumer: organisms that obtain their energy by eating other organisms.

Continentality: the influence of being close to or far away from the sea. Inland areas well away from the coast have a continental climate.

Convection currents: the circular currents of heat in the mantle.

Convectional rainfall: intense rainfall often in the form of thunderstorms resulting from very high temperatures and rapidly rising and cooling air.

Corrasion: the effect of rocks being flung at the cliff by powerful waves.

Corrie: a deep depression on a hillside with a steep back wall, often containing a lake.

Counter-urbanisation: the process of people leaving towns and cities to live in more rural areas such as the rural–urban fringe.

Country of origin: the country from which a migration starts.

Crest: the top of a wave.

Crop rotation: changing the use of a field regularly to help maintain soil fertility.

Cross profile: a line that represents what it would be like to walk from one side of a valley, across the channel and up the other side.

Crust: the outer layer of the earth.

Curtain: a broad deposit of calcite usually formed when water emerges along a crack in a cavern.

D

Death rate (DR): the number of deaths per 1,000 people per year.

Debt: money owed to others, to a bank or to a global organisation such as the World Bank.

Debt relief: forgiving a debt in part or in total, i.e. writing it off.

Debt relief: many poorer countries are in debt, having borrowed money from developed countries to support their economic development. There is strong international pressure for the developed countries to clear these debts – this is debt relief.

Decomposers: organisms such as bacteria that break down plant and animal material.

Deforestation: the removal of trees and undergrowth.

De-industrialisation: a process of decline in certain types of manufacturing industry, which continues over a long period of time. It results in fewer people being employed in this sector and falling production.

Dependency ratio: the balance between people who are independent (work and pay tax) and those who depend on them. Ideally, the fewer dependents for each independent person, the better off economically a country is. Here is the formula (figures can be in numbers or percentages):

$$\frac{\text{number of dependent people}}{\text{number of independent people}} \times 100$$

Depression: an area of low atmospheric pressure.

Destination: the country where a migrant settles.

Destructive wave: a wave formed by a local storm that crashes down onto a beach and has a powerful backwash.

Development measure: statistics used to show the level of development, which allows countries to be compared.

Discharge: the volume of water passing a given point in a river at any moment in time.

Disposal of waste: safely getting rid of unwanted items such as solid waste.

Diversification: moving into new activities to try to make a better living, e.g. a farmer offering tourist accommodation.

Donor country: a country giving aid to another country.

Drainage basin: area from which a river gets its water. The boundary is marked by an imaginary line of highland known as a watershed.

Drumlin: an egg-shaped hill found on the floor of a glacial trough.

Dry valley: a valley formed by a river during a wetter period in the past but now without a river.

E

Earthquake: a sudden and often violent shift in the rocks forming the earth's crust, which is felt at the surface.

Economic: this relates to costs and finances at a variety of levels, from individuals to government.

Economic core: the centre of a country or region economically, where businesses thrive, people have opportunities and are relatively wealthy. A highly developed area.

Economic migrant: someone trying to improve their standard of living, who moves voluntarily.

Economic periphery: the edge of a country or region in terms of economics. It may not physically be the edge, but is a more remote, difficult area where people tend to be poorer and have fewer opportunities. A less well developed area.

Ecosystem: the living and non-living components of an environment and the interrelationships that exist between them.

Ecotourism: tourism that focuses on protecting the environment and the local way of life. Also known as green tourism.

El Niño effect: a periodic 'blip' in the usual global climatic characteristics caused by a short-term reduction in the intensity of the cold ocean current that normally exists off the west coast of South America. It results in unusual patterns of temperature and rainfall and can lead to droughts and floods in certain parts of the world.

Emigrant: someone leaving their country of residence to move to another country.

Environmental: this is the impact on our surroundings, including the land, water and air as well as features of the built-up areas.

Epicentre: the point at the earth's surface directly above the focus of an earthquake.

Escarpment/cuesta: an outcrop of chalk comprising a steep scarp slope and a more gentle dip slope.

European Union (EU): a group of countries across Europe that work towards a single market, i.e. they trade as if they were one country, without any trade barriers.

Eutrophication: pollution of fresh water from agricultural waste or excess fertiliser run-off.

Exfoliation: flaking of the outer surface of rocks mainly caused by repeated cycles of hot and cold.

Exponential growth: a pattern where the growth rate constantly increases – often shown as a J-curve graph.

Extreme environments: locations with particularly difficult environments where the development of tourism has only recently occurred due to a niche market demand for somewhere different with physical challenges.

Extreme weather: a weather event such as a flash flood or severe snowstorm that is significantly different from the average.

Eye: the centre of a hurricane where sinking air creates clear conditions.

Eye wall: a high bank of cloud either side of the eye of a hurricane where wind speeds are high and heavy rain falls.

F

Fair trade: a system whereby agricultural producers in countries at lesser stages of development are paid a decent price for their produce. This helps them to attain a reasonable standard of living.

Fallow: land that has been left unseeded to recover its fertility.

Fetch: the distance of open water over which the wind can blow.

Fissures: extended openings along a line of weakness that allow magma to escape.

Flashy: a hydrograph that responds quickly to a period of rain so that it characteristically has a high peak and a short lag time.

Flood or storm hydrograph: a line graph drawn to show the discharge in a river in the aftermath of a period of rain, which is shown as a bar graph.

Floodplain: the flat area adjacent to the river channel, especially in the lower part of the course. This is created as a natural area for water to spill onto when the river reaches the top of its banks.

Floodplain zoning: controlling what is built on the floodplain so that areas that are at risk of flooding have low-value land uses.

Floods: these occur when a river carries so much water that it cannot be contained by its banks and so it overflows onto surrounding land – its floodplain.

Focus: the point in the earth's crust where an earthquake originates.

Fog: water that has condensed close to the ground to form a dense low cloud with poor visibility.

Fold mountains: large mountain ranges where rock layers have been crumpled as they have been forced together.

Food chain: a line of linkages between producers and consumers.

Food miles: the distance that food items travel from where they are grown to where they are eaten.

Food web: a diagram that shows all the linkages between producers and consumers in an ecosystem.

Fragile environment: an environment that is easily unbalanced and damaged by natural or human factors.

Freeze–thaw weathering: weathering involving repeated cycles of freezing and thawing.

Front: a boundary between warm and cold air.

Frost: frozen water resulting from the temperature of the ground or the air dropping below 0°C.

Function: the purpose of a particular area, e.g. for residential use, recreation or shopping.

G

Gender structure: the balance between males and females in a population. Small differences can tell us a great deal about a country or city.

Genetically modified (GM) crops: involves putting genes from other species (sometimes animals) into a crop to give it certain characteristics that increase yield.

Geological timescale: the period of geological time since life became abundant 542 million years ago, which geologists have divided into eras and periods.

Geothermal: water that is heated beneath the ground, which comes to the surface in a variety of ways.

Geyser: a geothermal feature in which water erupts into the air under pressure.

Glacial period: a period of ice advance associated with falling temperatures.

Glacial trough: a wide, steep-sided valley eroded by a glacier.

Glacier: a finger of ice usually extending downhill from an ice cap and occupying a valley.

Glacier budget: the balance between the inputs (accumulation) and the outputs (ablation) of a glacier.

Global warming: an increase in world temperatures as a result of the increase in greenhouse gases (carbon dioxide, methane, CFCs and nitrous oxide) in the atmosphere brought about by the burning of fossil fuels, for example.

Globalisation: the increasing links between different countries throughout the world and the greater interdependence that results from this.

GNI: Gross National Income – a measure of a country's wealth.

GNP: Gross National Product – a measure of a wealth that does not take account of some business taxes.

Gorge: steep-sided deep valley that may be formed by cavern collapse.

Green belt: land on the edge of the built-up area, where restrictions are placed on building to prevent the outward expansion of towns and cities and to protect the natural environment.

Greenfield sites: land that has not been built on before, usually in the countryside on the edge of the built-up area.

Greenhouse effect: the blanketing effect of the atmosphere in retaining heat given off from the earth's surface.

Greenhouse gases: gases such as carbon dioxide and methane, which are effective at absorbing heat given off from the earth.

Gross domestic product (GDP) per capita: the total value of goods and services produced by a country in one year divided by its total population. Foreign income is not included.

H

Habitat: the home to a community of plants and animals.

Hanging valley: a tributary glacial trough perched up on the side of a main valley, often marked by a waterfall.

Hard engineering: building artificial structures such as sea walls aimed at controlling natural processes.

Hard engineering: this strategy involves the use of much technology in order to try to control rivers.

Hazard: an event that occurs where people's lives and property are threatened and deaths and/or damage result.

HDI: Human Development Index – an index based on three variables: life expectancy at birth; level of education, including both literacy rate and years spent in school; income adjusted for purchasing power (how much it will buy). Maximum HDI = 1. Wealthy countries like Japan have an HDI of over 0.9, whereas poor countries are around half that figure or less. HDI concentrates on people's experience rather than economic measures.

Headland: a promontory of land jutting out into the sea.

High-access location: an area with excellent transport infrastructure, making it easy to reach for people and goods.

Honeypot site: a location attracting a large number of tourists who, due to their numbers, place pressure on the environment and people.

Hot deserts: regions of the world with rainfall less than 250 mm per year.

Hot spot: a section of the earth's crust where plumes of magma rise, weakening the crust. These are away from plate boundaries.

Household: a person living alone, or two or more people living at the same address, sharing a living room.

Hummock: a small area of raised ground, rather like a large molehill.

Hunter-gatherers: people who carry out a basic form of subsistence farming involving hunting animals and gathering fruit and nuts.

Hurricane: a powerful tropical storm with sustained winds of over 120 kph (75 mph). Also known as a tropical cyclone, a cyclone and a typhoon.

Hydraulic action: the power of the volume of water moving in the river.

Hydraulic power: the sheer power of the waves.

Hydroelectric power: the use of flowing water to turn turbines to generate electricity.

I

Ice cap: a smaller body of ice (less than 50,000 km²) usually found in mountainous regions.

Ice sheet: a large body of ice over 50,000 km² in extent.

Igneous rocks: rocks formed from the cooling of molten magma.

Immediate responses: how people react as a disaster happens and in the immediate aftermath.

Immigrant: someone entering a new country with the intention of living there.

Impermeable: rock that does not allow water to soak into it.

Incineration: getting rid of waste by burning it on a large scale at selected sites.

Industrialisation: a process in which an increasing proportion of the population are employed in the manufacturing sector of the economy.

Infant mortality: the number of babies that die under a year of age, per 1,000 live births.

Informal sector: that part of the economy where jobs are created by people to try to get an income (e.g. taking in washing, mending bicycles) and which are not recognised on official figures.

Inner city: the area around the CBD – usually built before 1918 in the UK.

Inputs: anything entering the farm system, e.g. climate, soil, seed, labour.

Insolation: energy from the sun used to heat up the earth's surface.

Intensive farming: high inputs of capital and/or labour to achieve maximum productivity.

Interdependence: the relationship between two or more countries, usually in terms of trade.

Interglacial: a period of ice retreat associated with rising temperatures.

Irrigation: artificial watering of the land.

J

Joints: cracks that may run vertically or horizontally through rock.

K

Kyoto Protocol: an international agreement aimed at reducing carbon emissions from industrialised countries.

L

Lahar: these secondary effects of a volcano are mudflows resulting from ash mixing with melting ice or water.

Land use: the type of buildings or other features that are found in the area, e.g. terraced housing, banks, industrial estates, roads, parks.

Landfill: a means of disposing of waste by digging a large hole in the ground and lining it before filling it with rubbish.

Lateral moraine: a ridge of frost-shattered sediment running along the edge of a glacier where it meets the valley side.

Leaching: the dissolving and removal of nutrients from the soil, typically very effective in tropical rainforests on account of the heavy rainfall.

Leakage: where profits made by the company are taken out of the country to the country of origin and so do not benefit the host country.

Levees: raised banks along the course of a river in its lower course. They are formed naturally but can be artificially increased in height.

Life expectancy: the number of years a person is expected to live, usually taken from birth.

Limestone pavement: a bare rocky surface, with distinctive blocks (clints) and enlarged joints (grikes).

Literacy rate: the percentage of adults in a country who can read and write sufficiently to function fully in work and society.

Livestock farming: rearing animals.

Living standards: people's quality of life, mostly measured economically but also socially, culturally and environmentally.

Load: material of any size carried by the river, from dissolved and small such as clay to very large boulders.

Long profile: a line representing the course of the river from its source (relatively high up) to its mouth where it ends, usually in a lake or the sea, and the changes in height along its course.

Long profile (glacier): this shows the changes in height and shape along the length of a glacier, from its source high in the mountains to its snout.

Longshore drift: the transport of sediment along a stretch of coastline caused by waves approaching the beach at an angle.

Long-term aid: aid given over a significant period, which aims to promote economic development.

Long-term responses: later reactions that occur in the weeks, months and years after an event.

Loose snow avalanche: a powdery avalanche usually originating from a single point.

M

Managed retreat: allowing controlled flooding of low-lying coastal areas or cliff collapse in areas where the value of the land is low.

Mantle: the dense, mostly solid layer between the outer core and the crust.

Marginal land: land that is only just good enough to be worth farming. It may be dry, wet, cool, stony or steep.

Maritime influence: the influence of the sea on climate.

Mass movement: the downhill movement of material under the influence of gravity.

Mass tourism: tourism on a large scale to one country or region. This equates to the Development and Consolidation phases of the Butler tourist resort life-cycle model.

Meander: a bend or curve in the river channel, often becoming sinuous where the loops are exaggerated.

Mechanical weathering: weathering that does not involve chemical change.

Medial moraine: a ridge of sediment running down the centre of a glacier formed when two lateral moraines merge.

Mercalli scale: a means of measuring earthquakes by describing and comparing the damage done on a scale of I to XII.

Metamorphic rocks: rocks that have undergone a change in their chemistry and texture as a result of heating and/or pressure.

Migration: the movement of people from one permanent home to another, with the intention of staying at least a year. This move may be within a country (national migration) or between countries (international migration).

Mixed farming: farming both crops and animals.

Moraine: sediment carried and deposited by ice.

Multilateral aid: countries at further stages of development give money to international organisations such as the World Bank, the International Monetary Fund (IMF) or the United Nations (UN), which then redistribute it to development projects in countries at lesser stages of development.

Multiplier: where initial investment and jobs lead to a knock-on effect, creating further jobs and providing money to generate services.

Multi-purpose project: a large-scale venture with more than one aim. Many water projects relate to flood control, water supply, irrigation and navigation.

N

National Park: an area usually designated by law where development is limited and planning controlled. The landscape is regarded as unusual and valuable and therefore worth preserving.

Natural change: the difference between birth rate and death rate, expressed as a percentage.

Natural decrease: death rate minus birth rate, expressed as a percentage.

Natural hazard: an occurrence over which people have little control, which poses a threat to people's lives and possessions. This is different from a natural event as volcanoes can erupt in unpopulated areas without being a hazard.

Natural increase (NI): the birth rate exceeds the death rate.

Newly industrialising countries (NICs): these include the Asian 'tigers' as well as other emerging industrial nations such as Malaysia, the Philippines and China.

Non-governmental organisation (NGO): an organisation that collects money and distributes it to needy causes, e.g. Oxfam, ActionAid and WaterAid.

North Atlantic Drift: a warm ocean current from the South Atlantic, which brings warm conditions to the west coast of the UK.

Nutrient cycling: the recycling of nutrients between living organisms and the environment.

O

Occluded front: a front formed when the cold front catches up with the warm front.

Ocean trenches: deep sections of the ocean, where an oceanic plate is sinking below a continental plate.

Organic farm: a farm that does not use chemicals in the production of crops or livestock.

Outer city or suburbs: the area on the edge of the city. Many suburbs were built after 1945 and get newer as they reach the edge of the city.

Outputs: products leaving the farm system, usually for sale.

Oxbow lake: a horseshoe or semicircular area that represents the former course of the meander. Oxbow lakes are cut off from a supply of water and so will eventually become dry.

P

Park-and-ride scheme: a bus service run to key places from car parks located on the edges of busy areas in order to reduce traffic flows and congestion in the city centre. Costs are low to encourage people to use the system – they are cheaper than fuel and car parking charges in the centre.

Permeable rock: a rock that allows water to pass through it.

Pervious: rock that allows water to soak into it via clear pathways of vertical joints and horizontal bedding planes.

Physical quality of life index (PQLI): the average of three social indicators: literacy rate, life expectancy and infant mortality.

Pillar: a calcite feature stretching from floor to ceiling in a cavern.

Pioneer plant: the first plant species to colonise an area that is well adapted to living in a harsh environment.

Plate: a section of the earth's crust.

Plate margin: the boundary where two plates meet.

Pleistocene period: a geological time period lasting from about 2 million years ago until 10,000 years ago. Sometimes this period is referred to as the Ice Age.

Plucking: a process of glacial erosion where individual rocks are plucked from the valley floor or sides as water freezes them to the glacier.

Pollarding: cutting off trees at about shoulder height to encourage new growth.

Pores: holes in rock.

Porous: rock that allows water to soak into it via spaces between particles.

Prediction: attempts to forecast an event – where and when it will happen – based on current knowledge.

Preparation: organising activities and drills so that people know what to do in the event of an earthquake.

Prevailing winds: the dominant wind, south-westerly in the case of the UK.

Primary effects: the immediate effects of an event, e.g. a volcanic eruption, caused directly by it.

Primary (virgin) rainforest: rainforest that represents the natural vegetation in the region unaffected by the actions of people.

Processes: jobs done on the farm to produce outputs.

Producers: organisms that obtain their energy from a primary source such as the sun.

Protection: constructing buildings so that they are safe to live in and will not collapse.

Push–pull factors: push factors are the negative aspects of a place that encourage people to move away. Pull factors are the attractions and opportunities of a place that encourage people to move there.

Pyramidal peak: a sharp-edged mountain peak.

Q

Quality of life: how good a person's life is as measured by such things as quality of housing and environment, access to education, health care, how secure people feel and how contented and satisfied they are with their lifestyle.

Quarry restoration: restoring or improving the environmental quality of a quarry, either during its operation or afterwards.

R

Receiving country: a country receiving aid from another country.

Recycling: collection and subsequent reprocessing of products such as paper, aluminium cans, plastic containers and mobile phones, instead of throwing them away.

Regeneration: improving an area.

Regional shopping centre: a major indoor shopping centre with a large car parking area, located close to a large urban area at a high-access point, such as a motorway junction, so having millions of customers within two hours' driving time.

Relief: the height and slope of the land.

Replacement rate: a birth rate high enough for a generation to be the same size as the one before it.

Responsible tourism: the idea of encouraging a balance between the demands of tourism and the need to protect the environment.

Resurgence: a stream that emerges from underground.

Retail parks: large warehouse-style shops often grouped together on the edge of a town or city, aiming to serve as many people as possible.

Retirement migration: migration to an area for retirement.

Ribbon lake: a long narrow lake in the bottom of a glacial trough.

Richter scale: a logarithmic scale ranging from 0 to 10 used for measuring earthquakes, based on scientific recordings of the amount of movement.

Rock cycle: connections between the three rock types, shown in the form of a diagram.

Rockfall: the collapse of a cliff face or the fall of individual rocks from a cliff.

Rotational slip: slippage of ice along a curved surface.

Rural depopulation: people leaving a rural area to live elsewhere, usually in an urban district.

Rural-to-urban migration: moving home from a rural area to settle in a city.

Rural–urban continuum: a graduation from rural to urban areas.

Rural–urban fringe: an area around a town or city where urban and rural land uses mix and compete.

S

Salinisation: the deposition of solid salts on the ground surface following the evaporation of water.

Salt marsh: low-lying coastal wetland mostly extending between high and low tide.

Saltation: a hopping movement of pebbles along the seabed.

Saltation (rivers): the bouncing movement of small stones and grains of sand along the river bed.

Scavengers: organisms that consume dead animals or plants.

Scree: deposits of angular rock fragments found at the foot of rock outcrops.

Second home: a home bought to stay in only at weekends or for holidays.

Secondary effects: the after-effects that occur as an indirect effect of an event, e.g. a volcanic eruption, on a longer timescale.

Sedimentary rocks: most commonly, rocks formed from the accumulation of sediment on the sea floor.

Segregation: occurs where people of a particular ethnic group choose to live with others from the same ethnic group, separate from other groups.

Selective logging: the cutting down of selected trees, leaving most of the trees intact.

Selective Management System: a form of sustainable forestry management adopted in Malaysia.

Self-help: sometimes known as assisted self-help (ASH), this is where local authorities help the squatter settlement residents to improve their homes by offering finance in the form of loans or grants and often installing water, sanitation, etc.

Shield volcano: a broad volcano that is mostly made up of lava.

Shock waves: seismic waves generated by an earthquake that pass through the earth's crust.

Shoreline Management Plan (SMP): an integrated coastal management plan for a stretch of coastline in England and Wales.

Short-term aid: aid given to relieve a disaster situation, e.g. people who have been made homeless and are starving after a serious flood.

Site and service: occurs where land is divided into individual plots and water, sanitation, electricity and basic track layout are supplied before any building by residents begins.

Slab avalanche: a large-scale avalanche formed when a slab of ice and snow breaks away from the main ice pack.

Slash and burn: a form of subsistence farming practised in tropical rainforests involving selective felling of trees and clearance of land by burning to enable food crops to be planted.

Snout: the front of a glacier.

Social: this category refers to people's health, their lifestyle, community, etc.

Soft engineering: a sustainable approach to managing the coast without using artificial structures.

Soft engineering (rivers): this option tries to work within the constraints of the natural river system and involves

avoiding building on areas especially likely to flood, warning people of an impending flood and planting trees to increase lag time.

Soil erosion: the removal of the layer of soil above the rock where plants grow.

Solution: the dissolving of certain types of rock such as chalk and limestone by rainwater. This is a means of transportation as well as an erosion process.

Solution (coastal transportation): the transport of dissolved chemicals.

Spit: a finger of new land made of sand or shingle, jutting out into the sea from the coast.

Spring: water re-emerging from the rock onto the ground surface. Springs often occur as a line of springs (springline) at the base of a scarp slope.

Squatter settlements: areas of cities (usually on the outskirts) that are built by people of any materials they can find on land that does not belong to them. Such settlements have different names in different parts of the world (e.g. *favela* in Brazil) and are often known as shanty towns.

Stack: an isolated pinnacle of rock sticking out of the sea.

Stalactite: an icicle-like calcite feature hanging down from a cavern roof.

Stalagmite: a stumpy calcite feature formed on a cavern floor.

Stewardship: the personal responsibility for looking after things, in this case the environment. No one should damage the present or future environment.

Straightening meanders: these occur when the natural curve in a river's course is left as the river follows an artificially more direct course that has been created for it, speeding up its flow out of an area.

Stratification: layering of forests, particularly evident in temperate deciduous forests and tropical rainforests.

Strikes: periods of time when large numbers of employees refuse to work due to disagreements over pay or other grievances.

Subduction: the sinking of oceanic crust at a destructive margin.

Subsistence farming: farming to produce food for the farmer and his/her family only.

Suburbanised village: a village within commuting range of a large urban area much in demand. Housing estates attached to the village edges aim to fulfil this demand.

Supervolcano: a mega colossal volcano that erupts at least 1,000 km^3 of material.

Suspension: lighter particles carried (suspended) within the water.

Suspension (rivers): small material carried within the river as it floats throughout the depth of it.

Sustainability: development that preserves future resources, standards of living and the needs of future generations.

Sustainable: ensuring that the provision of water is long term and that supplies can be maintained without harming the environment.

Sustainable city: an urban area where residents have a way of life that will last a long time. The environment is not damaged and the economic and social fabric, due to local involvement, are able to stand the test of time.

Sustainable community: community (offering housing, employment and recreation opportunities) that is broadly in balance with the environment and offers people a good quality of life.

Sustainable development: this allows economic growth to occur, which can continue over a long period of time and will not harm the environment. It benefits people alive today but does not compromise future generations.

Sustainable management: a management approach that conserves the environment for future generations to enjoy as it is today.

Swallow hole: an enlarged joint into which water falls.

Swash: the forward movement of a wave up a beach.

T

Tariffs: government taxes on imported or exported goods.

Temperate deciduous forest: forests comprising broad-leaved trees such as oak that drop their leaves in the autumn.

Terminal moraine: a high ridge running across the valley representing the maximum advance of a glacier.

Terraces: steps cut into hillsides to create areas of flat land.

The three Ps: the collective term for prediction, protection and preparation.

Top-down aid: aid used so that governments can run more efficiently or to build infrastructure such as roads and bridges.

Tor: an isolated outcrop of rock on a hilltop, typically found in granite landscapes.

Track: the path or course of a hurricane.

Traction: heavy particles rolled along the seabed.

Traction (rivers): the rolling along of the largest rocks and boulders.

Transnational corporations (TNCs): companies that spread their operations around the world in an attempt to reduce costs.

Tropical rainforests: the natural vegetation found in the tropics, well suited to the high temperatures and heavy rainfall associated with these latitudes.

Truncated spur: an eroded interlocking spur characterised by having a very steep cliff.

Tsunami: a special type of wave where the entire depth of the sea or ocean is set in motion by an event, often an earthquake, which displaces the water above it.

U

Urban Development Corporations (UDCs): set up in the 1980s and 1990s using public funding to buy land and improve inner areas of cities, partly by attracting private investment.

Urban sprawl: the uncontrolled outward expansion of the built-up area of a town or city.

Urbanisation: the increase in the proportion of people living in cities, resulting in their growth.

V

Vale: in the landscape, a flat plain typically formed on clay.

Vegetation succession: a sequence of vegetation species colonising an environment.

Vent: the opening – usually central and single – in a volcano, from which magma is emitted.

W

Warm front: a boundary with cold air ahead of warm air.

Warm sector: an area of warm air in between a warm front and a cold front.

Water pollution: putting poisonous substances into water courses, such as sewage, industrial effluent and harmful chemicals.

Water stress: this occurs when the amount of water available does not meet that required. This may be due to an inadequate supply at a particular time or it may relate to water quality.

Water table: the upper surface of underground water.

Waterfall: the sudden, and often vertical, drop of a river along its course.

Wave-cut notch: a small indentation (or notch) cut into a cliff roughly at the level of high tide caused by concentrated marine erosion at this level.

Wave-cut platform: a wide, gently sloping rocky surface at the foot of a cliff.

Weather: the day-to-day conditions of the atmosphere involving, for example, a description of temperature, cloud cover and wind direction.

Weathering: the breakup or decay of rocks in their original place at or close to the earth's surface.

Z

Zero growth: a population in balance. Birth rate is equal to death rate, so there is no growth or decrease.

Index

Acknowledgements

The author and publisher would like to thank the following for permission to reproduce material:

Source texts:

p48 Extract from Dartmoor National Park Authority. © Dartmoor National Park Authority. Reproduced with the kind permission of the Dartmoor National Park Authority. p63 Quote from BBC news story 'Is extreme weather due to Climate change?' by Paul Rincon, 23 August 2005. © bbc.co.uk/news. p111 Extract from 'Cost of deluge will be millions' from the Sheffield Star Special Edition. © 2007 by Johnston Press. Reprinted with permission from The Editor of The Sheffield Star, http://www. thestar.co.uk. p111 Quote from BBC article 'Why Bangladesh Floods are so bad' by Tracey Logan, 27 July, 2004. © bbc.co.uk/ news. p183 Extract from BBC News article 'Has China's one-child policy worked?', 20 September 2007. © bbc.co.uk/news. p187 Extract from BBC news article 'How Bad is the UK's Pension Crisis?' 23 September 2005. © bbc.co.uk/news. p187 Extract from 'The great betrayal: how the NHS fails the elderly', by Jeremy Laurance. The Independent, 27 March 2006. © 2006 by Newspaper Publishing plc. Reprinted by permission of The Independent. p189 Extract from BBC News 'Analysis: who gains from immigration?' by Steve Schifferes, 17 June 2002. © bbc.co.uk/news. p204 Quote taken from Leeds City Council website. © 2009 Leeds City Council. Reprinted with permission. p209 Extract from BBC News article 'Nairobi slum life: Into Kibera' by Andrew Harding, 4 October 2002. © bbc.co.uk/news. p224 Extract from BBC News article 'Gridlock fear for growing village', 3 June 2004. © bbc.co.uk/news. p260 Extract from BBC News article 'Hurricane Ivan blasts Caribbean'. 9 September 2004. © bbc.co.uk/news. p262 Extract from BBC News article 'G7 backs Africa debt relief plan'. 5 February 2005. © bbc.co.uk/news. p269 Extract from BBC Website 'Q&A: Common Agricultural Policy'. © bbc.co.uk/news. p277 © Toyota Motor Manufacturing (UK) Ltd. Reprinted with permission. p290 Quotes from Professor Gareth Edwards-Jones of Bangor University and Gareth Thomas, Minister for Trade and Development, taken from 'How the myth of food miles hurts the planet', by Robin McKie & Caroline Davies, The Observer, 23 March 2008. © 2008 by Guardian News & Media Ltd 2008. Reprinted with permission. p293 Quote from Gordon Ramsay, taken from 'Gordon Ramsay's war on out-of-season vegetables', By Caroline Gammell, 9 May 2008. © 2008 by Telegraph Media Group Limited. Reprinted with permission. p301 Extract from 'Tomorrow's tourism: A growth industry for the new Millennium', published by the Department for Culture, Media and Sport (extract from Foreword by Tony Blair), 26 February 1999. © Crown copyright, 1999. Crown copyright material is reproduced with the permission of the Controller of HMSO and the Queen's Printer for Scotland. p304 Extract from 'High stakes over Blackpool Supercasino', by Hugh Muir, The Guardian, 8 September 2006. © 2006 by Guardian News & Media Ltd 2008. Reprinted with permission.

Photographs courtesy of:

Alamy 14Aa, 60G, 60H, 73A, 84G, 86A, 87B, 95B, 106F, 106G, 110J, 121E, 138A, 144C, 154B, 162A, 162B, 184A, 193F, 195C, 198A (left and right), 205Ab, 206B, 213A, 214Ca, 214Cf, 216F, 216G, 227E, 232C, 237A, 244A, 250D, 265B, 278F, 299C, 299D, 306D, 307F, 308G, 308H, 309I, 313A, 313B, 313C, 315H, 317A, Chapter 11 header, Chapter 13 header; AP Photos 114D, 139B, 144A; Apex News & Pictures 64C; Art Directors: 207E, 239B; Bell Ingram Limited/ www.tarmac.co.uk 53B; Britain on View/ Flight Images LLP 165G; Collections/Graeme Peacock 104B; Corbis 15B, 28C, 30H, 67C, 90C, 118C, 182D, 210A, 213B, 241A (both), 244B, 283E, 285C, 306C, Chapter 8 header; Dave Dunford 51B; Defra Biodiversity Team 166I; FLPA/ Peter Wilson 228A; F E Mathes photo, courtesy of GNP archives 2008 - Lisa McKeon photo, USGS 127C; Fotolia 35B (Metamorphic), 171C, 203C, 203E, 231D, 238B, 247C, 292E; Geoscience Features Picture Library 110G, 133G; Getty Images 18C, 28B, 29F, 73B, 99G, 115F, 143C, 175C, 177C, 195C, 203D, 210 (top), 212B, 236C, 253C, 259A, 267C, 271A, 272C, 280A, 281C, 281D, 282E, 288C, 292E, 300A, Chapter 9 header (middle right and middle left); Hampshire & Isle of Wight Wildlife Trust/Peter Hutchings 167A, 196D; Ian West & Tonya West 2008 160A; iStockphoto 44D, 48B, 85H, 171B, 175B, 186A, 187B, 190B, 198Aii, 199Cb, 199Cc, 214Cb, c and e, 227C, 234A, 296B, 297C, 297D, 306B, 310A, 310B, 313D, Chapter 2 header, Chapter 4 header, Chapter 10 header, Chapter 12 header; Jason Hawkes 146A, 157B; Lonely Planet Images/Woods Wheatcroft 14Aa; Matt Below Photography 85H; Nick Bainton, West Sussex County Council 52A; PA Photos 19G, 29G, 76D, 107K, 141A, 164E, Chapter 1 header, Chapter 3 header; Panos/Stefan Boness 246A; Peter Cragg/Harting Stores 230C; Peter Smith Photography 146B; Photolibrary 85H, 134I, 136D, 152B; Photoshot 135B, 202G, 291B, 205Aa, Chapter 9 header (right); Reuters 29E, 182C; Rex Features 62A, 210A (Computers), 215Cd, 220B, 261A, 303E, Chapter 7 header; Rhett Butler 92B; RSPB Images 93C; Satellite Receiving Station, University of Dundee: 57B, 58A, 60J, 61K; Science Photo Library 18D (both), 20K; Simon Lewis 41C; Skishoot 140C; Skyscan.co.uk 155D; Skyscan/Jefferson Air 201E; South American Pictures/Tony Morrison 242D; Sprague Photo Stock 185C; Still Pictures 89B, 97D, 205Ac, 209 (both), 240C, 242C, Chapter 9 header (left); Tandridge District Council 229B; Tim Shuttlewood 296B; Tony Howell/tonyhowell.co.uk 41B; Tony Waltham/Geophotos 35B (Igneous and Sedimentary), 37A, 45B, 82C, 110H, 110I, 130A, 148A, 154A; Topfoto 303D; USGS/Mike Poland and Dan Dzurisin 20J; Visions of Tomorrow Inc, Science Faction 142B; Webb Aviation 114B, Chapter 5 header.

Ordnance Survey maps (43C, 49C, 103F, 105C, 105E, 107J, 114C, 133H, 153C, 156A, 166H, 197B and 278E) reproduced by permission of Ordnance Survey on behalf of HMSO. © Crown Copyright 2009. All rights reserved. Ordnance Survey Licence Number 100017284.

Cover photograph: courtesy of Getty/Gazimal

Photo research by Sue Sharp